David L. Andrews (Ed.)

Perspectives in Modern Chemical Spectroscopy

With Contributions by
D. L. Andrews, P. S. Belton, C. S. Creaser, M. E. A. Cudby,
S. F. A. Kettle, M. R. S. McCoustra, F. A. Mellon, D. B. Powell,
N. Sheppard, J. R. Sodeau, A. J. Thomson, D. J. Williamson
H. A. Willis and R. H. Wilson

Springer-Verlag
Berlin Heidelberg New York
London Paris Tokyo Hong Kong

David L. Andrews, B. Sc., Ph. D., C. Chem., F. R. S. C.
School of Chemical Sciences
University of East Anglia
Norwich NR 4 7 TJ, Great Britain

ISBN 3-540-52218-2 Springer-Verlag Berlin Heidelberg New York
ISBN 0-387-52218-2 Springer-Verlag New York Berlin Heidelberg

Library of Congress Cataloging-in-Publication Data
Perspectives in modern chemical spectroscopy / David L. Andrews. (ed.);
with contributions by D.L. Andrews ... [et al.].
Includes bibliographical references.
ISBN 0-387-52218-2 (U.S. : alk. paper)
1. Spectrum analysis. I. Andrews. David L.
QD95.P28 1990
543'.0858—dc20

Typesetting: Thomson Press Ltd., India
Offsetprinting: Color-Druck Dorfi GmbH, Berlin; Bookbinding: Lüderitz & Bauer, Berlin
2152/3020-543210 Printed on acid-free paper

List of Contributors

Andrews D. L.
School of Chemical Sciences, University of East Anglia,
Norwich NR4 7TJ, England

Belton, P. S.
AFRC Institute of Food Research, Norwich Laboratory,
Colney Lane, Norwich NR4 7UA, England

Creaser, C. S.
School of Chemical Sciences, University of East Anglia,
Norwich NR4 7TJ, England

Cudby, M. E. A.
School of Chemical Sciences, University of East Anglia,
Norwich NR4 7TJ, England

Kettle, S. F. A.
School of Chemical Sciences, University of East Anglia,
Norwich NR4 7TJ, England

McCoustra, M. R. S.
School of Chemical Sciences, University of East Anglia,
Norwich NR4 7TJ, England

Mellon, F. A.
AFRC Institute of Food Research, Norwich Laboratory,
Colney Lane, Norwich NR4 7UA, England

Powell, D. B.
School of Chemical Sciences, University of East Anglia,
Norwich NR4 7TJ, England

Sheppard, N.
School of Chemical Sciences, University of East Anglia,
Norwich NR4 7TJ, England

Sodeau, J. R.
School of Chemical Sciences, University of East Anglia,
Norwich NR4 7TJ, England

Thomson, A. J.
School of Chemical Sciences, University of East Anglia,
Norwich NR4 7TJ, England

Williamson, D. J.
School of Chemical Sciences, University of East Anglia,
Norwich NR4 7TJ, England

Willis, H. A.
4 Sherrards Park Avenue, Welwyn Garden City, Herts AL8 7JP,
England

Wilson, R. H.
AFRC Institute of Food Research, Norwich Laboratory,
Colney Lane, Norwich NR4 7UA, England

Foreword

A course entitled *Perspectives in Modern Chemical Spectroscopy* has been annually held at the University of East Anglia (UEA) since 1979. It is a course designed principally for chemists in industry, offering broad coverage of all the major spectroscopic methods, together with an introduction to some of the most recently available techniques. The week-long course is now one of the longest running of its kind, and reflects a commitment to the development of spectroscopic technique which has from the earliest days characterised much of the research at UEA's School of Chemical Sciences. As research links with the nearby Institute of Food Research Norwich (IFRN) have strengthened over the years, several members of the Institute have also become involved in the teaching. With the tenth anniversary of the course, the notion of publishing a book essentially based on the lectures was conceived, and the end result is the present volume.

It has been a great pleasure to undertake the editing of a book involving so many of my friends and colleagues at UEA and IFRN, and I am indebted to them for all the new spectroscopy I have learnt in the process. I must also record my thanks to the staff of Springer-Verlag, and particularly to Rainer Stumpe for his unswerving enthusiasm for this project. It is our hope that the book will prove both useful and interesting to its readers, and encourage further developments and still wider applications of modern chemical spectroscopy.

Norwich, April 1990   David L. Andrews

Contents

Chapter 4
Electronic Absorption Spectroscopy: Theory and Practice
(M. R. S. McCoustra).. 87

Chapter 5
Luminescence Spectroscopy (C. S. Creaser and J. R. Sodeau) 103

Chapter 6
An Introduction to Nuclear Magnetic Resonance in Fluids
(P. S. Belton) 137

Chapter 7
Multinuclear High-Resolution NMR in Solids
(M. E. A. Cudby and D. J. Williamson)................................ 155

CHAPTER 1

Chemical Applications of Molecular Spectroscopy – A Developing Perspective

N. Sheppard

1 Introduction

Chemistry is primarily concerned with the structure and transformation of matter at the molecular level. Over the past few decades the rate of progress in evaluating the results of chemical reactions has been increasingly determined by the availability of a wide range of physical methods. With major help from these techniques, whole new frontier areas of inorganic, organic and biological chemistry have been opened up with great efficiency.

The majority of the physical methods of determining molecular structure can be classified as members of the large families of diffraction and spectroscopic techniques. Within organic chemistry the determination of molecular structures of the products of a reaction is nowadays usually made by spectroscopic methods, e.g., by the use of ultraviolet/visible (UV/VIS), infrared (IR), nuclear magnetic resonance (NMR) and mass (MS) spectroscopies [1]. For larger biomolecules (where longer-range stereochemical relationships are important) and for the molecules of inorganic chemistry (where it is frequently difficult to predict the likely structures resulting from reactions involving a multiplicity of metallic and nonmetallic elements) the spectroscopic methods may prove inadequate. In such circumstances, the geometrically more powerful and direct (but also much more time-consuming) diffraction methods are additionally brought into play.

Spectroscopic methods are also of major importance in analytical chemistry, where they can be used – often with high specificity – to identify the components in a mixture, and to determine their relative proportions. In this context spectroscopies are also widely used to measure the kinetic rates of transformation from reactants to products during chemical reactions. The measured reaction-rates in turn are of importance in the evaluation of chemical reaction mechanisms.

The purpose of this introductory chapter is to give a panoramic perspective on the major uses of spectroscopy in chemistry. The more detailed applications of the more widely applicable techniques will be explored in subsequent chapters.

2   A Historical Perspective

The great scale of present-day contributions of the spectroscopies to chemistry can be put into perspective by noting that 50 years ago, shortly before the onset of the Second World War, only ultraviolet/visible molecular spectra were in routine use for structural or quantitative chemical analysis. They were recorded by photographic methods. As will be explained below, these electronic spectroscopies found applications to the characterisation of conjugated double-bond systems in organic chemistry, and of the bonding of transition metals in inorganic and biologically-important molecules.

Although by this time (the late 1930s) the potentialities of infrared vibrational spectra for structural analysis were well appreciated in principle – following pioneering studies such as those of Coblentz in the 1910s [2], and Lecomte from the 1920s [3], – the experimental methods were considered to be too difficult for routine application. Most published infrared work in the 1930s was concerned with the vibration–rotation spectra of small molecules in the gas phase as studied more by physicists than by chemists. Nevertheless by the end of the 1930s a few academic and industrial spectroscopists were beginning to construct their own infrared spectrometers for structural analytical purposes.

The Raman Effect was discovered in 1928, and received much attention from chemists and physicists in the 1930s. Indeed, because of the comparative ease of photographic recording of such spectra in the visible region, it was seen by many as a more promising area of vibrational spectroscopy than the infrared. Kohlrausch and his colleagues in particular had recorded, by 1939, the Raman spectra of series of most of the common types of organic molecules and had identified the group-characteristic wavenumbers that could be used for structural analysis [4]. Nevertheless the comparative weakness of Raman spectra, the requirements for considerable-sized samples, and the propensity for Raman spectra to be obscured by the much stronger fluorescence spectra from small quantities of impurities, all militated against the wide use of this technique for general analytical purposes.

By the end of the War, in the mid-1940s, developments in electronics had provided for substantial potential improvements in all the above spectroscopies. In the ultraviolet/visible region used for electronic absorption or Raman spectroscopy, photoelectric detection was beginning to replace the photographic plate and was leading to more accurate quantitative measurements. In the infrared region semiconductor-based thermocouples gave higher sensitivities, and more rugged electronic amplifiers were enabling the elimination of the previously-used high-sensitivity galvanometers. The latter had been very in-conveniently susceptible to mechanical vibration. The electronic methods were also making possible the construction of automatic double-beam spectrometers to compensate for the very wavenumber-dependent backgrounds in infrared spectra. Within another few years this latter development enabled infrared spectroscopy to develop into a very versatile routine technique for chemical analysis.

From then on advances have been headlong. Totally new spectroscopic methods, such as nuclear magnetic resonance (NMR) and electron spin resonance (ESR), have been developed and very advantageously exploited, and mass spectrometry has grown greatly in practical importance. Furthermore, existing spectroscopies, such as in the infrared region, have continued to be developed to ever-higher degrees of sensitivity and sophistication. An important advance since the early 1970s has been the introduction of dedicated mini- or micro-computers into spectroscopic apparatus which have enabled the increasingly widespread use of Fourier-transform methods, described below in Sect. 3, and has led to a general sensitivity revolution during the past 20 years. To give a flavour of the continuing rapid rate of developments, which has implications for future prospects, we shall trace in outline the historical developments of the more widely applicable spectroscopies as they are discussed individually below. For this purpose, the dates indicated will refer to estimates of when a particular development led to widespread usage, rather than to the onset of earlier pioneering work.

3 An Introduction to the Use of Fourier-Transform (FT) Methods in Chemical Spectroscopy

Over the past 2 or 3 decades the use of Fourier-transform (FT) methods has brought about sensitivity-improvements of at least one to two orders of magnitude in a number of important branches of spectroscopy, including the infrared, NMR, and more recently the mass and Raman spectroscopic fields. The FT approach is likely to penetrate other branches in the near future. The experimental methods used in FT spectrometers vary greatly from one branch of spectroscopy to another. However the great improvements in sensitivity have in common the exploitation of the principle of the Multiplex Advantage that can be gained by FT methods. The Multiplex Advantage is measured relative to the sensitivity that can be attained when a spectrum is scanned, element by element, from one end to the other. In NMR this latter procedure is termed the continuous-wave (CW) method of recording a spectrum; in the optical spectroscopies it is termed the dispersive method. As its name implies, the Multiplex Advantage arises from the possibility of measuring simultaneously signals from a multiplicity of positions in the spectrum by analysis of a single combined signal from the detector.

The first application of Multiplex-FT methods in chemical spectroscopy was made in the far-infrared region, where the sources of radiation are weak and where there had long been a struggle to measure spectra with acceptable

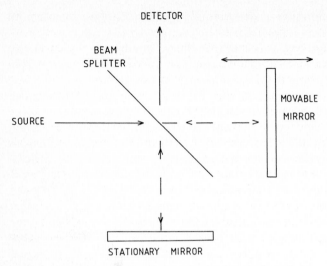

Fig. 1. A simplified schematic diagram of a Michelson interferometer

signal/noise at a reasonable resolution. This was achieved in the far-infrared, and subsequently in the mid-infrared regions, by replacing the diffraction grating (the rotation of which led to spectral scanning across an exit slit) by a Michelson interferometer. The latter (Fig. 1) consists of a beam-splitter at $45°$ to the radiation beam which divides this into two equal and mutually-perpendicular half-beams. One of these is returned to the beam-splitter by normal reflection off a fixed mirror, and the other by reflection off a mirror that can be varied in position. When the movable mirror is at the same distance from the centre of the beam-splitter as the fixed one, the returning beams will be in phase, whatever the wavelength or wavenumber of the radiation, and a large combined signal is passed to the detector. This gives what is termed the 'centre-burst'. As the mobile mirror gradually moves away from the equidistant position, radiation of a particular wavelength λ_i will go out of phase, and transmit a zero signal to the detector, when there are path-differences between the two beams of odd-numbers of half-wavelengths. Because of the to-and-fro passage of the light beam this corresponds to the same number of quarter-wavelengths of mirror-movement. An interferogram is collected by measuring the radiation reaching the detector at a series of closely spaced mirror distances on one or both sides of the position of the centreburst.

    With a continuum source, such as is common in the infrared region, the interferogram based on the output of the detector represents the sum of series of cosine waves of different periodicities, one for each wavelength, all of which are in phase at the centreburst position. Such an interferogram can be analysed into its component cosine functions, with amplitudes carrying intensity information, by the mathematical procedure of Fourier-transformation. In its simplest form this consists of multiplying the interferogram with another cosine function of the same periodicity as the contribution to the interferogram by radiation of the chosen wavelength. The $\cos^2 x$ form at that wavelength ensures that an always positive and gradually increasing quantity will be recorded on integration over the length of the interferogram. Other wavelengths will give rise to functions of the type $(\cos x \cos y)$ which periodically cancel to zero through the integration of positive and negative values. The greater the length of the recorded interferogram the less will be the 'leakage' of signal from one wavelength to the next, i.e. the higher will be the spectral resolution. If certain experimental conditions are met, it can be shown that an interferogram recorded over a path-difference of x, will give a resolution of $(1/x)$ i.e. of $1\,cm^{-1}$ from a path-difference of $1\,cm$. With modern computer techniques these numerous cosine-multiplications of the interferogram can be carried out very rapidly and in practice digital computer techniques also lead to more efficient computational procedures than would be provided by the straightforward multiple-cosine method.

The number of distinguishable spectral elements, N, in a spectrum is the ratio of the total wavenumber range to be scanned $(\tilde{v}_{max} - \tilde{v}_{min})$ to the resolution interval, $\Delta\tilde{v}$ i.e. $N = (\tilde{v}_{max} - \tilde{v}_{min})/\Delta\tilde{v}$. In the interferometric experiment, radiation from all spectral elements is measured throughout the time spent to scan the interferogram (or to co-add short-duration scans) whereas in the dispersion experiment a single spectral element is only measured for $(1/N)$ of the total scan time. Hence, since the signal/noise ratio (S/N) of a signal with random noise is proportional to $\sqrt{time}$, then when the time for accumulating the interferogram and for scanning the dispersion spectrum is equal, we have

$$\frac{(S/N)\,\text{interfer}}{(S/N)\,\text{disp}} = \sqrt{N}$$

This is the Multiplex Advantage assuming that all other factors are equal in the two experiments. The latter is rarely strictly the case, but nevertheless large gains in sensitivity can be made by this method.

Because of the long wavelengths in the far-infrared region, the initially required precision for driving the movable mirror was not difficult to achieve. Furthermore a maximum of a few hundred spectral elements had to be analysed from the interferogram and this was carried out by transferring the digitised interferometer data to the relatively slow main-frame computers of the 1960s. The development of dedicated minicomputers that could be incorporated as part of the interferometric spectrometer increased both the speed and convenience of the experiment and, with parallel developments in precision of mirror-drive etc., enabled the much more complex mid-infrared region to be measured by this method.

For technical reasons, it has proved more convenient in the mid-infrared region to record the intensities at different points in the interferogram while the mirror is continually scanning. If v is the mirror-velocity, then radiation of wavenumber $\tilde{v}$ will give a signal with frequency $2v\tilde{v}$, where the factor of 2 allows for the to-and-fro motion of the light beam. Hence each position in the spectrum makes a contribution to the interferogram with its own characteristic frequency. For example, using a KBr-backed germanium beam-splitter, the mid-infrared wavenumber range from 4000 to $400\,\mathrm{cm}^{-1}$ studied at $1\,\mathrm{cm}^{-1}$ resolution (3600 spectral elements) should give a theoretical Multiplex Advantage of $\sqrt{3600} = 60$. In practice values of 20 or more are achieved. Nowadays medium-priced infrared spectrometers incorporating microcomputers are available with this type of performance.

It is noteworthy that N and hence the Multiplex Advantage can be increased either by increasing the wavenumber range to be investigated at a given resolution, or by decreasing $\Delta\tilde{v}$ for a given wavenumber range. Conversely, for spectra of moderate resolution scanned over limited wavenumber ranges, the relative advantage of an FT over a dispersive spectrometer is much less. The Multiplex Advantage calculated above only applies as long as the signal/noise is independent of signal strength, as in the infrared region where it arises from the detector. In the visible region, where high-efficiency photomultipliers can pulse-count the number of photons arriving, the noise is proportional to $\sqrt{\text{signal}}$ and this precisely cancels out the Multiplex Advantage. However interferometers can have other energy-throughout advantages over slit-aperture-limited grating spectrometers giving the same resolution. This has meant that the Fourier-transform methods can still provide useful, but less spectacular, performances in relation to a dispersion instrument in the UV/VIS region. Recently it has been found advantageous (because of a consequent reduction of fluorescence problems) to use near-infrared laser excitation in Raman spectroscopy. In this region, the noise is also detector-determined and FT methods once again become advantageous.

It should be emphasised that the FT analysis of an interferogram is not the only method of obtaining a Multiplex Advantage. This can alternatively be achieved by recording the spectrum with a photographic plate, which also records many wavelengths simultaneously, by using an array of small detectors in the focal plane of a dispersion spectrometer or by alternative Hadamard transform techniques. However, when applicable, the FT method can more conveniently give higher Multiplex Advantages, and can also be used to provide higher and variable resolution.

In nuclear magnetic resonance spectroscopy, NMR, the replacement of CW scanning method by high-intensity pulses of radiation has also led to the use of FT methods. From FT considerations a short pulse simultaneously excites all the resonances over a frequency interval given by the reciprocal of the pulse-duration, centred at the 'carrier' frequency of the original pulse. The combined effect of these frequency components over the period of the 'free-induction decay' (FID) of the combined signal can be analysed back into the separate intensity vs frequency components by the same FT methods described above for optical spectroscopy. Once again this leads to very large gains in sensitivity, and to major advances in chemical applications as outlined below in Sect. 4.11 and described in more detail in Chapter 6.

Mass spectra can also be recorded in frequency terms by using the phenomenon of ion cyclotron resonance. In this method the unit-charged ions perform circular paths around a magnetic field with frequencies that decrease with the increasing mass of the ion. Detection of the combined ion-beam once again gives a signal with overlapping frequency components that can be separated by the same FT mathematical methods. This procedure is to be contrasted with the conventional scanning of a mass spectrum whereby one mass is measured at a time. Once again the number of spectral elements (resolvable masses) can be large with a correspondingly advantageous Multiplex Advantage.

4 The Electromagnetic Spectrum (The Photon Spectroscopies)

With the single notable exception of mass spectrometry, until comparatively recently all the commonly-used chemical spectroscopies involved the interaction of the photons of electromagnetic radiation with matter. The different regions of the electromagnetic spectrum are set out in Table 1 and are labelled either according to the wavelength/wavenumber range used, or according to the types of molecular energy levels involved e.g. ultraviolet (electronic) spectra, infrared (vibrational) spectra or radiofrequency (NMR) spectra. The types of chemical information provided by the different principal regions are outlined below, starting at the short-wavelength/high-frequency end.

Table 1. The electromagnetic (photon) spectroscopies

Region	Wavelength range[a]	Wavenumber range[a]	Type of molecular spectrum
γ-rays	< 10 pm	$> 10^9$ cm^{-1}	Mössbauer (excited states of nuclei)
X-rays	10 pm to 10 nm	10^9 to 10^6 cm^{-1}	Electronic (core orbitals)
Vacuum-UV	10–200 nm	10^6 to 5×10^4 cm^{-1}	Electronic
Near-UV	200 to 380 nm	$(5$ to $2.6) \times 10^4$ cm^{-1}	(valence orbitals)
Visible	380 to 780 nm	$(2.6$ to $1.3) \times 10^4$ cm^{-1}	
Near-IR	780 nm to 2.5 μm	$(13$ to $4) \times 10^3$ cm^{-1}	Vibrational (overtones)
Mid-IR	2.5 to 50 μm	4000 to 200 cm^{-1}	Vibrational fundamentals
Far-IR	50 μm to 1 mm	200 to 10 cm^{-1}	Vibrational fundamentals and rotational
Microwave	1 to 100 mm	10 to 0.1 cm^{-1} (3×10^{11} to 3×10^9 Hz)	Rotational; ESR/EPR
Radiofrequency	> 100 mm	$< 3 \times 10^9$ Hz	NMR; NQR

[a]These ranges are schematic and not sharply defined. UV-ultraviolet; IR-infrared; ESR-electron spin resonance; EPR-electron paramagnetic resonance; NMR-nuclear magnetic resonance; NQR-nuclear quadrupole resonance

4.1 Mössbauer (γ-ray-Absorption Spectroscopy) [5]

This involves the resonant absorption of γ-rays emitted from excited states of certain nuclei, usually produced by radioactive decay, e.g. from ^{57}Co to an excited state of ^{57}Fe. Resolution is very high in a narrow region about the frequency of the source because of the long lifetime of the excited state, and resonance frequency-matching with the same type of nucleus in another chemical environment is brought about by changing the in-line velocity of the sample, making use of the Doppler effect. Because relatively few elements have suitable radioactive nuclei, e.g. Fe, Sn, Ni, Au and I, chemical applications are specialised, and mainly applied to the investigation of inorganic coordination compounds or biomolecules containing the element in question. The position of the resonance and its hyperfine structure (see the discussion of electron spin resonance, ESR, spectra below) provide information respectively about the magnitude of the electron density (which can often be correlated with oxidation state), and the field gradient at the nucleus in question. The method is only readily applicable to solid samples.

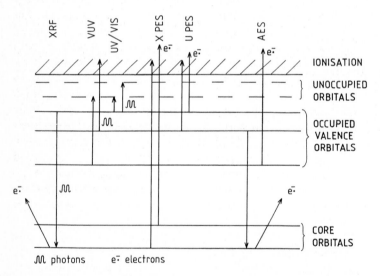

Fig. 2. A schematic molecular energy level diagram showing transitions that lead to various types of spectra
XRF-X-ray fluorescence, following ejection of an electron from a core orbital
VUV-vacuum-ultraviolet absorption involving strongly bound valence orbitals
UV/VIS-ultraviolet/visible absorption involving outer valence orbitals
XPES-X-ray photoelectron emission from a core orbital
UPES-ultraviolet photoelectron emission from a valence orbital
AES-the coupled transitions leading to the ejection of an Auger electron following the removal of an electron from a core orbital

4.2   X-ray Spectroscopy [6]

The most widely-used form is X-ray fluorescence (XRF) spectroscopy. Excitation of the sample by an incident electron beam leads to ejection of an electron from an inner strongly-bound (core) electron shell. This is followed by fluorescence as electronic transitions occur from higher occupied (valence) energy levels to the vacant inner core orbital. By this means many of the elements from F to higher atomic weight can be detected in most liquid or solid samples. Figure 2 is a schematic representation of electronic energy levels on which XRF transitions are indicated.

More recently, using the continuously wavelength-variable X-ray radiation source provided by synchrotron radiation, X-ray absorption spectroscopy has become of importance. This will be discussed below in the form of Extended X-ray Absorption Fine Structure (EXAFS) which is a hybrid technique involving spectroscopic and diffraction processes (Sect. 6.6.1).

4.3   Vacuum-Ultraviolet Spectroscopy (VUV)

The absorption bands in this region arise from electronic transitions from energy levels associated with valence orbitals to those of vacant bound orbitals of higher energy or within the continuum corresponding to ionisation (Fig. 2). For small molecules in the gaseous state such spectra provide important information about electronic structure (the spacing of electronic energy levels), molecular force-constants (from vibrational fine-structure) and molecular geometry (from rotational fine-structure). For larger molecules in condensed states the rotational fine structure is normally fully quenched and the vibrational structure partially so. The positions of the electronic absorption bands can be related to molecular structure. However because most types of molecules, saturated or unsaturated, absorb in this region there is relatively poor chemical specificity. This, combined with the fact that an evacuated spectrometer has to be used because of absorption in this region by the atmospheric gases oxygen and nitrogen, militates against the widespread use of Vacuum UV spectra for routine chemical analysis of larger molecules.

Similar information about the same electronic energy levels is also more frequently and readily obtained using Photoelectron Spectroscopy as described in Sect. 6.1 below.

4.4   Near-Ultraviolet/Visible (UV/VIS) Spectroscopy [7]

The low-wavelength/high-wavenumber limit of this region occurs at 200 nm (5×10^4 cm^{-1}) which is the position of the onset of atmospheric absorption by oxygen. It is the region of electronic spectra that can be used without the necessity for evacuation of the optical path and for this reason finds wide routine application.

As in the Vacuum-UV region, the absorption bands arise from electronic transitions from bound states (in this case of outer valence orbitals) to excited electronic states. In organic chemistry these exceptionally low energy transitions are associated with conjugated double-bond, aromatic and related molecules with delocalised electronic orbitals. Lone double-bonds, involving C, N and O usually absorb strongly just inside the Vacuum − UV region. The strongest (UV/VIS) absorption bands are associated with $\pi \rightarrow \pi^*$ transitions, and weaker ones with $n \rightarrow \pi^*$ transitions, where π and π^* refer to bonding and antibonding p-type orbitals and n refers to lone-pair orbitals. The weaker $n \rightarrow \pi^*$ transitions can also be observed from isolated double-bonds containing lone pairs of electrons, i.e. N, O, S etc. For inorganic and bioinorganic molecules, other absorptions can occur in the UV/VIS region from $d \rightarrow d$ orbital transitions associated with transition metals. Table 2 provides an overview of the historical development of the techniques and principal applications in the (UV/VIS) region.

Because saturated molecules do not absorb in this region, quantitative absorption work on solutes is possible in a wide range of solvents, including

Table 2. Historical Development of ultraviolet/visible spectroscopy since 1940

	Techniques	Applications
1940	H$_2$ lamp, glass/quartz prisms, photographic recording	Conjugated/aromatic organic molecules; inorganic transition-metal compounds
1950	Photoelectric recording	Better quantitative analysis
	Flash lamps	Transient species
1960	Diffraction gratings	Improved resolution
	Double beam operation	Improved intensities
	Optical rotatory dispersion	Chiral molecules
	Circular dichroism	Absolute configurations
1970	Sample automation	Speedier analyses
	Commercial fluorimeters	Chemical kinetics
	CW lasers	Raman spectroscopy
	Diffuse reflection	Powdered solids
1980	Picosecond laser pulses	Very short-lived transients
	Photoacoustic detection	Irregular highly scattering solid samples High gas-phase sensitivity
1990	Microprocessor operation	Higher quantitative accuracy and ease of spectrometer operation
	Tunable dye lasers	Resonance Raman spectroscopy; laser-induced fluorescence

water. Many additional applications of electronic spectra in this region can be made using fluorescence/phosphorescence techniques. Laser-excited fluorescence finds useful applications in remote-sensing analytical work in the atmosphere. Gas-phase emission spectra provide information about diatomic or small polyatomic molecules present in flames and in the atmospheres of stars. Pulses of radiation from lasers, from nanosecond to picosecond duration, have enabled the detection of transient species in ultra-fast chemical reactions.

Qualitative UV/VIS spectra of powdered solid samples can be obtained through absorption-attenuation of diffuse reflectance. This method is discussed further in the infrared context (Sect. 4.7.3). Spectra resembling those of absorbance can be derived by transforming the observed diffuse-reflectance spectrum with the theoretically-derived Kubelka-Munk function. This assumes that efficiency of diffuse reflectance is constant across the spectrum, which is not always so (Sect. 4.7.1).

Molecular chirality can be studied by optical rotatory dispersion or circular dichroism techniques in the UV/VIS region, and these can be used to classify absolute configurations of enantiomers. Principal applications are to large organic or biomolecules. Magnetic circular dichroism (using the Faraday Effect generated by magnetic fields) also gives valuable information about the electronic structures of such molecules.

4.5   Raman Spectroscopy [8]

Raman spectra also occur in the visible or near-UV regions. They measure the change in wavenumber of re-emitted (scattered) radiation from molecules following irradiation by a strong monochromatic laser source, usually in the visible region. The changes in wavenumbers are associated with transitions between vibrational and/or rotational energy levels in the ground electronic state. The technique provides the same type of vibrational and rotational information as infrared and far-infrared spectroscopy (Sect. 4.7 and 4.8 below) but is complementary in the sense that many strong features in infrared spectra are weak in Raman spectra and vice versa. For vibrations of centrosymmetric molecules there is a strict Rule of Mutual Exclusion between infrared and Raman spectra. The different intensity relationships are caused by the fact that the intensities of Raman lines are determined by the electrical polarizability change during the vibration (the induced dipole moment that gives rise to the re-emitted radiation is proportional to the polarizability), whereas those in infrared spectra depend on the vibrational change in electrical dipole moment.

The most intense, and therefore the most useful, group-characteristic bands in the Raman spectra of organic molecules are caused by vibrations of the more-polarizable (and less polar) bonds or groups such as $C-H$, $C{\equiv}C$, $C{=}C$, $C-C$, aromatic rings etc. Groups incorporating electron-rich atoms such as sulphur, the halogens, or metals also give strong features. However Raman spectra have not found as wide applicability to organic chemistry as have infrared spectra. This is because the structural information provided by their group-characteristic features can nowadays often be more readily obtained by other means from ^{1}H and ^{13}C NMR spectra (Sect. 4.11). If NMR had not been invented, Raman spectroscopy would have been a much more important technique to the organic chemist. In inorganic chemistry, where heavy atoms are involved, Raman spectroscopy has more advantages because low and high wavenumber features can readily be measured in the same spectrum. In absorption the mid-infrared and more difficult far-infrared techniques have to be separately employed. On the other hand highly-coloured inorganic compounds are liable to thermal decomposition through absorption of energy from the laser beam used to excite the Raman spectrum. Of course, in organic or inorganic chemistry, both infrared and Raman spectra are needed in order to obtain comprehensive information on the vibrational frequencies of a molecule.

Water is a very good solvent for Raman spectroscopy as its own spectrum is weak. The use of glass sample-cells is also very convenient. A persistent problem in Raman spectroscopy of mixtures or of poorly-purified samples has been the simultaneous excitation of fluorescence spectra, by components or impurities which have absorption bands at the laser frequency. Because fluorescence spectra can be orders-of-magnitude stronger than Raman spectra, very small amounts of conjugated impurities can cause difficulty. Lasers of longer wavelength are less likely to find overlap with an electronic absorption in the sample but have the disadvantage that for a given intensity in the laser beam the strength of scattering varies as $1/\lambda^4$. However a recent promising development has been the use of a laser in the near-infrared region (a YAG laser at 1.064 μm) together with sensitivity enhancement by use of Fourier-transform techniques (see Sect. 3).

4.5.1   Resonance Raman Spectroscopy (RRS) [9]

The classical Raman Effect applies to compounds that do not have an electronic absorption band at the wavelength of the incident laser beam. If there is an overlap with the edge of an absorption, but electronic fluorescence is not yet excited, a much stronger Resonance Raman Spectrum may be obtained. Intensity-enhancements may be by several orders of magnitude, although these will vary for different Raman lines and depend on polarizabilities associated with the excited as well as the ground electronic states. As such spectra are best obtained by excitation near the edge of an electronic absorption, tunable dye lasers are very useful as monochromatic sources for this work. The high intensity of the effect has led to a number of analytical applications in inorganic chemistry and to coloured biomolecules. Of course absorption of energy from an exciting laser can lead to hazards of sample decomposition by photochemical or thermal means so that the intensity of the laser beam has often to be limited, but this difficulty is usually offset by the inherent high sensitivity of the technique.

4.5.2   Surface-Enhanced Raman Spectroscopy (SERS) [10]

This is another form of Raman Effect of high intensity which can give enhancement up to a factor of 10^6. It therefore has attractive analytical aspects for studying monolayers of adsorbed species on metal surfaces. It is however a very specific effect as it only readily occurs for roughened metal samples such as silver (*par excellence*), gold and copper or the alkali metals; other metals give limited enhancements. Despite great efforts, the mechanism of SERS remains controversial. The high specificity may arise from the requirement for two favourable factors. The first is that metal particles are present of the right size for the excitation of their plasma resonances in the visible region; roughened or cold-vapour deposited surfaces are particularly effective. Additionally a degree of electronic interaction between the adsorbed molecule and the metal appears to be necessary in order to give Resonance Raman effects.

A particularly attractive analytical application of SERS is to the identification of the structure of surface species on electrode surfaces of the above metals. Because water has a very weak Raman spectrum, it is an ideal solvent for such studies.

4.6   Near-Infrared Spectroscopy (NIR) [11]

Only a few special classes of molecules have electronic absorptions that extend into the near-infrared region. Other absorptions that do occur are intrinsically weak and result from overtones and combinations of the fundamental molecular vibration frequencies that are characteristic of the mid-infrared region. These overtones etc. are also strongest from the high-frequency bond-stretching fundamentals of hydrogen-containing bonds such as OH, NH, CH etc., perhaps in combination with hydrogenic angle-bending fundamentals associated with the same groupings. Although the absorptions are weak, for liquids they can readily be enhanced by increasing the pathlength of liquid samples. They can alternatively be measured effectively by absorption-attenuation of diffuse-reflectance from powdered solid samples.

The relative simplicity of these spectra lends itself to automatic computer-selected choice of analytical absorption bands based on measurements of spectra of series of reference compounds. Near-infrared spectra have found substantial analytical application in the food industry, e.g. for quantitatively determining the proportions of fatty, carbohydrate or protein constituents.

4.7   (Mid)-Infrared Spectroscopy [12–14]

The absorption bands in this region of the electromagnetic spectrum correspond to fundamental vibration frequencies of molecules, interspersed with the usually substantially weaker overtones and combinations involving lower frequency fundamentals.

For the smaller polyatomic molecules these spectra can be completely analysed in terms of the expected number of vibrational fundamentals. From the frequencies of these the force-constants that control bond-stretching and angle-bending motions can be calculated. For more general analytical or structural diagnosis work on larger molecules, the characteristic wavenumber ranges for the vibrations of particular functional groups, such as CH, NH, OH, $C\equiv N$, $C=O$, $C=C$, aromatic rings etc. are of great importance in organic chemistry[12]. In inorganic chemistry the characteristic spectra of symmetrical complex ions, such as NO_3^-, CO_3^{2-}, ClO_4^-, SO_4^{2-}, PO_4^{3-}, $[FHF]^-$ are of much use, and in organometallic chemistry particular metal-atom-bonded ligands such as CN, CO, NH_3, CH_3 etc. can be identified from their infrared absorptions [13]. Intensities in infrared spectra depend on dipole-changes occurring during the vibrations and so the polar bonds and groups frequently give the strongest absorptions, in contrast to the non-polar ones which often give prominent Raman features. The general range of applicability of infrared spectra is much greater than of UV/visible spectra because saturated as well as unsaturated, unconjugated as well as conjugated, molecules all have vibration frequencies.

As molecular vibration frequencies vary only slightly from one phase to another, a major advantage of infrared spectroscopy is that spectra can be obtained from small samples (from a few mg down to a few ng when special equipment is used) in the gaseous, liquid and solid phases. Table 3 chronicles some of the advances in techniques in the infrared region since the 1930s and indicates the corresponding general applications.

Compared with UV/VIS spectroscopy there are however difficulties over the choice of solvents for the infrared region because most of these have their own absorptions. For organic molecules with light atoms the heavy-atom solvents CCl_4, $CHCl_3$ (with $CDCl_3$), CS_2 and C_2Cl_4 etc. can be used. In the low frequency region hydrocarbons and liquid paraffin (Nujol), which mainly have high frequencies, can be used. Because of very strong absorption by the highly-polar water molecules, this solvent can only be used in thin layers, although modern high-sensitivity Fourier-transform spectrometers can more readily detect the relatively very weak absorptions from solutes.

During the 1950s and 1960s improvements in infrared sensitivity were brought about by replacing prisms by the more highly-dispersing diffraction gratings. However, as described in Sect. 3, the biggest improvement in performance came about when dispersive spectrometers were replaced by interferometers, in the late 1960s to mid-1970s, together with the necessary computing facilities for Fourier-transform conversion of the interferogram into a normal (intensity vs wavenumber) spectrum. At one end of the scale this has led to much higher resolution spectra at acceptable signal/noise; this is mostly exploited in the analysis of the detailed vibration/rotation spectra of smaller molecules in the gas phase. Such spectra have also been very useful in analytical atmospheric research.

Table 3. Historical development of infrared spectroscopy since 1940

	Techniques	Applications
1940		
	Single-beam spectrometers (NaCl and KBr prisms)	Functional-group analysis of organic molecules
	Nujol mulls	Powdered solids
1950		
	Double-beam spectrometers (optical null)	Routine organic structure determination
	KBr discs	Powdered solids
	CsBr/CsI prisms	Low frequencies; inorganic and organometallic chemistry
	Diffraction gratings	Better quality spectra
1960		
	ATR	Applications to many industrial samples
	FT Far-infrared interferometric spectroscopy	Inorganic heavy-atom molecules; intermolecular vibrations; torsional modes
	Beam condensers	Microsampling
1970		
	Minicomputers, FT spectrometers	High resolution, high sensitivity, and absorbance subtraction; applications to adsorbed species and catalysts.
	IR Diffuse Reflectance (DRIFT)	Neat powdered solids
	Microprocessors and ratio-recording dispersion spectrometers	Improved performance at moderate price
1980		
	Microprocessor FT spectrometers	High performance at moderate price-FT spectrometers application to difficult samples
	Reflectance and RAIRS	Monolayers on flat surfaces
	Photoacoustic spectroscopy	Neat powders, irregular solids
1990		

IR-infrared; FT-Fourier transform; DRIFT-diffuse-reflectance infrared Fourier-transform spectroscopy; RAIRS-reflection/absotption infrared spectroscopy; ATR-attenuated total internal reflection

At the other end the higher signal/noise performance at resolutions of $1 \, \mathrm{cm}^{-1}$ or greater has found many analytical applications to condensed phase samples, including the detection of very small amounts of impurities by use of computer-based transmission-ratioing (or absorbance subtraction) procedures.

The very high sensitivity brought about by FT methods has also led to the first effective use, or the further development, of a number of additional nontransmission techniques for obtaining spectra from difficult samples [14]. The special characteristics of these, in comparison with the transmission method, are discussed in the next few sections. The use of these various methods will be discussed in more detail in Chapters 2 and 3.

4.7.1   Absorption Spectra as Measured by Infrared Transmission

The transmission technique has always been the one most widely employed for obtaining infrared absorption spectra. For gases and dilute solutions the method is straightforward. With liquids the principal experimental problem has occurred when the absorptions are so strong that it is difficult to obtain a uniform layer of sufficiently small thickness. The transmission spectra of powders that scatter radiation are conveniently improved in quality by surrounding the particles with a non-absorbing medium of similar refractive index, such as liquid Nujol, or by pressing the powder in a KBr or CsI disc.

Such transmittance (fractional transmission) measurements, τ, have been used to obtain absorptance (fractional absorption), α, on the assumption that $\alpha = 1 - \tau$. Such spectra are also conventionally assumed to be 'reference' spectra against which are assessed those obtained by other experimental methods, some of which are described below. However the above relationship between α and τ are only holds if there are no reflection losses, or if such losses can be accurately compensated. This causes little problem for gases, and spectra of dilute solutions can be corrected for reflection losses by ratioing against the spectrum of the pure solvent in the same cell. However the above assumption no longer holds when there are uncompensated reflection losses, and particularly selective reflection peaks in the vicinity of intrinsically strong absorptions, such as can occur for neat liquids and solids. In principle all absorption bands are associated with local changes in the refractive index n, and selective reflection losses in the vicinity of a strong absorption depend on n and on the absorption index, k. The latter is a measure of the absorption per unit wavelength of thickness and the linear (decadic) absorption coefficient can be calculated from it. The effect of reflection losses, when neglected in this manner, is to increase the apparent strength of the absorption bands derived from transmission spectra and to slightly displace the peak wavenumber. Because n is always higher on the low wavenumber side and lower on the higher wavenumber side, the apparent absorption maximum as determined from $(1 - \tau)$ is slightly shifted to low wavenumbers. With organic molecules it is usually only the strongest absorption bands which are distorted in these ways but with ionic inorganic compounds the distortions can be very substantial. In reality the selective reflection *reduces* the absorptance, which is the fraction of the *incident* energy that is absorbed, because the reflected radiation does not enter into the sample. However the reflection losses enhance the *apparent absorptance* derived on the assumption that $\alpha = 1 - \tau$, because energy reflected is equivalent in transmission-loss terms to energy totally rather than partially absorbed. The quantities (decadic) absorbance, $A = \log_{10}(I_0/I)$, and linear (decadic) absorption coefficient $\alpha = (1/l)\log_{10}(I_0/I)$ where l is the path-length, only have a meaning when I_0 is the intensity that enters the sample and I is the intensity transmitted at the other side.

Whereas attempts to measure absorption from uncorrected transmittance spectra in fact measure $(\alpha + \rho)$, where ρ is the reflectance (fractional reflection), other means of obtaining absorption spectra depend on different functions of α and ρ as we shall see below. In such circumstances the transmission-derived apparent absorption spectrum is not a suitable model for judging the performance of the other methods.

Applications of transmission infrared spectroscopy are very many and have become even more widespread using the high sensitivity of FT spectrometers. An obvious area of application is to the obtaining of acceptable spectra from very difficult samples. These might be very small in size (FT methods have led to a revival of infrared microscopy) or very poorly transmitting, as in the case of studies of highly absorbing and/or highly scattering pressed discs of catalysts. FT methods have provided a very strong stimulus to studies and identifications of chemisorbed species on finely-divided metal catalysts of the oxide-supported type, or on heavy-metal oxide catalysts themselves. There are nowadays few heterogeneous catalyst systems that are not profitably studied by infrared methods.

Another major application has been the widespread use of FT infrared methods to identify the small amounts of separate gaseous components that are eluted from a gas chromatography (GC) column. The carrier-gases such as helium or nitrogen do not have interfering spectra. This is known as GC/IR 'on the fly' and the group-characteristic nature of the IR spectra provide valuable complementary information to that generated by mass spectrometric detection (GC/MS) to be discussed in Sect. 5. Substantial efforts have also been made to apply FTIR to the detection of components separated by liquid chromatography (LC/IR) but there are more substantial difficulties to be overcome because of infrared absorption by the carrier solvents.

4.7.2 Attenuated Total Internal Reflection (ATR) [14]

In this well-established method the sample has to be capable of making good optical contact with a suitably-shaped plate of a high-refractive-index material, such as KRS-5 (thallium bromide/iodide) or germanium. Liquids cause no problem but solids have to be soft enough to be pressed into contact. The total internal reflection within the high refractive-index material is attenuated by absorption within a thin layer (of the order of a wavelength in thickness) of the sample in contact with the plate. This method can be very helpful in obtaining qualitative absorption spectra from pastes and plastics, and from the top layer of a composite material. It is also of substantial use, using cylindrical CIRCLE cells, for obtained spectra from thin layers of strongly absorbing liquids or solvents such as water. ATR materials of different refractive index, or variation of the angle of internal incidence for a given material, can give different penetration depths of the sample in contact. Absorption sensitivity can also be enhanced by multiple internal reflections in the same plate. ATR has proved to be an extremely versatile means of dealing with a wide range of samples of the types that are studied in industrial analytical laboratories.

4.7.3 Diffuse Reflectance (DRIFT) [14]

As mentioned earlier, (Sect. 4.4) diffuse reflectance from the surfaces of powders has been used for a considerable period in the UV/visible region where the detectors are very efficient. In the infrared region the availability of high-sensitivity FT techniques has turned the method from being one of extreme difficulty to one of routine feasibility. Hence, although it is not totally logical, it is customary to include the letters FT in the abbreviation.

The light scattered from the bulk of the powdered sample is partly attenuated through absorption processes during passage within the particles. In regions of strong intrinsic absorption, entry into particles is however reduced by selective reflection and the absorptance α is correspondingly reduced. Additional partial reflection losses, including selective contributions, occur at the front-surface of the powder. Hence the actual attenuation of the scattered light by absorption processes is $(\alpha - x\rho)$ where $0 < x < 1$, to be compared with $(\alpha + \rho)$ from apparent transmission measurements. Compared with transmission spectra, the strongest absorption features in the spectra of organic molecules are therefore relative reduced in intensity by these selective reflection processes. For the intrinsically very strong absorptions of ionic inorganic compounds the effect can be dramatic, with absorption intensities drastically reduced by reflected light that reaches the detector after little or no absorption. The relative importance of selective reflection and absorption can be dependent on particle sizes, with smaller particles reducing both the bulk and front-surface reflection contributions. Front-surface reflection losses can also be further reduced by diluting the powder in non-absorbing powdered KBr, with the latter by itself being used to provide the reference spectrum of the output from the infrared source.

The great practical advantage of the DRIFT method is that powdered samples can be studied with little or no pretreatment. Furthermore the weaker absorptions can be very well recorded with little distortion by DRIFT, and these are often more than adequate for structure and identification purposes. An attractive future possibility for the method is the study of the spectra of powdered catalysts *in situ* while reactions are taking place.

4.7.4 Photoacoustic Spectroscopy (PAS)

This is another technique first applied in the UV/VIS region that is now feasible in the infrared region, given the higher sensitivity of FT spectrometers.

In this method the infrared radiation from a modulated beam is absorbed by a sample (usually a solid, often in powdered form) giving rise to correspondingly fluctuating temperatures at the surface of the sample. An overlying gas is thence subject to modulated pressure decreases and increases leading to the generation of sound waves that are picked up by a sensitive microphone. The method is ideally suited to mid-infrared FT measurements because radiation of each wavenumber is modulated with a corresponding frequency by the moving mirror of the interferometer. Hence the same computer programs that are used to FT-analyse a transmission interferogram can also be used to analyse the output from the microphone. Below a certain depth the absorbed energy is dissipated into a general temperature-rise of the sample, i.e. heat no longer pulses back to the surface. In this sense the technique is a surface-sensitive one. Powdered carbon-black is a near-perfect absorber and is used to give a reference spectrum from the source.

A primary advantage of PAS is that a solid, in bulk or (as with DRIFT) in powdered form, can be studied without prior sample preparation. Finely powdered solids are nevertheless advantageous, both because of their higher surface areas and their reduced reflection losses. Clearly, as only absorbed radiation can give rise to a signal, PAS only measures the absorptance, α, which differs from $(\alpha - x\rho)$ for DRIFT. Compared with transmission, which measures $(\alpha + \rho)$, there will still be reductions in intensity of the strongest absorptions due to selective reflection losses that diminish the amount of energy that can be absorbed by the sample. However the latter effects are more pronounced in DRIFT.

4.7.5   Emission Infrared Spectra

The strong infrared emission spectra from high-temperature flames are readily used for identifying the gaseous molecules present. The infrared radiation from lower-temperature samples (gas, liquid or solid) is much weaker and biassed in intensity towards the lower wavenumber region. Once again, however, FT methods have made feasible the measurement of such spectra at modest temperatures. In principle such spectra can even be measured from room-temperature samples using a cooled detector, although at this temperature the total radiation emitted is small. In principle the emission method provides an alternative way of determining absorptance since by Kirchoff's Law the absorptance, α, is equal to emittance, ε, which is the fraction of energy emitted compared with that from a blackbody at the same temperature. Again, however, selective reflections reduce emission (in the same way that they reduce absorptance) in the vicinity of intrinsically strong absorption bands.

The emission method can be convenient for studying opaque samples, e.g. thin surface films on metal surfaces. As the metals have very high reflectivity in the infrared, they have very low emittance. The emission spectrum from a surface film can then be measured against the background of a very weak metal contribution. The method finds applicability to surface layers on rough or curved surfaces where the alternative specular reflection method (Sect. 4.7.6) cannot conveniently be used.

4.7.6   Specular Reflectance

For a thin layer on a reflecting flat metal surface, the exploration of its absorption spectrum can also be carried out by reflecting an infrared beam off the metal surface. This method is often more efficient than the study of emission because

(i) the sample can be studied at ambient temperature and (ii) the magnitude of energy absorbed from the beam from a normal high-intensity, high-temperature, infrared source will be much greater than that emitted from a sample heated to a moderate temperature. This method is called *reflection-absorption infrared spectroscopy* (RAIRS) and, with the help of FT-sensitivity, has recently been spectacularly successful in obtaining spectra from monolayers of adsorbates on metal surfaces, including single-crystal surfaces with known atomic arrangements of the metal atoms [15].

In principle, the wavenumber-dependence of the specular reflection off the flat surface of any bulk sample, metal or non-metal, when measured as a function of the angle of reflection, can be used to determine the wavenumber-dependence of the two intrinsic optical parameters, the refractive index, n, and the absorption index, k. The latter is related to the linear decadic absorption coefficient, α. The calculations are carried out by use of the Kramers–Kronig relationships between n and k and computer programs have been put together for this purpose. This is one of the best ways of determining the true absorption coefficients as a function of wavenumber. An alternative procedure is to make transmission measurements as a function of thickness and to correct for the fact that the reduction in transmission is partly due to wavenumber-dependent reflection losses as well as absorption, as described, for example by R.N. Jones and his colleagues (see for example Ref. [16]).

4.8   Far Infrared Spectroscopy [17]

This is essentially a continuation of mid-infrared spectroscopy to lower wavenumbers, although somewhat different experimental techniques (sources, diffraction gratings, beam-splitters) have to be used.

The lowest frequency vibrational fundamentals give absorption bands in this region, and these are particularly of use for characterising inorganic heavy-atom structures, or the torsional modes of lighter groupings such as CH_3. In addition absorptions can be observed due to reorientational motions in crystals or liquids and to intermolecular vibrations such as the stretching of hydrogen-bonds. In the gas-phase fine-structure spectra from rotational motions of dipolar small molecules can be analysed, but this is usually carried out with greater resolution and precision in the microwave region described in Sect. 4.9.

4.9   Microwave Spectroscopy of Gases [18]

With quite different experimental techniques, using klystrons as tunable monochromatic sources and waveguides as absorption cells, very high resolution pure rotation gas-phase spectra can be obtained providing the molecule in question has an electrical dipole moment. Although the analysis of the experimental data can be complex, except in the case of molecules with substantial symmetry, the results provide the most accurate information available on molecular dimensions in terms of bond-lengths and interbond angles. Readily resolved fine-structure splittings of rotation lines in applied electric fields (the Stark Effect) or magnetic fields (the Zeeman Effect) also provide additional information on the electrical and magnetic properties of molecules. Hyperfine splittings observed in the absence of applied fields arise from the presence of atoms with 'non-spherical' nuclei, such as the halogens, nitrogen, boron etc., which consequently have electrical quadrupole moments. The splittings are interpretable in terms of the product of the nuclear quadrupole moment and the electric field gradient at the site of the nucleus. The latter is particularly sensitive to p contributions to atomic orbitals in the vicinity of the nucleus.

Although in principle a low-symmetry gas-phase molecule gives a complex set of individual rotation lines which could be used as a molecular 'fingerprint' in analysing gaseous mixtures, the microwave method has in practice found rather limited use for general analytical purposes. It has however found application to analysing mixtures of isotopically-substituted molecules such as are produced in mechanistic studies of catalytic reactions.

4.10   Electron Paramagnetic Resonance (EPR) and Electron Spin Resonance (ESR) Spectroscopies [19]

This spectroscopic method is applicable to molecular species which have one or more unpaired electrons, such as organic free radicals or radical-ions, and transition-element compounds with incompletely filled electronic d-shells. Such molecules have magnetic moments arising from the spin of the unpaired electron(s) with sometimes additional contributions from orbital electronic motions. When both contributions are substantial the more general description Electron Paramagnetic Resonance is used. When the orbital contribution to the magnetic properties is negligible the technique is conventionally termed Electron Spin Resonance.

The spin angular momentum associated with an electron leads to a circulation of its charge distribution and hence to the production of a magnetic field, the magnitude of which can be symbolised by an effective magnetic dipole moment μ_S. The spin angular momentum has a magnitude which in an applied magnetic field takes up the familiar quantised values of $\pm\frac{1}{2}\hbar\,(=\pm\frac{1}{2}h/2\pi)$. Each of these is associated with a different energy level, the magnitude of the separation between the two being dependent on the interaction of the applied magnetic field B_0, and the magnetic dipole moment μ_S. The separation is proportional to $\mu_S.B_0$ and hence increases linearly with the magnitude of B_0. Transitions between the two energy levels can be brought about by a sinusoidally fluctuating magnetic field associated with electromagnetic radiation of the appropriate frequency. The requirement is that this should satisfy the Bohr relationship $\Delta\varepsilon = hv$, where $\Delta\varepsilon$ is the separation of the two energy levels, v is the frequency of the radiation and h is Planck's constant.

Typically magnetic fields used to excite ESR spectra cause absorption frequencies in the microwave region. When only electron-spin contributions to the molecular magnetic dipole moment are present, the ESR resonance frequency will be virtually independent of the structure of the free radical in question. To a high approximation this is the situation with organic free radicals or radical-ions where the orbital containing the unpaired electron is fixed in orientation with respect to the nuclear framework of the molecule. Where electron orbital contributions also contribute to the molecular magnetic dipole moment, the net angular momentum will be associated with a combined quantum number, which includes the electron spin contribution, and gives rise to an appropriate number of energy levels. The resonance position will then be dependent on the magnitude of the additional orbital contribution to the magnetic properties. Orbital contributions arise when the orbitals are free to reorient in the magnetic field, as is sometimes the case with single-electron d-orbitals of the transition elements.

In addition to the position of a resonance, another extremely important property of EPR/ESR spectra is the occurrence of so-called hyperfine structure. This arises through interaction of the much smaller magnetic moments of certain nuclei, such as ^1H, ^{13}C or ^{31}P, with the magnetic moment of the unpaired electron. Like the electron, a magnetic nucleus can also have spin angular momentum and therefore an effective magnetic moment, μ_N as will be described in more detail in the next section on NMR. Such a nucleus with an angular momentum quantum number, I (which can be integral or half-integral) can take up $(2I + 1)$ quantised orientations in a magnetic field, corresponding to angular momentum components from

$$+ I(\hbar), \quad (I - 1)(\hbar).....(1 - I)(\hbar), \quad - I(\hbar).$$

The interactions between the electron and nuclear magnets is a mutual one of magnitude depending on $\mu_S \cdot \mu_N$ and is, unlike the position of the centre of the resonance, independent of the applied field, B_0. Each quantised orientation of the magnetic nucleus will have a different magnitude of interaction with the magnetic electron. This is such that, as is the case in classical physics, the least stable arrangement corresponds to the two magnets being nearest to parallel and the most stable arrangement to them being closest to antiparallel. The actual result from interaction with a single magnetic nucleus is a set of I equally spaced hyperfine-structure lines replacing the single line of the electron resonance. As an example the hydrogen nucleus ^1H, has $I = \frac{1}{2}$ and so the interaction of this with an unpaired electron will cause the resonance of the latter to split into a doublet. ^{14}N has $I = 1$ and its interaction will cause a triplet of lines of equal intensity. These situations, together with the splitting patterns to be expected when 2 or 3 chemically equivalent $I = \frac{1}{2}$ nuclei are present, are shown in Fig. 3.

Let us consider as an example the ESR spectrum to be expected from the ethyl

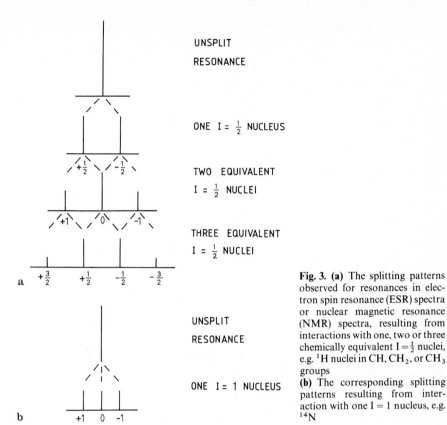

UNSPLIT
RESONANCE

ONE $I = \frac{1}{2}$ NUCLEUS

TWO EQUIVALENT
$I = \frac{1}{2}$ NUCLEI

THREE EQUIVALENT
$I = \frac{1}{2}$ NUCLEI

a $+\frac{3}{2}$ $+\frac{1}{2}$ $-\frac{1}{2}$ $-\frac{3}{2}$

UNSPLIT
RESONANCE

ONE $I = 1$ NUCLEUS

b $+1$ 0 -1

Fig. 3. (a) The splitting patterns observed for resonances in electron spin resonance (ESR) spectra or nuclear magnetic resonance (NMR) spectra, resulting from interactions with one, two or three chemically equivalent $I = \frac{1}{2}$ nuclei, e.g. ^{1}H nuclei in CH, CH_2, or CH_3 groups
(b) The corresponding splitting patterns resulting from interaction with one $I = 1$ nucleus, e.g. ^{14}N

radical $CH_3CH_3\cdot$. The presence of the two equivalent ^{1}H nuclei of the CH_2 group will split the electron resonance into a 1:2:1 triplet, whereas the 3 equivalent nuclei from the CH_3 group will cause an additional splitting of form 1:3:3:1. The overall pattern of a quartet of triplets hence provides an extremely powerful way of characterising the presence of this alkyl free radical. More generally, because ESR spectra of free radicals are intense and predictable in position, this spectroscopic technique has proved to be a very powerful one for studying the kinetic and species involved in free radical reactions, e.g. in polymerisation processes.

Such spectra are at their simplest in the gaseous or solution states where there is free orientation of the magnetic molecules with respect to B_0. In the solid state, where the orbital containing the unpaired electron can have a particular fixed direction, the resonance frequency will vary with respect to the orientation of the orbital with respect to B_0. For a powdered sample this leads to broadened resonances; single crystals spectra vary systematically with orientation. Nevertheless many useful applications of EPR have also been made in inorganic and catalyst research involving species with unpaired electrons in the solid state.

4.11   Nuclear Magnetic Resonance (NMR) Spectroscopy [20]

This is the same type of magnetic phenomenon that occurs in 'spin-only' ESR, as discussed in Sect. 4.10, except that it is the magnetic dipole moment of a nucleus that is involved. Typically this is about 3 orders of magnitude smaller than that of the electron. The resonances are correspondingly lower in frequency and fall in the radiofrequency region. The position of a particular nuclear resonance is once again linearly dependent on the strength of the magnetic field, B_0. Another extremely important difference from ESR is that there is no longer a restriction to molecules with unpaired electrons. Any molecule with a magnetic nucleus will give an NMR spectrum. Each type of magnetic nucleus (including isotopic nuclei) has its own frequency range for resonance in a particular magnetic field. Only very rarely does the resonance of one type of nucleus overlap that of another. Even so, because of the extremely high resolution attainable in NMR [< 1 Hz, contrasted with, say, $0.1 \, \text{cm}^{-1}$ ($\cong 3 \times 10^9 \, \text{Hz}$) in infrared spectra] the slightly different resonance frequencies for the same type of nucleus in different chemical environments can be easily separated. For a given type of nucleus these chemical shifts are conventionally measured in δ units of parts per million (ppm) to high frequency (low field) with respect to the resonance of a nucleus of the same type in a chosen reference environment. For example the resonances of the three types of magnetic nuclei, ^1H, ^{13}C and ^{29}Si in $Si(CH_3)_4$, tetramethyl silicon or TMS, are each chosen as the reference origin of the δ scale for the nucleus in question.

The applied magnetic field, B_0, which causes different energy level separations for different quantised orientations of a magnetic nucleus, also sets up currents in the surrounding electron orbitals. These tend to shield the nucleus from the applied magnetic field. Different electronic environments cause different shieldings, and this is the origin of the chemical shift. When measured in δ units of parts per million (ppm) the chemical shifts are independent of B_0. However for some purposes, described below, it is desirable to express the chemical shifts in the same units (Hz), as those of fine-structure splittings, the latter being independent of B_0. In units of Hz the chemical shift separations between adjacent resonances are proportional to B_0.

As described in Sect. 4.10 on ESR, the number of quantised energy levels, separated through interaction with B_0, is equal to $(2I + 1)$ where I is the spin quantum number for the nucleus in question. Some nuclei are non-magnetic and therefore do not give rise to NMR spectra. These have their relative atomic masses divisable by 4, e.g. ^4He, ^{12}C, ^{16}O, ^{24}Mg, ^{28}Si, ^{32}S, etc. However several of these elements to have less abundant isotopes that are magnetic, e.g. ^{13}C, ^{17}O, ^{29}Si etc. A fair number of nuclei have $I = \frac{1}{2}$ (with quantised orientations corresponding to $M_1 = +\frac{1}{2}$ and $M_1 = -\frac{1}{2}$). These include ^1H, ^{13}C, ^{15}N, ^{19}F, ^{29}Si, ^{31}P etc. Such nuclei have non-quadrupolar (spherical) nuclei and give particularly sharp resonances. Others have $I > \frac{1}{2}$, e.g. ^2D, ^{14}N ($I = 1$), $^{11}B(I = \frac{3}{2})$, $^{27}Al(I = \frac{5}{2})$ etc. Because of interactions of the electric quadrupoles of these nuclei with electric field gradients associated with the surrounding electron distribution, their resonances are broader – sometimes substantially so – and are therefore more difficult to observe.

For a given type of nucleus, whatever the value of I, the allowed transitions between adjacent energy levels of a given set are all equal in magnitude, leading to a single resonance frequency. However, as with hyperfine structure in ESR spectra, other adjacent magnetic nuclei can cause fine-structure splittings of a given resonance, with the same types of patterns as those described above for the ESR case. However, in frequency terms these are much smaller by about 3 orders of magnitude as the interaction is proportional to $\mu_{N1} \cdot \mu_{N2}$ (where μ_N represents the magnetic moment of a nucleus), compared with $\mu_S \cdot \mu_N$ in the ESR case. In this respect the description of the splittings in ESR spectra as hyperfine is unfortunate and historically related to the then best-resolved features in atomic electronic spectra caused by nuclear spin.

The fine-structure splitting patterns are usually simple and easy to recognise when different types of nuclei are involved e.g. ^1H and ^{31}P. This is also so when the chemical shift between two nuclei of the same type expressed in Hz is much greater than the coupling constant, also in Hz. Such spectra are said to be 'first order'. The coupling that gives rise to the fine structure is a mutual one between the two nuclei (it is transmitted by the intervening electron distribution) and, unlike the shielding, does not depend on B_0. When the coupling constant and chemical shift in Hz are comparable, deviations from simple patterns are observed to give 'second order' spectra. However these can still be analysed using standard quantum-mechanical procedures. When there is no chemical shift different between adjacent nuclei, e.g. between the three chemically identical ^1H nuclei in a CH_3 group, there is no observed mutual splitting of resonances and this is a very useful simplification. The magnetic nuclei still interact but all the allowed transitions have the same frequency. Further simplifications of spectra can be achieved by strongly irradiating one nucleus with the appropriate frequency while observing the resonance of another. It is then found that the fine structure of another resonance, of which the irradiated nucleus is the cause, will collapse. Sometimes all the nuclei of a different type can be decoupled simultaneously. For example, by 'noise-decoupling' it is possible to irradiate simultaneously all the ^1H spectra while observing ^{13}C spectra. This has the advantage that each of the ^{13}C resonances is reduced to a strong singlet.

Between them, the chemical shifts (which have group-characteristic values) and the fine structure patterns due to internuclear coupling can provide a very great deal of information about molecular structure. In the organic chemical context early work was carried out with ^1H nuclei which gives strong signals, and much progress was made. However, it was considered a great pity that the abundant nucleus of carbon, ^{12}C, was non-magnetic. The ^{13}C isotope is magnetic but, nucleus for nucleus, gives weaker resonances than ^1H and has only 1% natural abundance. It was clear that if ^{13}C resonances could also be measured, the total information provided by ^1H and ^{13}C spectra would be such as to enable structure determinations of very large organic molecules. As a result many attempts were made to measure the weak ^{13}C spectra, but there proved to be very formidable intensity problems. This was ultimately only successfully solved by changing from continuous CW scanning of the spectrum (by varying the frequency at constant B_0 or varying B_0 at constant frequency) to high-power pulse operation with Fourier-transformation of the resulting free induction decay (FID) into a normal (intensity vs frequency) spectrum as outlined earlier in Sect. 3. As ^{13}C resonances arise from an $I = \frac{1}{2}$ nucleus, they are very sharp and a typical CW spectrum might consist of 10^4 resolvable spectral elements. The Multiplex Advantage in signal/noise terms is then approx. 10^2. In the earlier 1970s the newly-introduced FT spectrometers first enabled ^{13}C spectra to be obtained with ease, and when these data were combined with those from ^1H resonances, NMR emerged as the most efficient single spectroscopic technique for determining the structures of organic molecules.

It is interesting to realise that if the abundant ^{12}C nucleus had been magnetic, the initially-studied ^1H spectra from organic molecules would have been so much more complex, for reasons of H—C and C—C couplings, that NMR could probably never have been developed as a tool for studying the structures of large organic or biomolecules. Once the ^{13}C intensity problem had been overcome, there were major advantages in a ^{13}C nucleus having a very few magnetic carbon neighbours. Nature is not always unkind! The main historical development in NMR techniques are shown in Table 4. The high sensitivity of the FT method subsequently led to the application of NMR spectroscopy to many previously inaccessible nuclei 'across the Periodic Table'. This opened the door to many NMR applications in inorganic chemistry [21].

Since the beginning of the era of commercial instrumentation for NMR in the mid-1950s, attempts have been made to increase the magnetic field that is available, while at the same time preserving resolution. An increase of field has two major advantages. One of these is that sensitivity increases approximately in proportion to B_0^2. The second is that an increased B_0 proportionately increases the separation between adjacent resonances when expressed in Hz. As the coupling constants are independent of B_0 this means that second-order spectra from the same type of nuclei frequently tend towards the more readily-interpretable first-order type.

The first commercially available NMR spectrometers in the 1950s had magnetic fields giving ^1H resonances at 40 MHz. Nowadays superconducting solenoid magnets enable such resonances to be measured at ten times that frequency. The greater separations of resonance and the even more greatly increased intensities have turned NMR into a technique that nowadays gives information about the structures and interactions of very large biomolecules, such as enzymes, in solution.

For several decades NMR was limited to work in solution. The high

Table 4. Historical development of nuclear magnetic resonance (NMR)

	Techniques	Applications
1940		
1950	1945-'Discovery' of NMR	
	Electromagnet spectrometers (30–40 MHz) Resolution 1 in 10^7 EM spectrometers 60 MHz (1 in 10^9)	$^1H(^{19}F)$ spectra from organic molecules for structure determination Routine ^1H spectra on printed charts
	Permanent magnets 60 MHz	
1960		
	Double resonance techniques	Very wide usage in organic laboratories
1970	100 MHz EM spectrometers (continuous wave)	Rate-processes and dynamic equilibria
	Fourier-transform NMR; ^1H noise decoupling; use of minicomputers	^{13}C NMR-major applications to organic structure determination
	Magic-angle spinning	
1980	New pulse sequences	High resolution solid-state NMR
	Superconducting magnets (200–500 MHz)	Most inorganic nuclei and other nuclei of low abundance e.g. ^{17}O
1990	2D-NMR	Large organic and biomolecules

EM–electromagnetic; 2D-NMR-two dimensional NMR

resolution obtainable in the liquid phase was usually catastrophically degraded when attempts were made to study the same samples as solids. There are several reasons for this. The first is the direct dipole-dipole interaction between adjacent magnetic nuclei. This is much greater in magnitude than the indirect electron-coupled interactions that give rise to fine structures of resonances in solution spectra. In the liquid phase the direct dipole-dipole coupling is precisely averaged to zero, due to rapid rotational and translational 'tumbling' motions. In a static situation the relative magnitude of the direct dipole-dipole interaction is proportional to $\mu_{N1} \cdot \mu_{N2} (1 - 3\cos^2 \theta) r^3$ where r is the internuclear distance between the two magnetic nuclei N_1 and N_2, and θ is the angle between the internuclear vector and the direction of B_0. The angle-dependence of this interaction led to the suggestion that a sufficiently rapid rotation of a solid sample about an axis at the 'magic angle' $\theta = \cos^{-1}(1/\sqrt{3})$ with respect to the direction of B_0 could average the direct dipole–dipole coupling to zero, as is brought about by molecular tumbling in solution. At the same time the chemical shift anisotropy – another complication in the solid state – might also be averaged to a single value. Additionally, where the two interacting nuclei are different in type, it should be possible to 'stir' the spins of one of them by strong double-irradiation to further average out the direct and indirect dipolar couplings with the other. These and other procedures, dependent on the use of sophisticated pulse-sequences, has led to very valuable high-resolution type spectra from, for example, ^{13}C nuclei in organic grouplings in the solid state, including crystals and polymers, and from ^{29}Si nuclei in aluminosilicate minerals. The study of dilute spin-systems, such as ^{13}C and ^{29}Si, keeps down the requirement for a very high rate of sample spinning, after other large direct couplings between ^1H and ^{13}C nuclei have been eliminated by strong double-irradiation of ^1H.

The continued advances in sensitivity and capability of NMR spectroscopy over three decades, as chronicled briefly in Table 4, has been a truly remarkable story and a great landmark in chemical spectroscopy.

4.12   Nuclear Quadrupole Resonance (NQR) Spectroscopy [22]

In connection with gas-phase microwave spectroscopy, ESR/EPR amd NMR, phenomena have been pointed out that depend on the interaction of non-spherical electrically quadrupolar nuclei with electric field gradients at the nucleus caused by the electronic environments. In fact the precessional motions of such nuclei about the principal direction of the field gradient can be measured directly in solid samples as resonant absorptions of electromagnetic radiation of the appropriate frequency. These resonances usually occur at lower radio frequencies that are used in NMR spectroscopy. In addition to the presence of a quadrupolar nucleus, such as ^{14}N or the abundant isotopes of the halogens, it is necessary for the nucleus to be in a non-spherically-symmetric environment in a solid sample. For example ^{14}N in NH_4^+ or ^{35}Cl in a Cl^- ion are located in zero field gradients for symmetry reasons and do not give NQR spectra. However in $H-C{\equiv}N$ or CH_3-Cl these nuclei do give NQR resonances. In fact, for a given type of nucleus, the resonant frequency is very variable dependent on the field gradient. Conversely the observed frequencies have been shown to be particularly useful in evaluating p-orbital contributions to the electron distribution surrounding the nucleus in question: s-orbital contributions give zero field gradients for symmetry reasons. Also the electron density, and hence field gradients, are weak for d orbitals in the vicinity of nuclei.

Although such experiments have found some useful applications to furthering the understanding of electron distributions in solids, this type of spectroscopy has found few general analytical applications, partly because of difficulties in predicting the field gradients and hence the frequencies of the resonances.

5   Mass Spectrometry* [23, 24]

Besides the photon spectroscopies such as IR and NMR, the other technique which has found major structural analytical applications is mass spectrometry.

---

*Editorial note. The Editor assumes full responsibility for insisting on the use of the term *mass spectrometry* in place of *mass spectroscopy* here and throughout the book.

The principle of the method is that molecules in the gas phase are converted into 'parent' (initial formula) and 'daughter' (fragment) positive (or less often negative) ions which are sorted into their mass numbers by passage through a combination of magnetic and/or electrical fields. The method has exceptionally high sensitivity because of the capability of counting individual ions as they transfer their charge to an electrode.

The earliest chemical analytical applications were developed in industrial laboratories of the oil companies in the 1950s. A standard excitation procedure, using electron beams accelerated by 70 volts, caused fragmentation of mixtures of volatile petroleum hydrocarbons to give characteristic spectral patterns from the positively-charged ions. A data-base was built up of the fragmentation patterns of different pure hydrocarbons which could be used to analyse multi-component mixtures. The fragmentation patterns, in terms of mass numbers and intensities, were often very usefully different for geometrical isomers of hydrocarbons of the

Table 5. Historical development of mass spectrometry from 1940

	Techniques	Applications
	Simple spectrographs; photographic recording	Isotope analysis
1940		
	Electron-beam ionisation (Resolution 1 m.u. at 200–300 m.u. max)	Quantitative analysis of hydrocarbon mixtures
1950		
	Double-focussing spectrometers Resolution 10^{-3} m.u.	For H, C, N, O; molecular formula obtainable from every peak → larger molecules
1960		
	High temperature sampling	Solids
	Time of flight spectrometers	Kinetic studies
	GC/MS	Major analytical application
1970		
	Minicomputers	Ease of operation; data handling
	Small quadrupole spectrometers	Easy analysis of low molecular weight mixtures
	Chemical and field ionisations	Better work on unstable compounds; parent peaks; simpler spectra
1980		
	FTMS (ICR)	Higher sensitivity
	Laser microprobe (LAMMA)	Spatial analysis of solids
	SIMS	Surface species on metals
	FAB	Large biomolecules; protein/polypeptide sequencing
1990		

ICR-ion cyclotron resonance; SIMS-secondary ion mass spectroscopy; FAB-fast atom bombardment; GC/MS-gas chromatography/mass spectrometry; FTMS-Fourier transform mass spectroscopy; LAMMA-laser microprobe mass analysis

same molecular formula. At that time resolutions were obtainable that could resolve integral masses up to several hundred mass units.

Nowadays the mixtures themselves are pre-separated by gas chromatography (GC) and the very high sensitivity of mass spectrometry enables the identification of the different components as they come off the column. The GC/MS combination is of major application and can for example be used to detect small quantities of large organic molecules, such as drugs used illegally to stimulate performance in horse-racing and athletics. These and other advances, discussed below, are presented in historical perspective in Table 5.

The next step was the gradual development of an understanding of the mechanistic fragmentation pathways of the parent ions, so that – in conjunction with other chemical spectroscopies – mass spectra could play a rôle in identifying the unknown structures of new compounds. Such applications are now widespread in organic chemistry, where the mechanistic pathways can be systematically identified, and in inorganic chemistry where the fragmentation of ligands of identifiable masses from coordination and cluster compounds provide much structural information.

The late 1950s and early 1960s saw the development of double-focussing mass spectrometers which greatly increased resolution, leading both to the opening-up of the higher mass-ranges, and to the measurement of low or medium masses with a precision of up to a thousandth of a mass unit. With this performance it became possible, for example, to distinguish between molecular formulae within which a contribution of 28 mass units was caused by two nitrogen atoms (14 + 14) or by an oxygen and a carbon atom (16 + 12). With the help of pre-calculated mass tables it became possible to write down the values of w, x, y and z in a molecular formula $C_w H_x N_y O_z$ for the parent and for each of the fragmentation peaks. Usually such precision was only needed for a few of the latter in order to make a successful structural diagnosis. The exact molecular formula obtained from the parent ion was of course of immediate assistance in interpreting the infrared and NMR spectra of the same compound.

The electron-beam ionisation method can lead to very weak parent-ion features in the mass spectra of certain classes of compounds. This led to the development of less vigorous ionisation methods which would give more intensity in the parent peaks. Chemical ionisation is an example of this, whereby a high voltage electron beam ionises a 'reagent' molecule, such as methane, which after equilibration provides a gentle means of ionising the larger 'sample' molecules by protonation or hydrogen-ion subtraction. This and other methods will be described in more detail in Chapter 11.

Another problem is related to the necessity for getting sufficient number of molecules from involatile samples into the gas-phase for mass spectroscopic analysis. However the gas-phase pressures needed are not high, and the use of heated probes could extend the method to covalent molecules of relatively high molecular weight. Where thermal heating leads to sample decomposition, sometimes field desorption from high-curvature metal tips can be used at moderate temperatures. For very high molecular-weight or refractory materials, it is only possible to eject molecular fragments into the gas phase before mass analysis. A variety of high energy probes, such as lasers, or beams of electrons, ions or atoms, have been used for this purpose.

Moderation by surrounding viscous liquids has, for example, enabled the recently-introduced fast-atom bombardment (FAB) method to produce relatively large polypeptide fragments from high molecular-weight proteins. The patterns of different fragments can often be pieced together to give detailed information about long sequences of amino-acids in the polypeptide chains.

High energy ion beams, usually of helium, have also been used to sputter fragments from surfaces in Secondary Ion Mass Spectroscopy (SIMS). This method has found considerable applications to identifying the formulae of species chemisorbed on metal surfaces, although the fragmentation process itself can lead to ambiguities.

Intense pulses of photons from a laser have also been used to vaporise portions of solid samples for analysis, using time-of-flight mass spectrometry. By use of an optical microprobe the vaporisation can be made to occur from very small parts of a heterogeneous sample. The technique is known as laser microprobe mass analysis (LAMMA) or laser-induced mass analysis (LIMA) and finds wide application in solid-state physics, chemistry and in geology.

Valuable developments in the techniques of mass spectrometry continue to be made. As mentioned in Sect. 3 of this Chapter, Fourier-transform methods based on ion-cyclotron detection promise further important advances in sensitivity in the near future. This and other techniques have recently opened up a new research field for studying the interaction of molecules in the gas-phase to form clusters that are readily identified by their mass values. Examples include clusters of carbon or of metal atoms, some of which turn out to have selective and interesting stabilities, and the interactions of rare-gas atoms or protons with different numbers of water molecules. In short, mass spectrometry is another of the major chemical spectroscopies that continues to grow from strength to strength in techniques and applications.

6   Some More Specialised Spectroscopies

During the last two decades a number of new spectroscopic methods have been developed which, although they do not find widespread usage on the scale of infrared, NMR or mass spectrometry, have important applications in specific areas. Some of these, with their principal applications, are listed in Table 6 and are described in more detail in the remainder of this Section.

6.1   Photoelectron Spectroscopy (PS or PES) [25]

This form of spectroscopy involves the use of monochromatic radiation in the Vacuum UV or X-ray regions to excite electrons from the valence or core electron

Table 6. More recent and additional spectroscopies

Ultraviolet Photoelectron Spectroscopy (UPS or UPES)
Applications to the measuring of binding-energies of outer valence orbitals in small gaseous molecules; investigating the structures of small chemisorbed species on metal surfaces.

X-ray Photoelectron Spectroscopy (XPS or XPES)
The identification of atoms within molecules, with chemical shifts related to the electronic environment; studying the nature of small chemisorbed species (oxidation states etc.). Also known as ESCA (Electron Spectroscopy for Chemical Analysis).

Auger Electron Spectroscopy (AES)
Surface analysis, the detection of elements present on surfaces; widely used for identifying residual impurities on metal surfaces; the pattern of chemical shifts can be used to empirically classify particular species on different metal surfaces. Surface scanning applications.

Electron Energy Loss Spectroscopy (EELS)
Electronic or vibrational spectra of surfaces or adsorbed species; important applications (high sensitivity, low resolution) for identification of structures of species adsorbed on metal and semiconductor surfaces through their vibration frequencies.

Inelastic Electron Tunnelling Spectroscopy (IETS)
Another form of vibrational spectroscopy; requires the sample to be adsorbed onto an oxide film between metal electrodes and measurements to be made at liquid helium temperatures.

Inelastic Neutron Scattering (INS)
Yet another form of vibrational spectroscopy of limited sensitivity and resolution with particular applications to lower wavenumber vibrations involving movements of hydrogen atoms; applications to the crystalline state and to adsorbed species.

Extended X-ray Absorption Fine Structure (EXAFS)
A spectroscopy/diffraction hybrid providing information, through diffraction of photoelectrons, about the distances and nature of atoms surrounding the element causing the X-ray absorption; applications to coordination patterns around metal atoms in disordered systems, biomolecules, and catalysts; most conveniently studied with tunable synchrotron radiation that can excite X-ray absorptions by different elements.

Near-Edge X-ray Absorption Fine Structure (NEXAFS)
Pre-ionization X-ray absorption with directional-dependent applications, using polarised synchrotron radiation to provide information about electronic structures and bond directions of adsorbed species relative to metal surfaces.

---

orbitals respectively to give ionisation (Fig. 2). The difference between the energy of the exciting photon and the kinetic energy of the emitted electron, as measured with wire grids at appropriate potentials, gives the binding energy of the orbital from which the electron is ejected. The sample can be a low pressure gas (high pressures interfere with the electron trajectories) or an evacuated condensed-phase sample.

The ultraviolet method of excitation (UPS or UPES), usually using emission lines from He discharges, has proved to be a very valuable means of measuring the patterns of energy levels of small organic or inorganic molecules and of correlating these with those predicted by molecular-orbital quantum-mechanical methods. Although, for example, transitions from lone-pair orbitals of halogen atoms have some degree of group specificity in photoelectron spectra in the valence UV region, this method cannot compete for this purpose with infrared or NMR spectroscopy. X-ray photoelectron spectroscopy (XPS or XPES), otherwise named historically, (but less fortunately) Electron Spectroscopy for Chemical Analysis (ESCA) identified elements by characteristic ionisations from core orbitals. Different chemical environments, e.g. different oxidation states of metals, can give systematically different peak-positions.

In surface science the UV and X-ray photoelectron spectroscopies have provided valuable information about the changes in structure in molecules such as CO, acetylene and ethylene that occur on chemisorption on metal surfaces. In general however the interpretation of the photoelectron spectra of surface species in terms of molecular structure tends to be more difficult than for vibrational spectra.

6.2   Auger Electron Spectroscopy (AES) [26]

As an alternative to X-ray fluorescence (Sect. 4.2), the energy released when an electron from an outer orbital makes the transition to a vacancy in a core orbital, may be passed to another electron leading to its ionisation as shown in Fig. 2. The energy of the resulting emitted Auger electron is measured as in photoelectron spectroscopy. It has a value characteristic of the atom from which the core electron has been removed, although dependent to a degree on the outer-electron molecular orbitals from which the two electrons involved make their respective transitions to the core vacancy and to vacuum. For analytical purposes the main application is to determining the lighter elements present on a metal surface, where the method has high sensitivity and specificity. The detailed shape of the Auger feature can be used to empirically characterise different forms of adsorption of a given surface species, when precautions are taken to minimise decomposition in the electron or X-ray beam that is used to create the core vacancies.

6.3   Electron Energy Loss Spectroscopy (EELS) [27]

This is an alternative form of electronic or vibrational spectroscopy in which excitations to higher electronic or vibrational energy levels are measured as the corresponding energy losses recorded by an incident beam of monoenergetic electrons. The principal analytical applications are to the study of the vibrational spectra of monolayers of molecules chemisorbed on metal or semiconductor surfaces. As with other forms of electron spectroscopy, the signals from insulating samples can be complicated by charge accumulation by the sample. Prior to the application of FT methods in reflection-absorption-infrared spectroscopy, the EELS method was the most sensitive form of vibrational spectroscopy that could be used to determine the structure of chemisorbed species on metal surfaces, and made a major impact on this area of research.

For electrons reflected close to the specular direction (when the angle of reflection equals the angle of incidence), the same mechanism of excitation of spectra applies to infrared and EELS spectra. This is dependent on a pulsed electrical field from the incident electron exciting the vibrational dipole motions of the molecule. Over a metal surface the intensity depends on the degree to which the vibration of the adsorbed species causes dipole changes perpendicular to the surface. Off-specular electrons, which involve substantial momentum transfer parallel to the surface, carry information about vibrational modes parallel to the surface. These are excited by the so-called impact mechanism. There is no equivalent of this mechanism in infrared spectroscopy and hence EELS provides valuable supplementary information. However for the perpendicular modes the infrared resolution of approx. 1 cm^{-1} is much higher than is possible with EELS (resolution approx. 40 cm^{-1}). Also, unlike with electron spectroscopies in general, infrared spectra can be obtained in the presence of considerable pressures of gas over the surface. This much enhances the infrared applications to catalyst research.

6.4   Inelastic Electron Tunnelling Spectroscopy (IETS) [28]

This is another form of vibrational spectroscopy involving the tunnelling of electrons across a metal/insulator/metal sandwich. On varying the voltage between the metal electrodes tunnelling occurs selectively when vibrational energy transitions of the insulating sample bridge the energy gap. The method gives good resolution (~ 10 cm^{-1}) and vibrational frequencies are recorded that can appear in both infrared and Raman spectra. The equipment required is rather straightforward, and the samples needed are small, but a disadvantage is that the experiments have to be carried out at liquid helium temperatures. The sample in question is usually deposited on alumina as the insulating oxide. The alumina itself has a relatively weak vibrational spectrum.

6.5   Inelastic Neutron Scattering (INS) [29]

In this case, neutrons rather than electrons cause transitions between energy levels of the sample. The usual application is to vibrational spectroscopy in the lower wavenumber range up to 1000 or 1500 cm^{-1}. Symmetry-based selection-rules do not apply to this form of spectroscopy so that in principle all modes can give spectral features. However, in practice, the ^{1}H hydrogen nucleus has a particularly large cross-section for incoherent nuclear scattering. Hence where the molecule contains this element its motions are responsible for most of the intensity of the stronger features; ^{2}D has a much weaker cross-section leading to valuable differences in the isotopic spectra. Not only modes involving internal vibrations of hydrogen-containing groups, but also those involving motions of heavier atoms on which the hydrogens 'ride', give a considerable intensity of neutron scattering. An advantage of INS, not applicable to infrared or Raman spectroscopy, is that the intensities of the various vibrational features can in principle be calculated from the amplitudes of motion of the various atoms. Because angle-dependent measurements of scattering also record momentum as well as energy exchanges it is possible to record complete frequency dispersions of the lattice modes by INS, rather than only the in-phase vibrations of unit cells which are accessible to the infrared and Raman spectroscopies.

The practical disadvantages of the technique are that experiments have to be carried out at special institutes where high-flux reactors provide the neutron beam, and that rather large samples are needed (10 g upwards). Nevertheless valuable vibrational spectroscopic studies have been made by this technique, for example, of the low frequency modes of vibrations of hydrogen-containing crystals and the adsorption of hydrogen-containing species on finely-divided metal or oxide catalysts.

6.6   The Use of Synchrotron Radiation [30]

When electrons are circulated in a large-scale storage ring at relativistic velocities, these lead to the lateral emission of polarised electromagnetic radiation. This is in the form of a continuum of which the intensity/wavelength distribution can be calculated from the dimensions of the storage ring and the speed of the electrons. This is termed synchrotron radiation and such sources are now available at institutes in a number of countries. They produce the strongest and best quality continua of radiation across the vacuum UV and X-ray regions, and they also improve on conventional sources in the far infrared region.

6.6.1   Extended X-ray Absorption Fine Structure (EXAFS) and Near-Edge X-ray Absorption Fine Structure (NEXAFS) [31]

With such a continuum source it is, for example, possible to study X-ray absorption spectra that arise from different elements in a compound. When this is done, the sharp absorption edge is found to be followed by decreasing absorption on which are superimposed a series of intensity-fluctuations. This is termed Extended X-ray Absorption Fine Structure (EXAFS) [31]. The phenomenon is a spectroscopy/diffraction hybrid which has some very useful chemical applications. As the radiation is scanned with frequencies above the X-ray absorption edge, the excess energy produces photoelectrons of gradually increasing energy and shorter wavelengths. These are diffracted by adjacent atoms and hence give periodic interference with newly emerging electrons. This results in a series of overlapping interference fringes which in principle can be Fourier-analysed to give the distances and diffraction amplitudes of nearby atoms. EXAFS therefore provides a very good way of exploring coordination patterns around, for example, transition metal atoms in coordination compounds, biomolecules and metal catalyst particles. Because of double-diffraction effects some corrections have to be made, the magnitudes of which can best be assessed by EXAFS measurements on molecules of known structure containing the same elements. Coordination-patterns deduced from EXAFS for one type of atom can be checked for consistency with that from the EXAFS spectra of an adjacent diffracting element, with the synchrotron radiation continuum providing the different X-ray frequencies required.

Near-edge fine structure to the absorption edge itself (NEXAFS) otherwise described as X-ray Absorption Near-Edge Structure (XANES) studied with the polarised radiation can also determine the orientations of some of the orbitals (and associated chemical bonds) which are subject to pre-ionisation excitation by X-ray absorption. For example, using X-ray absorption by carbon atoms it is possible to tell whether a hydrocarbon such as ethylene or acetylene adsorbed on a metal surface has its CC bonds oriented parallel or perpendicular to the metal surface.

Samples for EXAFS or NEXAFS work need not be large, but they do have to be taken to the institute housing the synchrotron source, and they have to be held in a good vacuum during measurement.

6.7   Some Other Laser-Induced Spectroscopies [32]

The development of the laser, which provides a thin parallel beam of coherent radiation that can be, when desired, of very high intensity in the form of very short-duration pulses, continues to lead to many advances in molecular spectroscopy. Some of the better-established of these that are of analytical importance have been mentioned earlier in this Chapter. Thus monochromatic, fixed-frequency, lasers have been used as powerful sources for Raman spectroscopy in the visible and near-infrared regions or, in pulsed operation, for the induction or monitoring of extremely rapid chemical reactions in the visible/near UV region (Sect. 4.5 and 4.4 respectively). Tunable dye lasers in the visible region have likewise been used to excite Resonance Raman Spectra or to explore wavelength-dependent fluorescence spectra (Sect. 4.5.1 and 4.4 respectively). In mass spectrometry pulsed lasers have been used to vaporise solids for compositional and spatial analysis of heterogeneous solids (Sect. 5).

Another rapidly-growing analytical application of laser-induced light scattering or fluorescence is to the determination of concentration/distance profiles for pollutants or aerosols in the atmosphere such as are emitted from smoke-stacks. The laser beam is pulsed at regular intervals and the back-scattered or fluorescent radiation is collected by a telescope. The distance associated with particular returning photons is determined by the time-lapse from the exciting pulse. The acronym lidar stands for light detection and ranging. The technique is an optical equivalent of radar, and is molecule-specific.

There is now a considerable number of lasers that provide powerful monochromatic radiation at different wavelengths in the visible and near infrared (see Chapter 8). Frequency-mixing or frequency-doubling brought about by interaction of strong pulses with certain types of optical materials, further add to the number of wavelengths available. A particular monochromatic source can be used to 'pump' a dye laser which can itself emit over a continuum to longer wavelength than that of the pump. By use of several such pumps of different wavelengths, in conjunction with a dye laser, it is now feasible to obtain coherent laser radiation at wavelengths of choice between the near-ultraviolet and near-infrared regions. Many applications ensue, from the measurement of very high resolution spectra of gases through the use of resonant multiphoton processes, to selective photoionisation prior to mass-spectrometric detection.

Unfortunately, widely wavelength/wavenumber tunable radiation is much more difficult to achieve in the 'chemistry-rich' infrared region. Continuously tunable radiation down to about $2000 \, cm^{-1}$ can be obtained by means of difference-frequency generation employing the tunable dye-lasers in the visible region, or optical parametric oscillators whereby a monochromatic visible beam is subdivided into an infrared component and a visible beam of longer wavelength Gas lasers involving CO or CO_2 give sets of discrete laser-lines from vibration-rotation transitions in the 2000 or $1000 \, cm^{-1}$ regions respectively over several hundred cm^{-1}; a high-pressure CO_2 laser can give a continuum. With semiconductor diode lasers very high resolution radiation can be obtained, but tuned over only a few 10's of cm^{-1} at a time; these can be used to produce excellent vibration-rotation spectra of gases. However much work in the infrared and far-infrared regions still involves choosing the molecular species to suit the laser rather than vice-versa.

A laser giving a tunable output across most of the infrared region would find very many applications. Thus a thin parallel infrared beam would be ideal, for example, for infrared reflection/absorption (RAIRS) studies of adsorbed mono-layers on metal single-crystal surfaces. Such work is at present limited by sensitivity considerations even when Fourier-transform methods are used with conventional continuum sources. There would also be substantial promise for work with monolayers at gas/liquid or liquid/liquid or liquid/solid interfaces as shown by a recent experiment in which a fixed-wavenumber visible and a tunable infrared beam were caused to interact at a liquid/solid surface [33]. The resulting sum-frequency spectrum in the visible region, which was primarily generated at the interface, gave information about the chemical nature and orientation of the molecules in an adsorbed monolayer.

Various forms of stimulated Raman spectra have also been generated by multiphoton or mixed-photon interactions involving powerful laser beams, and these find a number of specialised applications. An example is Coherent AntiStokes Raman Spectroscopy (CARS) which has found applications to analysing molecular processes in combustion and other high temperature gaseous systems. Although space does not allow the description of these more specialised techniques, Ref. [32] provides information.

7   Sampling Considerations and Conclusions Across the Spectroscopies

In the previous sections of this Chapter, we have considered, one by one, what the well-established spectroscopies have to offer the chemist in terms of structural or quantitative analytical information, and in terms of their experimental capabilities. In this last Section we change the viewpoint to that of the chemist who regularly prepares or is presented with, particular types of sample, and who wishes to obtain information about them by efficient choice amongst the various spectroscopies available. For this purpose, for the most part, we restrict attention to the more widely applicable chemical spectroscopies, i.e. to the UV/visible, infrared, NMR, and mass spectroscopies. We also indicate how the capabilities relate to the costs of different types of spectroscopic equipment.

Tables 7 and 8 collect together, and supplement, the information provided by the sections devoted to the individual spectroscopies. They show that infrared has an advantage for chemical structural applications in that it can be applied to all types of sample (large or small, solid, liquid or gas) and, in the form of dispersive spectrometers, it can be relatively cheap. NMR spectroscopy is relatively more expensive but is an outstanding analytical technique for work in solution, with increasing (but costly) applications to solids. Mass spectrometry has the principal

Table 7. Availability of various spectroscopic techniques

Chemical Nature of Sample	Cheap	Moderately Priced	Expensive
Moderate-sized organic molecules	UV/visible; dispersive IR; ^{1}H NMR (CW).	^{13}C and ^{31}P FT NMR; Single-focussing Mass spec.	Double focussing Mass spec.
Large organic and biomole-cules	UV/visible; dispersive IR; ^{1}H NMR (CW);	^{13}C and ^{31}P FT NMR	High field (superconducting) NMR FAB Mass Spec.
Inorganic molecules	UV/visible; dispersive IR; CW NMR	FT infrared.; Raman; FT NMR; Mass Spec.	High field NMR Double-focussing Mass Spec.
Catalysts and adsorbed species	Computer-controlled dispersive IR	FTIR	SIMS and electron-spectroscopy techniques

Physical Nature of sample			
Solid polymers	Dispersive IR	FT Infrared; ^{13}C FT NMR; MS/Pyrolysis	High resolution solid state NMR
Volatile mixtures	GC/IR	GC/IR	GC/MS
Involatile mixtures	LC/IR	LC/FTIR	LC/MS
Surface films	Dispersive IR	FTIR	Electron spectroscopies (Auger/UPES/XPES/ Electron Energy Loss). SIMS.

IR-infrared; UV-ultraviolet; NMR-nuclear magnetic resonance; MS-mass spectrometry; FT-Fourier transform; GC-gas chromatography; LC-liquid chromatography; FAB-fast atom bombardment; SIMS-secondary ion mass spectroscopy; UPES-ultraviolet photoelectron spectroscopy; XPES X-ray photoelectron spectroscopy.

Table 8. Sampling requirements for different spectroscopic techniques

Spectroscopic Technique	Gas	Liquid	Solid	Solution (solvents)	Poly-mers	With GC	With LC	Amount of Sample needed in mg.
UV/Visible	$\sqrt{}$	$\sqrt{}$a	[$\sqrt{}$]b	H_2O, EtOH saturated solvents	[$\sqrt{}$]c	—	$\sqrt{}$	1–10
Infrared	$\sqrt{}$	$\sqrt{}$	$\sqrt{}$	CS_2, CCl_4 $CHCl_3/CDCl_3$	$\sqrt{}$	$\sqrt{}$	$\sqrt{}$	$1-10[10^{-2}-10^{-3}]$ $[10^{-6}]$d
NMR ^{1}H	—	$\sqrt{}$	—	Many solvents	—	—	—	10–100
^{13}C	—	$\sqrt{}$	[$\sqrt{}$]e		[$\sqrt{}$]e	—	—	100+
Other nuclei	—	$\sqrt{}$	[$\sqrt{}$]e		[$\sqrt{}$]e	—	—	Variable (100–500)
Mass Spec.	$\sqrt{}$	$\sqrt{}$	[$\sqrt{}$]g	—	[$\sqrt{}$]f	$\sqrt{}\sqrt{}$	$\sqrt{}$	1–10 or less

[]—with special accessories: [[]]—with special spectrometers: a)—if absorption not strong; b)—diffuse-reflectance; c)—thin sheets only; d)—FT infrared; needed; e)—high resolution solid-state NMR needed; f)—FAB; g)—sampling in the gas-phase; multiple techniques are available for achieving this-even from involatile compounds. UV—ultraviolet; NMR-nuclear magnetic resonance; FT—Fourier transform; GC—gas chromatography; LC—liquid chromatography.

advantage of very high sensitivity which enables it to be used for detecting and analysing very small amounts of substances in mixtures, particularly when used in the GC/MS combination.

The spectroscopic literature on structure determination in organic chemistry is well integrated [1], and such a chemist has available to him moderately-priced equipment for the UV/visible, infrared, ^{1}H and ^{13}C NMR, and mass spectroscopies. These should be used in conjunction. The biochemist or pharmaceutical chemist, who deals with large organic or biomolecules, should be able to make economically-worthwhile use of even the expensive high-field (superconducting) NMR facilities.

Inorganic chemists, with a particularly wide range of structural problems to solve, can make profitable use of most spectroscopic facilities that can be afforded. FTIR, FTNMR and higher-mass mass spectrometries all find important applications. Also for the inorganic chemist the spectroscopic facilities can often be very profitably combined with X-ray crystallography.

For the analytical chemist also, only cost is likely to limit the spectroscopic facilities that are desirable. Computer-based UV/visible and infrared spectrometers have greatly helped in the analysis of mixtures with few components and the detection and identification of minor impurities. If complex mixtures have to be analysed then GC/IR and the even more expensive GC/MS systems have much to offer. Modern FTIR facilities also provide the capacity to obtain spectra from solid samples with the minimum of (or no) sample preparation.

The industrial chemist working in the plastics industry will find the versatility for sample-handling, and the sensitivity, of FTIR to be particularly valuable. Also ^{13}C NMR spectroscopy (including the more expensive high-resolution solid-state techniques) can give much direct and valuable information on the microstructures of polymers and copolymers. Also, of importance in industry, the catalytic chemist can profitably study and characterise the heterogeneous catalysts themselves, or the molecules chemisorbed on them, by means of FTIR spectroscopy.

Nowadays the purchasing and servicing of spectroscopic facilities make up a high proportion of the financial resources needed by an active chemical research or analytical laboratory. However, over the past few decades, these more sophisticated facilities have added whole new dimensions to the scope and applications of even the well-known spectroscopic methods, enabling them to be applied to an ever-widening range of chemical problems.

8   References

1. Williams DH, Fleming I (1987) Spectroscopic methods in organic chemistry 4th edn, McGraw-Hill, London
2. Coblentz WW (1962) Investigations of infra-red spectra, The Coblentz Society, Washington (reissue of Publication No. 45 of the Carnegie Institute of Washington, Washington, 1905)
3. Lecomte J (1928) Le Spectre Infrarouge, les Presses Universitaires, Paris.
4. Kohlrausch KWF (1972) Ramanspektren, Heyden, London (a reprint of the edition by Becker and Erler, Leipzig, 1943)
5. Long GJ (ed) (1984) Mössbauer spectroscopy applied to inorganic chemistry, vol. 1, Plenum, New York
6. Tertian R, Claisse F (1982) Principles of quantitative X-ray fluorescence analysis, Heyden, London
7. Rao CNR (1961) Ultraviolet and visible spectroscopy, Butterworths, London
8. Baranska H, Labudzinska A, Terpinski J (1987) Laser Raman spectroscopy analytical applications, Horwood, Chichester, UK
9. Clark RJH, Hester RE (eds) (1983–85); Advances in infrared and Raman spectroscopy, Wiley, Chichester, UK and Heyden, London: articles on Resonance Raman Spectroscopy in vols 11–13
10. Creighton JA, Cotton TM (1988) in: Creaser CS, Davies AMC (eds) Spectroscopy of surfaces, Wiley, Chichester, UK (Advances in Spectroscopy, vol 16)
11. Creaser CS, Davies AMC (eds) (1988) Analytical applications of spectroscopy, Royal Society of Chemistry, London (several articles).
12. Bellamy LJ (1975) The infrared spectra of complex molecules, 3rd edn, Chapman and Hall, London
13. Nakamoto K (1986) Infrared and Raman spectra of inorganic and coordination compounds, 4th edn, Wiley-Interscience, New York
14. Willis HA, Van der Maas JH, Miller RGJ (1988) Laboratory methods in vibrational spectroscopy, Wiley, Chichester, UK
15. Chesters MA (1986) J. Electron Spectros. Relat. Phenom. 38:123
16. Young RP, Jones RN (1971) Chem. Rev. 71:219
17. Finch A, Gates PN, Radcliffe K, Dickson FN, Bentley FF (1970) Chemical applications of far infrared spectroscopy, Academic, London
18. Sugden TM, Kenney CN (1965) Microwave spectroscopy of gases, Van Nostrand, London
19. Symons M (1978) Electron spin resonance spectroscopy, Van Nostrand Reinhold, New York
20. Akitt JW (1983), NMR and chemistry, 2nd edn, Chapman and Hall, London
21. Harris RK, Mann BE (1978) NMR and the periodic table, Academic, London
22. Smith JAS (1986) Chem. Soc. Rev. 15:225
23. Howe I, Williams DH, Bowen RD (1981) Mass spectrometry, principles and applications, 2nd edn, McGraw-Hill, New York
24. Hill HC (1978) Introduction to mass spectrometry, Academic, New York
25. Brundle CR, Baker AD (eds) (1977–78) Electron spectroscopy-Theory, techniques and applications. vols 1 and 2, Academic, London

26. Briggs D, Seah MP (1983) Practical surface analysis by Auger and X-ray photoelectron spectroscopy, Wiley, Chichester UK
27. Ibach H, Mills DL (1982) Electron energy loss spectroscopy and surface vibrations, Academic, New York
28. Hansma PK (ed) (1982) Tunnelling spectroscopy, Plenum, New York
29. Willis BTM (ed) (1973) Chemical applications of thermal neutron scattering, Oxford Univ. Press, Oxford
30. Winick H, Doniach S (eds) (1980) Synchrotron radiation research, Plenum, New York
31. Teo BK, Joy DC (1981) EXAFS spectroscopy, Plenum, New York
32. Andrews DL (1986) Lasers in chemistry, Springer, Berlin, Heidelberg, New York
33. Guyot-Sionnest P, Hunt JH, Shen YR (1987) Phys. Rev. Letts 59:1597

CHAPTER 2
Recent Advances in Vibrational Spectroscopy

H.A. Willis and D.B. Powell

1 Introduction

The principal techniques of vibrational spectroscopy, infrared absorption and Raman scattering have been practised for many years. They are general methods, giving qualitative and quantitative information directly on solids, liquids and gases, and on both the physical and chemical nature of the samples.

Many spectroscopic methods have been developed subsequently, which have the ability to produce more information on particular samples than either IR or Raman, but at the expense of the range of applicability. Thus, mass spectrometry has higher sensitivity but it can be applied only to the vapour phase, and involatile samples must be physically or chemically modified before examination. XPS (X-ray photoelectron spectroscopy) and related techniques also have higher sensitivity but depend upon decomposing the sample before examination; the information content of the spectrum is low and indeed it is often supplemented by mass spectrometry, as in Secondary Ion Mass Spectrometry (SIMS). NMR is usually able to give far more detailed structural information than IR or Raman, but it generally requires larger samples, and the analysis of the spectra of material in the solid phase presents considerable difficulty.

The survival of IR and Raman spectroscopy depends upon their wide application range and the richness of the spectra they produce. Sensitivity improvements (i.e. improved signal/noise ratio) in these techniques in recent years have extended this range, so ensuring that vibrational spectroscopy maintains its position as an essential laboratory technique. This increased sensitivity has been particularly welcome in dealing with inorganic materials and has been exploited in a number of ways.

a) *Poor IR transmission.* In IR spectroscopy, samples with very poor transmission properties (i.e. $\leqslant 1\%$ transmission) can yield useful spectra. This applies generally to the newer ratio-recording instruments. FTIR (Fourier Transform IR) instruments are even more powerful in dealing with low radiation levels, but with reservations. The FTIR instrument works well on those samples in which heavy light loss occurs throughout the spectrum, i.e. where the beam is attenuated by obscuration, scattering by particulate fillers, or by featureless absorbers such as carbon black. It does not apply to samples which are good transmitters in the part of the spectral region observed and poor transmitters in other regions. This is a consequence of the interferometric measurement technique used. Energy from all the regions observed is measured simultaneously. The large signals reaching the detector from regions of high transparency swamp the weak signals from regions of high attenuation, leading to substantial distortion of the final spectrum. This problem can be overcome by optically attenuating the signal in transparent regions, e.g. with an optical filter.

b) *FT Raman spectroscopy*. Raman spectroscopy with a visible radiation source has been difficult to apply to fluorescent samples, because the weak Raman signal is difficult to detect in the presence of the large fluorescence signal. The ratio of Raman signal to the interfering fluorescence signal can, in general, be improved by increasing the wavelength of the exciting line, e.g. by use of a neodymium/YAG laser emitting at $1.06 \mu m$. Unfortunately, this reduces the intensity of the Raman signal, but this energy loss can be recovered by using a near-infrared interferometer rather than a dispersive spectrometer such as has been used conventionally for recording Raman spectra. Raman spectroscopy is described in detail in Chapter 9.

c) *Data handling and library search*. With improved quality of spectral data, it is possible to manipulate the data to improve its presentation. Examples are background correction, smoothing, enhancement of resolution to separate overlapped features in the spectrum, and difference spectroscopy to reveal the spectra of minor components present in the spectra of mixtures. These processes can be useful both qualitatively and quantitatively. The identification of substances from their IR or Raman spectra depends upon high precision in the recording of the spectral data both in frequency (horizontal scale) and intensity (vertical scale). The great improvement in the quality of the data in databanks has put identification of unknown from their spectra on a much more secure and reliable basis.

d) *Reflectance and photoacoustic spectroscopy*. Useful IR spectra can now be measured by reflectance and photoacoustic spectroscopy. These methods in which the signal level is low can be very convenient for the characterisation of samples such as aqueous solutions, bulk solids and powders.

e) *Quantitative analysis*. Quantitative precision in IR spectroscopy has improved to the extent that it is now favoured in laboratory and process control in many cirumstances over conventional analytical procedures.

f) *Fast spectroscopy*. Spectral information can now be obtained very rapidly. Both Raman and IR can measure reactions and processes occurring in microseconds. In Raman spectroscopy, sub-nanosecond time resolution has been achieved.

g) *Microsampling*. Information can now be obtained, by both IR and Raman spectroscopy, on very small samples down to microgram, or in some cases picogram, quantities.

In this Chapter recent developments in the IR technique are described; corresponding developments in Raman spectroscopy are considered in Chapter 9. We detail each of the points listed above with examples, with the exception of diffuse reflectance, multiple internal reflectance and photoacoustic spectroscopy described in Chapter 3. An area of growing interest, specular reflectance, is additionally dealt with below.

2   Data Handling and Library Search

2.1   Data Handling

Most modern data stations in the infrared use suitable PC's of good quality on which the appropriate programs are installed. Hard disks are now generally employed because of their greater capacity, but some systems with twin floppy disks are also used. It is highly desirable that the system incorporates a colour monitor as this makes spectrum comparison and manipulation much easier. Computer controlled systems are essential for FT instruments and there are few dispersive instruments now sold without data stations. The main functions of the data handling programs are control of the spectrometer scanning conditions, spectrum manipulation, storage of spectra obtained and of reference spectra, and library search programs to assist in spectral identification.

2.1.1   Spectrometer Scanning Conditions

The conditions used for obtaining a spectrum can be set using the data station— e.g. scan time, slit program, noise suppression etc. Two advantages arise from this control.

a) Spectrometer conditions can be varied and the resulting spectrum viewed on screen before plotting the spectrum. When a satisfactory spectrum is obtained it can then be plotted in various formats using a plotter or printer.

b) The conditions used can be stored and instantly recalled and used for future samples under exactly the same conditions.

2.1.2   Spectrum Manipulation

Once a spectrum has been obtained it can be displayed over various ranges and formats. It can be converted to absorbance from % transmission or displayed as various order derivative spectra. Weak spectra can be expanded, or strong spectra reduced, and the spectrum can be smoothed or flattened to remove excessive noise or correct a sloping background arising from light scattering. While these latter changes are largely cosmetic, they can be of value when spectral comparisons are being made. At any stage in this process the modified spectrum can be stored or plotted. As in many cases the spectrum is not plotted on printed chart paper, one useful facility is to be able to locate peak positions and absorbances and obtain a printed record of the exact values. It is important to be able to distinguish peaks from noise and the best systems only record peaks above a chosen noise level.

Another valuable facility is to obtain quantitative data using a selected peak or peaks. The data system will automatically calculate peak heights and areas of a series of samples and relate these to a standard of known concentration, giving a direct read-out of concentrations.

2.1.3 Storage of Spectra

Storage of spectra is greatly facilitated by use of computers. Complete spectra can be stored on a hard disk system, for convenience in a series of directories or libraries. These can be dumped in convenient groups on to floppy disks. With twin floppy disk computers the spectra are saved directly on to floppy disks. With all these systems any spectra can be directly accessed and displayed for comparison with the spectra of unknown samples. Sets of reference spectra can be stored and, for example, in solution work there is no need to scan the solvent with each sample, as reference solvent spectra can be loaded into the computer and the solvent bands subtracted from the solution, as described in section 2.3.

2.1.4 Library Search

An important development arising from the storage of reference spectra on floppy disks is the development of library search programs. These are programs which will automatically search through a set of library spectra and find the best match with a given sample spectrum.

Design of the programs for library search has many pitfalls. The essential feature of any such program is to match peak frequencies and relative absorbances in the sample spectrum with those of the reference spectra. Various algorithms can be used to obtain a match, e.g. simple subtraction, least squares fit, subtraction of first derivatives etc. The result of the computer matching will come out as a hit-list table with the most probable match at the top. A simple match of peak frequencies and absorbances often gives problems, particularly when the comparison and sample spectra differ in quality. Some systems allow for the 'blanking out' of uncharacteristic parts of the spectrum, e.g. the hydrocarbon peaks in Nujol mull spectra. It is also possible to de-resolve spectra to the rather low resolution of most spectra libraries before making a match. Low resolution spectra are used in reference libraries as the smaller number of data points required permits holding more spectra.

Every practical spectroscopist knows that in comparing spectra in order to obtain a match, band shapes are as important as frequencies and absorbances. Some of the more sophisticated search programs take such features as bandwidths into the search algorithm. The problem with all search programs is to select criteria which are narrow enough to correctly define a spectrum and yet broad enough to allow for spectral differences arising from instrumental and sampling differences. In all search procedures, final selection is always made by visually comparing the sample spectrum with spectra from the search library using the overlay procedure on the screen. Most search libraries are centred around compounds which are commonly met in the area of work concerned. These libraries can be, for example, from paints, plastics or drugs etc., or from chemical types of compound, e.g. heterocyclics, aromatics or halogen derivatives. However, the range may be broader and in some cases the complete catalogue of a chemical supplier is stored as a library.

Library search can be extremely valuable but must be used with discretion as sometimes the most unlikely matches come top of the hit-list. The problem is that the computer may not recognise as important a relatively weak feature that is in a particularly characteristic part of the spectrum. For example, a relatively weak but sharp band at 2250 cm^{-1} would be instantly recognised by the polymer spectroscopist as being due to the nitrile group, whereas the computer may not 'understand' its significance. In general, library search works best when the library is based on a limited range of compounds related to the systems being studied.

2.2   Resolution Enhancement

IR spectra are frequently measured on liquids or amorphous solids. Gases at low pressure and ordered solids show very narrow absorption bands. Because of the disorder in liquids and amorphous solids absorption bandwidths increase, often to the extent that information within the spectrum is concealed. There are a number of mathematical procedures [1] by which the bandwidth may be reduced and, in appropriate circumstances, useful information may then be obtained.

An example given by Maddams and Willis [1], is shown in Fig.1. These are IR spectra in the carbonyl region of oxidised high-density and low-density polyethylene. The broad carbonyl band in the spectrum as measured in each case is shown by resolution enhancement to consist of the same four peaks (1724, 1719, 1715 and 1697 cm^{-1}), but of considerably different relative intensities in the two cases, demonstrating that the oxidation process depends upon polymer structure. Many other examples are given in the reference quoted [1].

Resolution enhancement, correctly applied, does not affect the quantitative validity of the spectrum. Thus Willis et al. [2], have determined the relative concentrations of different ethylene sequences in propylene/ethylene copolymers working on the resolution enhanced IR spectrum.

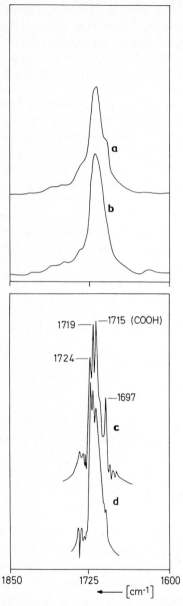

1719 —— —1715 (COOH)

1724—

—1697

c

d

1850   1725   1600
                    ◄—— [cm⁻¹]

Fig. 1. Detection of various carbonyl containing species in oxidised polyethylene by resolution enhancement of the IR spectrum in the C═O region.
a) IR spectrum of oxidised high-density polyethylene in the $1700 \, \text{cm}^{-1}$ (C═O) region.
b) As a) but for low-density polyethylene
c) Spectrum a), resolution enhanced by Fourier self-deconvolution.
d) Spectrum b), resolution enhanced by Fourier self-deconvolution, using the same parameters as (c) so as to be strictly comparable.
Data from Ref. [1] reproduced with permission.

2.3   Difference Spectroscopy

In difference spectroscopy, the object is to exploit the very high signal/noise ratio to observe both qualitatively and quantitatively weak features due to minor components in the presence of substantial levels of absorption by a major component.

An example is the identification of a plasticiser in PVC film. A portion of the film is placed in methanol, which extracts some of the plasticiser. The spectra of the unextracted and extracted films are recorded, (Fig. 2). Differences in the spectra due to sample thickness are accommodated by converting both spectra from a transmittance to an absorbance ordinate. The ordinates can then be scaled so that the bands of one component exactly cancel; in this case the clear bands due to PVC near $650 \mathrm{cm}^{-1}$ are cancelled: the bands remaining in the difference spectrum are due to the plasticiser, which can now be identified without difficulty [3]. When the amount of minor component is small, there is usually a problem in selecting the scaling factor to cancel precisely the major component spectrum.

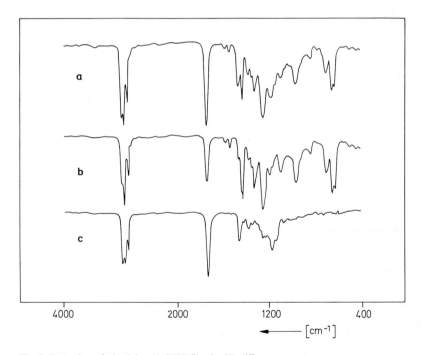

Fig. 2. Detection of plasticiser in PVC film by IR difference spectroscopy
a) Spectrum of film with high plasticiser content.
b) Spectrum of film with low plasticiser content.
c) Difference spectrum with PVC bands eliminated.
Data from Ref. [3] reproduced with permission

This arises through minor but significant changes which appear in the spectrum due to refractive index effects and component interactions such as hydrogen-bonding. Also negative bands appear if the major component is over-compensated. This may be overcome by varying the scaling factor systematically and recording the sequence of spectra. Hannah [4] shows the practical effect of this (Fig. 3). The bands marked A, B, C and D in the spectra, which remain effectively the same while the rest of the spectrum changes, arise from an impurity in one of the samples.

Difference spectroscopy is also appropriate for the quantitative determination of minor components in materials. Figure 4 shows an example, the determination of the methyl content of polyethylene. The band at $1380\,\mathrm{cm}^{-1}$ arises from methyl groups; those at $1355\,\mathrm{cm}^{-1}$ and $1368\,\mathrm{cm}^{-1}$ arise from methylene groups in a gauche conformation, and the intensity of these latter bands varies substantially in the spectrum of a particular polyethylene according to the thermal history of the sample examined. This problem can be overcome by subtracting the spectrum of a polyethylene free of methyl groups, and scaling the spectra appropriately to remove the methylene bands exactly. Examination of a series of samples of known methyl concentration gives an excellent linear relation between methyl content and absorbance per mm thickness of sample measured on the difference spectra at $1380\,\mathrm{cm}^{-1}$.

The use of a computer to obtain difference spectra means that in solution work it is no longer necessary to obtain an exact balance with the pure solvent since the spectra of the solution and solvent can be displayed together on the screen and a subtraction of the solvent peaks made, either automatically or visually with the operator choosing the peaks to be cancelled.

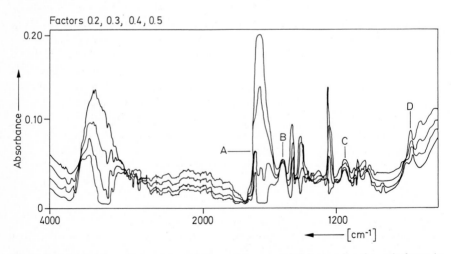

Fig. 3. Real bands of a minor component in a mixture revealed by varying the scale factor in difference measurements. Reproduced from Ref. [4] with permission

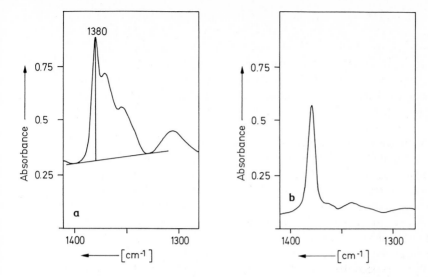

Fig. 4.
a) IR spectrum of polyethylene in the region of the band at $1380\,\mathrm{cm}^{-1}$ due to the methyl group.
b) IR spectrum of polyethylene in the $1380\,\mathrm{cm}^{-1}$ region after removal of bands due to $-CH_2-$ groups by subtracting the spectrum of a sample of polyethylene of very low methyl content

3   Reflectance Methods

Reflectance methods have been developed which will deal with a number of difficult condensed phase samples (Chapter 3), but hard solids with specularly reflecting surfaces (e.g. inorganic crystals, polymer fabrications) have presented difficulties. Recently programs have been written [5] using the Kramers-Kronig dispersion relations which permit absorbance spectra to be calculated from specular reflectance spectra. Although this spectrum is weak, the signal/noise ratio is high enough to permit calculation from the refractive index (n) and absorbance (k) spectra. Spectra from polymethyl methacrylate sheet[5], namely the measured specular reflectance spectrum and the calculated n and k spectra, are given in Fig. 5. Note the close similarity of the calculated spectrum to the absorbance spectrum measured in the usual way. The method appears to deal satisfactorily with polymer fabrications heavily filled with carbon black.

4   Quantitative Analysis

IR quantitative analysis has been practised for many years, but until recently it has not been the usual method of choice, in spite of its relative speed and simplicity, because of its relatively poor precision. Recent improvements in

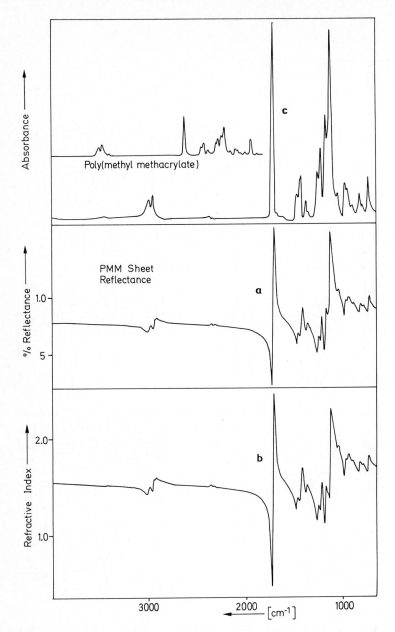

Fig. 5. Absorbance and refractive index spectra calculated from the specular reflectance spectrum of polymethyl methacrylate sheet
a) Specular reflectance spectrum.
b) Calculated refractive index spectrum.
c) Calculated absorbance spectrum. (Insert: Library reference spectrum of polymethyl methacrylate).
Data from Ref. [5] reproduced with permission

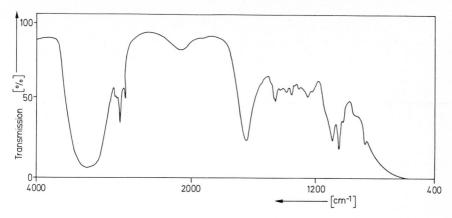

Fig. 6. Spectrum of a liquid laundry detergent, measured by multiple internal reflection

instrumentation have completely changed this picture resulting in improvements in both precision and detection limits.

The sensitivity of FTIR instruments makes it possible to analyse multicomponent mixtures in aqueous solution. An example is the analysis of a liquid laundry detergent with the spectrum shown in Fig. 6. By reference to the spectra of known mixtures, the concentration of individual constituents in the mixture can be determined by solving a series of simultaneous equations by a matrix method. The results in two cases shown in Table 1 demonstrate that such a mixture can be analysed to high precision.

Another important area is the analysis of samples which are not fully characterised chemically, but which can be characterised in terms of other standard tests and properties. The implication is that running a relatively rapid IR spectrum can be used as a substitute for a whole series of time-consuming tests. An example of this approach is the analysis of coal [6]. A large number of samples

Table 1. Analytical results obtained from the spectra of liquid detergent solutions

Sample being evaluated: mix 02-liquid detergent		
Component	Conc. in %	Actual
1) Sulphonate	9.73	9.71
2) Non-ionic	12.6	12.61
3 Water	66.94	66.98
4) Ethanol	10.38	10.37

Sample being evaluated mix 04-liquid detergent		
Component	Conc. in %	Actual
1) Sulphonate	10.68	10.78
2) Non-ionic	11.25	11.22
3) Water	67.56	67.88
4) Ethanol	9.73	9.74

Table 2. Correlations for 43 coking coals using factor analysis of the 2100–300 cm^{-1} region (26 factors) [6].

Property	Range of values	Coefficient of determination (R^2)	Standard deviation
Carbon (wt %)	70.9 –86.9	0.990	0.39
Organic hydrogen (wt %)	4.20– 5.73	0.936	0.12
Total sulphur (wt %)	0.25– 0.233	0.868	0.10
Ash (wt %)	2.5 –16.2	0.985	0.42
Specific energy (MJ/kg)	29.58–35.10	0.935	0.39
Fluidity	0– 4	0.251	0.93
Hardgrove grindability index	49–99	0.847	6.0
Vitrinite (vol %)	36.6 –81.4	0.905	4.1
Exinite (vol %)	0.0 –16.8	0.935	1.4

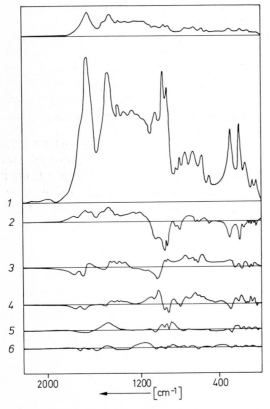

No.	Factor	Loading
1	1	0.1450
2	2	0.2101
3	3	0.0214
4	4	0.0634
5	5	-0.1240
6	6	0.1197

2000   1200 [cm^{-1}]   400

Fig. 7. Factors composing the IR spectrum of coal, measured by diffuse reflectance. Reproduced from Ref. [6] with permission

(in this case 43) were examined by a series of tests (Table 2); the samples were also finely ground and their IR spectra measured by diffuse reflectance. The analytical method is demonstrated in Fig. 7. The top part of this figure is the spectrum measured on one of the samples. By examining the spectra of all the 43 samples, a series of synthetic spectra were generated. The addition of these spectra in particular proportions constitutes the infrared spectrum of one of the unknowns. The synthetic spectra, which can have both positive and negative features, are called factors, and the proportion of each factor which when summed will generate the spectrum of an unknown is described as a 'loading'. The 'loadings' for the factors which make up the spectrum at the top of Fig. 7 are as shown in the figure.

In this case, a total of 26 significant factors were found (Table 2), and in terms of this analysis, the precision with which other properties could be predicted is shown. Clearly the IR spectrum is excellent in predicting some properties (% carbon, % hydrogen, % ash) but gives no useful information on some other properties. This is a very important development because the analysis is achieved without a complete chemical characterisation of the material, or by recourse to the preparation of synthetic mixtures. We may expect that Factor Analysis, and related computational methods, will significantly increase the range of samples which may be analysed by quantitative IR methods.

FTIR spectroscopy is now making an important contribution in process control applications. This is at the moment applied to high value products to justify the comparatively high cost of the equipment. An example of a widely used analysis is concerned with silicon wafers for semiconductor applications. These

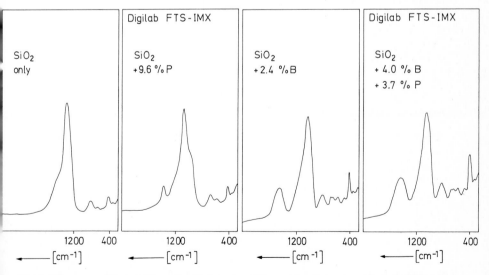

Fig. 8. IR spectra of doped silicon dioxide layers on silicon chips. Reproduced from Ref. [7] with permission

wafers are coated with silicon dioxide doped with phosphorus and boron. These latter elements may be determined by measuring the transmission spectra of the silicon wafers by FTIR [7]. Spectra of some wafers are shown in Fig. 8. Phosphorus is measured by means of a band near $1330\,\mathrm{cm}^{-1}$ and boron from the band at $1400\,\mathrm{cm}^{-1}$.

5 Fast Spectroscopy

It is possible to follow fast reactions of physical changes relatively easily by Raman spectroscopy. This is discussed in Chapters 8 and 9. In infrared studies, where an FTIR instrument is normally used, it is necessary to observe one complete interferogram before any data can be recovered, giving a minimum response time of some hundreds of milliseconds. However, it is possible to reduce this time interval substantially by utilising a form of stroboscopic measurement, with a reaction which can be repeated at short time intervals, and with precise synchronisation between reaction initiation and interferometric scan initiation [8]. This is demonstrated diagrammatically in Fig. 9 where it is presumed that the event cycle is such that reaction complete within the time period required to record 5 data points.

Interferograms are recorded in a succession of files, as shown; the experiment is arranged so that data collection is offset by one point for each successive scan. Individual files do not contain useful data, since the points represent data measured at a large number of different time intervals following the initiation of the experiment. However, it will be seen that by combining the first point from the first file with the second point from the second file, and so forth, a new series of

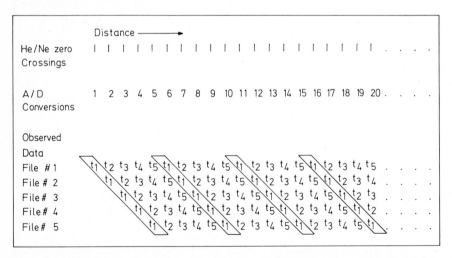

Fig. 9. Diagrammatic representation of 'stroboscopic' FTIR measurements. Reproduced from Ref. [8] with permission

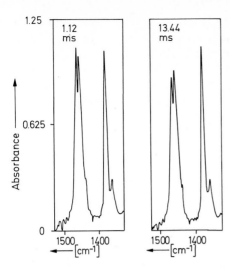

Fig. 10. Spectra measured on polypropylene film while being stretched. Reproduced from Ref. [8] with permission

files can be constructed representing complete interferograms for times 1, 2, 3 and so on. At least in principle, interferograms can be obtained at repetition times which represent the time interval between successive points in the interferogram, rather than the time intervals between successive interferograms, and a time resolution in the millisecond range is possible.

There are severe practical limitations to this apparently admirable scheme because any timing errors, source intensity fluctuations or other variations in the experiment can have a catastrophic effect on the reconstituted interferograms. Nevertheless, Molis, MacKnight and Hsu [9] have conducted experiments with a time resolution of about 100 milliseconds.

One problem is to find a system which can be repeatedly cycled; a popular system has been to extend and relax a polymer specimen and observe the corresponding changes in molecular conformation. Graham, Hammaker and Fateley [8] have published time-resolved spectra measured during the extension of polypropylene film with a time resolution of a few milliseconds. Fig. 10 shows spectra measured after 1.12 and 13.44 milliseconds, which are clearly significantly different. It may be concluded that IR spectroscopy with millisecond resolution is now perfectly feasible.

6 Microsampling

Microsampling by IR spectroscopy has been the subject of study for many years [10], although early measurements with insensitive apparatus were not particularly successful. They did, however, demonstrate the general principle of microsampling systems, which is to reduce the beam area in the sample plane by a mirror system, (Figs. 11a and 11b), and this kind of device has proved very valuable in both ratio-recording and FTIR instruments.

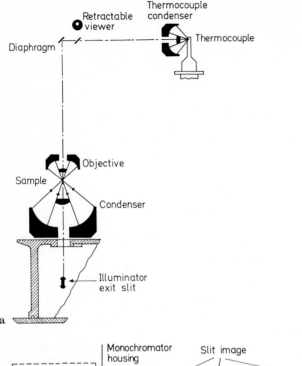

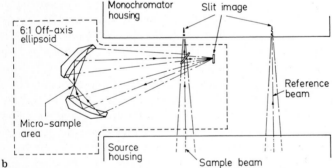

Fig 11. Beam condensers for reducing the size of the focal spot of the IR beam.
a) Cassegrain beam condenser. (Reproduced with permission from J. Coates et al., (1953) J. Opt. Soc. Am. 43:984
b) Off-axis ellipsoid condenser. (Perkin-Elmer Corp)

6.1   Microsampling in Chromatography and Thermogravimetry

This approach is now used in the direct (on-line) examination of fractions eluting from a gas chromatography column [11]. The cross sectional area of the gas cell is reduced to one or two square millimetres (Fig. 12) and in this way with a given quantity of gas a significant increase in path length is possible. In an FTIR

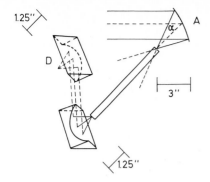

Fig. 12. Light pipe and associated optics for direct measurement of gas chromatography fractions. Collimated radiation from an interferometer is focused by mirror A($\alpha = 45°$) onto the light pipe. The emerging radiation is collected and re-focused by a paraboloid onto the detector D. Reproduced from Ref. [11] with permission

instrument, a beam which is circular in cross section at the sample focus can be used entirely, while with a dispersive instrument only a slit-shaped light patch can be used, the rest being blocked off by the slit jaws. This factor, combined with the greatly enhanced sensitivity of FTIR instruments, means that these are almost exclusively used in this work. The cells used are of light pipe form, i.e. the walls are internally gold plated to reflect the light beam forward through a cell of between 10 cm and 30 cm length. Considerable effort has been made to optimise the design [12]. An 'IR chromatogram' is obtained by transforming all the interferograms and plotting the total area of the successive spectra against time. This chromatogram is very similar to that measured with the conventional flame ionisation column detector. This work has now reached the stage at which GC apparatus can be purchased with a dedicated FTIR detector [13]. Sensitivities for components leaving the GC column are in the nanogram range.

An alternative approach to GCIR is the Mattson Cryolect (Fig. 13) [14]. Gas from the GC column is directed from a fine jet towards a reflecting metal drum at 12 K. The gas immediately solidifies as a thin layer which is examined by 'transflectance' [13]. The sample is thus confined to an extremely small area and, because it is examined at low temperature rather than at the elevated temperature required in the conventional GC/FTIR apparatus, the bandwidth of the bands is decreased, and the height increased, thus giving a further sensitivity improvement. Detection limits comparable with those of GC/MS are claimed.

Attempts have been made to use FTIR as a detector for liquid chromatography, but flow-through methods as applied in GC/FTIR suffer from severely reduced sensitivity because of absorption by the liquid solvent [15]. The situation is more satisfactory in supercritical fluid chromatography/FTIR (SFC/FTIR) because mobile phases can be chosen, particularly carbon dioxide, which have relatively useful regions of transparency in the mid-IR spectrum [15]. There are, however, technical problems with the construction and maintenance of IR cells which will resist the relatively high pressure at room temperature of the supercritical fluid. There has therefore been more effort directed to collecting fractions from LC and SFC with evaporation of the mobile phase. Again SFC

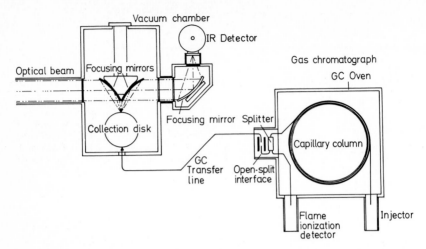

Fig. 13. The 'Cryolect' system for measuring IR spectra of gas chromatography fractions as solids at 4 K. Reproduced from Ref. [14] with permission

proves more satisfactory because the mobile phase is easily removed. By good design of the evaporator, the sample may be deposited on a very small area of a transparent plate, to be examined on an IR microscope [16] as described below.

The detection of microquantities of gas has been developed for the on-line examination of gases evolved in thermogravimetric analysis (TGA) [17]. In the TGA/FTIR technique the loss in weight of a sample on heating is measured as a function of heating time either at constant temperature or, more usually, with a constantly increasing temperature, e.g. 20 °C increase per minute. The weight loss curves obtained are highly characteristic of the sample, but it is also very useful to be able to identify the volatile material. In TGA/FTIR, the TGA unit is modified so that the volatiles are collected for transfer to an IR gas call. An 'IR chromatogram' is obtained similar to that in GC/FTIR, which is remarkably similar in character to the first derivative of the TGA weight-loss curve. Suggested uses of the technique include characterisation of polymer compositions [18] and the qualitative and quantitative analysis of residual solvents in pharmaceutical preparations [19].

6.2   Infrared Microscopy

In the examination of small samples with devices such as those shown in Fig. 11a and b, the area of the sample at the beam focus may be smaller than the area of the beam and radiation may bypass the sample, travelling straight through the instrument to the detector. This problem has been overcome in practice by mounting the sample over, or in, a pin-hole or slit in a metal foil [20]. Although a

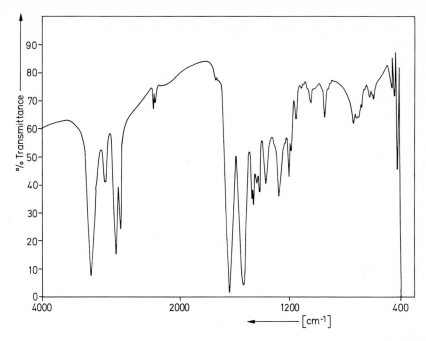

Fig. 14. IR spectrum of a nylon 66 filament measured in a slit aperture. Reproduced from Ref. [20] with permission

slit is not the optimum shape for an FTIR instrument, it is particularly useful for mounting short fibre samples, as is a requirement in forensic science studies. Figure 14 provides an example.

An extension of this technique is to replace the beam condenser by a reflecting microscope (Fig. 15). Here the sample is mounted on a stage with X, Y and Z motions. A small rooftop prism can be pushed into the beam to permit the sample to be aligned by visual observation. The problem of restricting the field to prevent radiation bypassing the sample is overcome by introducing an additional beam focus between the sample and the detector. Four moveable slit jaws are mounted at this focus to obscure radiation passing through the object plane but not through the sample.

There is still a problem with the diffraction limit, i.e. when the least dimension of the sample normal to the radiation propagation direction falls below the wavelength of the radiation being used. At one time it was supposed that measurement would be impossible if the wavelength exceeded this dimension. This is certainly not true in practice, but the distribution of radiation in the sample plane is changed by diffraction, and far more radiation bypasses a minute sample. A complete correction for diffraction is obtained in the pin-hole method but the 'second focus' method is less effective. A useful improvement is obtained

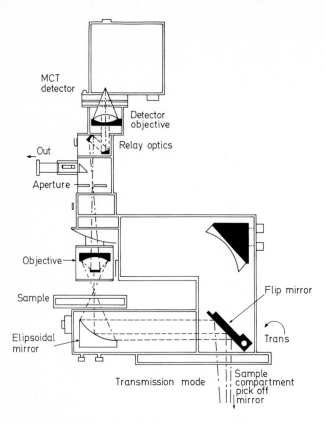

Fig. 15 Schematic diagram of IR microscope. Reproduced with permission from P.R. Griffiths, European Spectroscopy News, No. 64, Feb/March, 1986, p 8

by introducing yet another focal point, this time between the light source and sample, and placing another set of field-restricting slit jaws at this point. This is usually called a 'redundant aperture'.

The IR microscope is arranged so that both transmission and reflection studies are possible. This apparatus has introduced a new dimension into IR microsampling since spectra can be recorded at a series of pre-determined points arranged to fall on a straight line so that in effect a cross-section view is obtained [21]; the output can be presented as a stack plot as in Fig. 16 which represents the cross section across a film which contains a small included impurity. The successive spectra represent points 2.5 μm apart. It is noted that while all views have the characteristics of the polyester film, a band at about 3400 cm^{-1} grows and then declines. This indicates that the impurity is an amide, probably a crystal of slip agent. A further advantage of the IR microscope over the beam condenser is that visible radiation may be passed through the system and a microphotograph obtained from which the precise part of the sample viewed by IR may be deduced. An interesting example occurred in the examination of a sample of

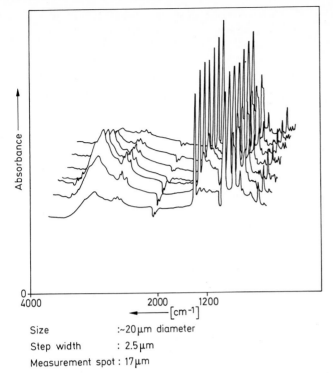

Size   : ~20 μm diameter
Step width   : 2.5 μm
Measurement spot : 17 μm

Fig. 16. Stack plot spectra of polyester film measured in an IR microscope. Reproduced from Ref. [21] with permission

Fig. 17. Microphotograph of mixed crystalline solids. Reproduced from Ref. [22] with permission

phenobarbitol [22] which from the microphotograph evidently contains two crystalline substances (Fig. 17). By closing the apertures the IR spectra of the square crystals (Fig. 18a) and the feathery crystals (Fig. 18b) may be obtained. Reference to standard reference spectra shows that these are both phenobarbitol, but in different polymorphic forms. It appears, therefore, that IR microscopy is a very significant development, and is due to become a major application of FTIR.

7   Inorganic Applications

Although in principle of technique inorganic compounds are no different from organic compounds, there are some particular problems in the use of infrared with inorganic materials. There are three main points; (a) the presence of heavy atoms increases the need for low-frequency data; (b) there is often a need to handle solid samples, since many inorganic compounds are insoluble in the usual infrared solvents; (c) the study of aqueous solutions presents further problems. Point (a) is a question of spectrometer design, but (b) and (c) depend on sampling methods.

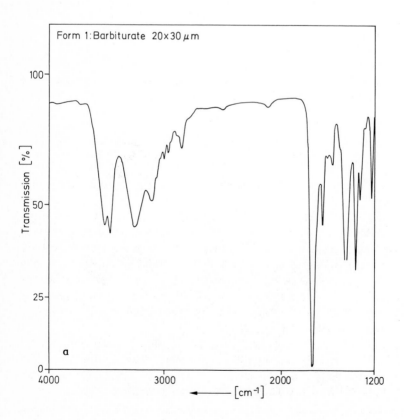

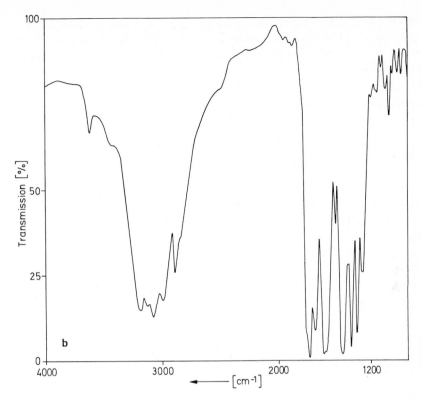

Fig. 18. IR spectra of parts of the sample shown in Fig. 17.
a) Square crystals.
b) Feathery crystals.
Reproduced from Ref. [22] with permission

Most inorganic substances can be handled using the usual halide disc technique, but special care has to be taken to avoid ion exchange with ionic samples. The choice of compound for the disc matrix is important. Many inorganic compounds have high refractive indexes and with such samples the best spectra are obtained using CsBr or CsI rather than the more usual KBr or KCl. Another useful compound is TlBr, which gives excellent spectra but is highly toxic. Aqueous solutions can be studied in transmission, using a cell with CaF_2 or BaF_2 windows. With modern FT instruments, solvent subtraction is relatively easy, giving good spectra of the solute. However, the technique is still limited to 'windows' in the water (or D_2O) spectrum.

Solvent absorption problems are, however, avoided with attenuated total reflectance (ATR) sampling. Several cells have been developed for use with FTIR instruments which give excellent spectra of aqueous solutions and enable quantitative measurements to be carried out. These cells nearly all use a zinc selenide crystal, which is not water sensitive. The three main commercially available cells are (1) the circle, (2) horizontal ATR and (3) square Col cells. In all these cells the solution is simply poured in and the spectrum taken. There are no alignment or solvent cancellation problems except in regions of very strong water absorption above $3000\,\text{cm}^{-1}$ and after use the cell is simply washed with water.

8 Future Developments

The next major advance in IR spectroscopy will depend upon the introduction of cheap tunable lasers. These should lead to at least an order of magnitude improvement in spectral resolving power over the finest FTIR systems, and the greatly increased intensity, if accompanied by good stability, should yield a further substantial improvement in signal/noise ratio.

9 References

1. Maddams WF, Willis HA (1988) Proceedings SPIE 917: 35
2. Willis HA, Chalmers JM, Bunn A, Thorne C, Spragg R (1988) In: Creaser CS, Davies AMC (eds) Analytical applications of spectroscopy, Royal Society of Chemistry, London, p 188
3. Spragg RA (1984) Perkin-Elmer Infrared Bulletin No. 100, Perkin-Elmer, Beaconsfied, Bucks, UK
4. Hannah RW (1988) Mackenzie MW (ed), Advanced in applied Fourier transform, infrared spectroscopy, Wiley, Chichester, p 29
5. Krishnan K (1987) FTS/IR Notes No.51, Biorad Digilab Division, Cambridge, MA
6. Fredericks PM, Lee JB, Osborn PR, Swinkels DAJ (1985) Applied Spectroscopy 39: 303
7. Krishnan K, Kuehl D (1985) Biorad Semiconductor Notes No. 105, Biorad Inc., Cambridge, MA
8. Graham JA, Hammaker RM, Fateley WC (1985) In: Durig JR (ed) Chemical, biological and industrial applications of infrared spectroscopy, Wiley, Chichester, p 301
9. Molis SE, Macknight WJ, Hsu SL (1984) Applied Spectroscopy 38: 529
10. Cole ARH, Jones RN (1952) J. Opt. Soc. Am. 42: 348
11. Yang PWJ, Griffiths PR (1984) Applied Spectroscopy 38: 816
12. Yang PWJ, Ethridge EL, Lane JL, Griffiths PR (1984) Applied Spectroscopy 38: 813
13. The Hewlett Packard HP 5965A infrared detector. Hewlett Packard Ltd., Uxbridge, UK
14. Bourne S   The Cryolect GC/IR system, Mattson Instruments Inc., WI
15. Griffiths PR (1988) In: Creaser CS, Davies AMC (eds) Analytical applications of spectroscopy, Royal Society of Chemistry, London, p 173
16. Pentovey SL Jr, Shafer KN, Griffiths PR, Fuoco R (1986) Journal of High Resolution Chromatography: Chromatographic Communications 9: 168
17. Compton DA (1987) International Labmate, June, p 37
18. Weiboldt RC, Adams GE, Lowry SR, Rosenthal RJ (1988) American Laboratory 20: 70
19. Johnson DJ, Compton DA (1988) Spectroscopy 3: 47
20. Curry CJ, Whitehouse MJ, Chalmers JM (1985) Applied Spectroscopy 39: 174
21. Zachmann G (1986) Bruker Report No.1, p 2
22. FTIR Sampling Technology, Product Update Spring 1988, Spectra-Tech. Inc., Stamford, CT, p 11

CHAPTER 3
Infrared Sampling Methods

P.S. Belton and R.H. Wilson

1 Introduction

Infrared spectroscopy has long been recognised as a powerful technique for obtaining structural information about molecules and for analysis. The traditional limitation of the technique was that samples had to be presented in a manner suitable for transmission of an infrared light beam, so that highly scattering or opaque samples were unsuitable. Thus a very wide range of scientifically and commercially important materials were precluded from study. Whilst transmission methods are still important today, and indeed are used by many spectroscopists as reference methods, newer techniques have become available which have widened the range of possible applications considerably. In this chapter three of these new methods: diffuse reflectance, attenuated total reflectance and photoacoustics are considered in addition to transmission methods. Together these methods represent those most widely used by spectroscopists and can, between them, produce spectra from samples in almost any physical form.

2 Transmission Methods

Before the advent of Fourier transform infrared spectroscopy, transmission methods constituted the most common class of sample presentation techniques. It is true to say that even today they are still important and their use is widespread. Transmission methods can be divided into those for solids involving the use of pellets and mulls, and those for fluids, be they liquids, solutions or gases. The underlying principle is the same for each of these methods. Light is passed through the sample and the difference in intensity between incident and transmitted light is compared. The sample is contained in a cell either neat or in solution, or dispersed in an inert matrix (for pellets) or a liquid such as nujol for mulls. In all cases infrared radiation of intensity I_0 is incident on the sample (Fig. 1). However, not all this energy passes into the sample since there will

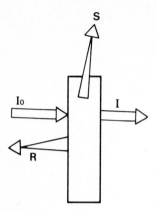

Fig. 1. Illustration of sources of energy loss at a sample. Incident light, intensity I_0, is reduced by reflection R, scattering S and absorption so that detector sees an intensity I

always be a reflected component, R and a scattered component S. The energy measured by the detector, I, therefore depends not only on the absorption of the sample but also the reflectance and scattering power as well. Thus where the Beer-Lambert law applies $I = (I_0 - R - S) 10^{-\varepsilon Cl}$, with ε the molar absorptivity, l the optical path length and C the concentration.

2.1 Solid Samples

Pellet techniques are the most widely used for the examination of solid samples. Typically the procedure involves the grinding of a small amount (approx. 1%) of the sample with an infrared transparent matrix such as potassium bromide to form an intimate dispersion. A quantity of the material is placed in a die and subjected to a large force, typically about 10 tonnes. During this time it is usual to apply a vacuum to the die. The resultant pellet or disc is mounted for analysis in a special holder in the infrared beam. There are several pitfalls with this technique. Firstly, the sample must be dry and the disc must not be allowed to come into contact with water vapour or it will fog. In practice, this can occur quite rapidly in a normal laboratory atmosphere. Secondly, the pressed disc must be fairly uniform with no areas of residual unpressed powder or artifacts will appear in the spectrum. Pellet techniques can also be extremely difficult to prepare for quantitative analysis since the concentration of sample in the matrix and the thickness must be both known and reproducible. This is clearly quite difficult and requires a posteriori measurement.

An alternative method for quantitative analysis involves the preparation of a mull. The solid is dispersed in a material such as nujol, a paraffin, and this is sandwiched between two salt plates. However, the spectrum exhibits strong additional bands arising from the dispersing medium, so that the available spectral range is limited.

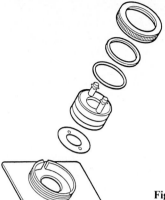

Fig. 2. Exploded view of a transmission cell. Courtesy of Spectra-Tech Europe Ltd.

2.2 Liquid Samples

In general a liquid cell consists of two windows separated by a spacer of fixed or variable length (Fig. 2). The precise design may vary. If the spacer is of fixed thickness and the cell is cemented together then a sealed cell results. Alternatively all the components may be detachable so that the cell can be cleaned and different spacers used. These cells are known as demountable cells. The third type of cell is one where the spacing of windows can be varied continuously using a moving piston arrangement.

Filling in nearly all cells is accomplished by injection via luer-type fittings; liquid cells are usually used for the analysis of solutions and reasonable quantification can be achieved. However the main problem is ensuring that the cell pathlength remains constant throughout a series of analyses. This is particularly important when the solvent is water. The strong absorption of water in the infrared necessitates the use of very short ($< 10\,\mu$m) pathlengths which are not easy to reproduce. Even at this pathlength the water will substantially mask a large part of the infrared spectrum. Aqueous work also involves the use of more exotic window materials such as zinc selenide or silver chloride since most salt windows are water soluble.

Neat liquids may be qualitatively sampled by direct compression between two windows in a similar fashion to the mull technique-however this procedure does not lend itself to quantitative analysis.

2.3 Variable Temperature Studies

It is possible to obtain commercial variable temperature cells for solid and liquid samples. For work above ambient temperature, simple cells are available with hot water jackets or electrical heaters. For sub-ambient work the cell is normally contained in a cryostat which can be cooled with liquid nitrogen. Intermediate (77 K–273 K) temperatures are obtained by electrically heating against the cooling effect of the liquid nitrogen. Condensation on the windows is prevented by evacuation of the cryostat and the use of heated windows in the cryostat chamber containing the cell.

2.4 Gases

Gases are usually analysed in long pathlength (10 cm) or special multipass cells in order to provide the necessary pathlength (approx. 10 m) required at low concentrations. The pressure of the gas is critical, particularly for quantitative work where pressure broadening of lines can result in large deviations from the Beer-Lambert law.

2.5 Disadvantages

Transmission methods generally demand a certain amount of sample handling and preparation. Liquid cells require the dissolution of the sample in solvent and rely on the maintenance of a fixed pathlength for quantitative work. The problems associated with such cells include:

(a) The appearance of interference fringes, particularly in cases where bubbles are inadvertently present in the cell. These fringes may mask bands of interest.
(b) Solvent bands may also be present, obscuring large regions of the spectrum. These may only be digitally subtracted in cases where the pathlength of the cell is very short ($< 10\,\mu$ m), but such pathlengths are not easy to reproduce.
(c) Wedging errors induced by the cell windows not being plane parallel.

Mulls exhibit similar problems but also present the further complication of uncertainty in concentration. This difficulty is also experienced with pellet techniques. Pellets have the added disadvantage of unreproducible thickness and the possible appearance of Christiansen effects [1]. The latter arise when the variation in refractive index at a strong absorption band causes the refractive index of the sample to be more similar to the matrix so that the scattering component, S, is reduced and more energy is seen by the detector than expected. Scattering can be particularly severe in poorly pressed and powdery pellets.

3   Diffuse Reflectance

The diffuse reflectance technique has become widely used for the analysis of solids and particulates and is generally considered to be a technique requiring minimal or no sample preparation. However, the spectra produced can sometimes exhibit

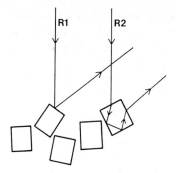

Fig. 3. Illustration of diffuse reflectance phenomenon. Ray R1 is specularly reflected whereas ray R2 enters sample and is diffusely reflected

unusual features when compared to the corresponding transmission spectra. In certain cases the diffuse reflectance spectrum can be so distorted as to be unidentifiable, and frequently diffuse reflectance spectra cannot be used for library searching databases composed of transmission spectra.

In order to understand these effects it will be necessary to consider the origin of the diffuse reflectance phenomenon. Consider a ray of light arriving at the surface of particulate medium (Fig. 3). At the interface there is a random arrangement of crystal faces so that there will be a range of angles of incidence with the crystal faces. Some rays (e.g. Rl) will arrive at such an angle that reflection off the surface will occur. This gives rise to a specular or surface reflected component. This component has no absorptive interaction with the sample. Another ray (R2), on the other hand, may arrive at such an angle that it is not reflected but is refracted and enters a crystallite. It may then emerge or become internally reflected a number of times, and in general, will undergo refraction and reflection through many crystallites before emerging from the solid. The angle of emergence for this ray depends on the ray path and can take any value. Such rays, having passed through the sample, have undergone interaction with the sample and contain spectral information.

An important point to bear in mind is that although specular reflection is

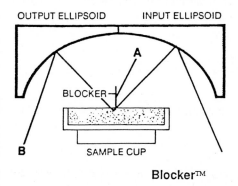

Fig. 4. Diagram of a typical diffuse reflectance accessory. The use of the Blocker™ device is illustrated for the elimination of specular reflectance

reflection where the angle of incidence equals the angle of reflection, these angles defined with respect to the normal perpendicular to the reflecting surface. In general there will be a distribution of normals and thus a distribution of the angles of incident and specularly reflected rays. Thus diffuse and specular reflection are always mixed and cannot be separated spatially. The energy arriving at a detector is therefore a sum of surface reflected radiation and diffusely reflected radiation from the sample. The presence of specular effects has important consequences in the spectrum as will be discussed later, but it should be noted that diffuse reflectance accessories that rely on spatial filtering or mechanical beam stops are not likely to eliminate specular reflection in most particulate samples. However under certain circumstances where flat surfaces are examined they can lead to spectral improvement (Fig. 5). Diffuse reflectance differs from transmission in that the surface reflected light is collected and not returned towards the source, and the observation of a spectrum depends on scattering of the light by the sample. Furthermore, the light path through the sample is less clearly defined and characterised than in a transmission experiment. A typical sampling arrangement for diffuse reflectance is shown in (Fig. 4).

Diffuse reflectance spectra are not presented in transmittance or absorbance units. Instead it is usual, particularly in the mid-infrared, to use Kubelka-Munk units [2]. These reflect the origin of the spectrum as it results from the absorbance of scattered light. The Kubelka-Munk relationship may be expressed as:

$$R = (1 - R_{\infty})^2 / 2R_{\infty} = k/s$$

for an infinitely thick sample – i.e. one in which no light reaches the bottom of the sample. Here k is a molar extinction coefficient and s a scattering coefficient; R_{∞} is the absolute reflectance, defined as:

$$R_{\infty} = J/I_o$$

where J is the total reflected radiation from a surface illuminated with light of intensity I_o.

It is not usually possible to measure the absolute reflectance R_{∞} of a sample, and instead the relative reflectance R'_{∞} is measured. This is the reflectance of the sample measured with respect to a reference material. The latter is usually an inorganic salt such as potassium bromide, and

$$R'_{\infty} = R_{\infty} \text{ (sample)}/R_{\infty} \text{ (standard)}$$

Since k is defined as:

$$k = 2.303 \, \varepsilon C$$

where ε is the molar (decadic) absorption coefficient and C is the concentration we have

$$(1 - R'_{\infty})^2 / 2R'_{\infty} = C/k'$$

where $k' = s/2.303\varepsilon$ for a given chromophore. Thus at any particular wavelength a plot of the Kubelka-Munk function versus C should be linear if s remains

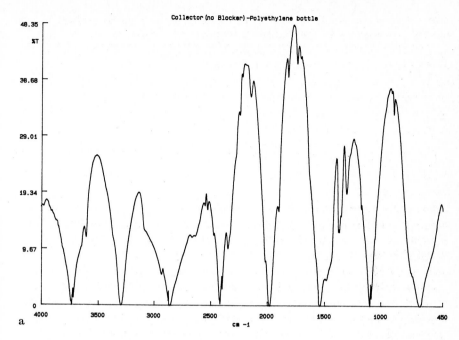

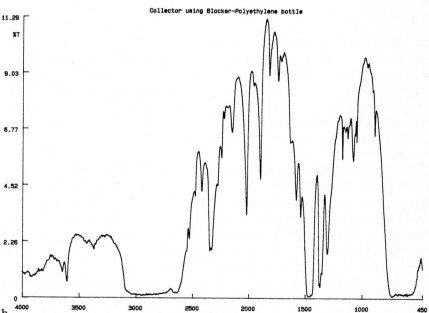

Fig. 5. Diffuse reflectance spectra of polyethylene bottle acquired **a)** without Blocker™ and **b)** with Blocker™ showing improvements that may be achieved under certain circumstances. Courtesy of Spectra-Tech Europe Ltd.

Table 1. Relative band intensities predicted for two bands of transmittance 0.001 and 0.1 with increasing specular component predicted from the Kubelka-Munk equation

% specular component	Ratio of band intensities in Kubelka-Munk units
0	123
5	3.5
10	2.3
15	1.9
20	1.7
25	1.6

constant. This equation has been well proven in many systems and is widely used for the quantification of diffuse reflectance data. However when $(1 - R'_\infty)^2$ is plotted versus wavelength to give a spectrum, apparent intensities can be distorted by specular effects [3].

Consider for example a compound which has scattering coefficients and molar absorptivities such that two absorption bands generate diffuse reflectances of 0.001 and 0.1 respectively. In the absence of surface reflectance the light intensities arriving at the detector are thus 0.001 and 0.1 for an incident light intensity, I_o, of unity. When transformed to Kubelka-Munk units these correspond to 499 and 4.05 respectively giving a ratio of approximately 123.

However, it is not possible in practice to eliminate the specular component which is always present and collected with the diffused component. If there is a 5% specular component, then the effective incident radiation is reduced to 95% of its intrinsic value. The transmitted portions are then 0.00095 and 0.095 respectively for the two bands. The reflected radiation is then the sum of the 5% surface reflectance plus the transmitted (diffuse) radiation so that the intensities become 8.84 and 2.52, and the ratio in band intensities is now 3.5. The band ratio can be calculated for a given specular component for these two bands; the result is shown in Table 1.

Clearly when a spectrum is plotted in Kubelka-Munk units significant compression can occur when there is specular reflectance. However even if the spectrum were plotted in terms of intensity of reflected light compression would still be apparent and indeed would be slightly worse. This is because when most of the light is absorbed within the sample, the largest contribution to signal intensity comes from the specular component. Since this contributes evenly to spectral intensity, the absorption spectrum then only makes a small perturbation to this spectral component. These effects are illustrated by the diffuse reflectance spectra of sucrose shown in Fig. 6. The main difference between the diffuse reflectance spectrum of Fig. 6a and the corresponding KBr pellet spectrum (Fig. 7), is the loss of spectral detail in the region 1800–900 cm−1 where there appears to be a compression of peaks leading to a loss of spectral discrimination and resolution.

The usual method of overcoming band suppression is to dilute the sample in

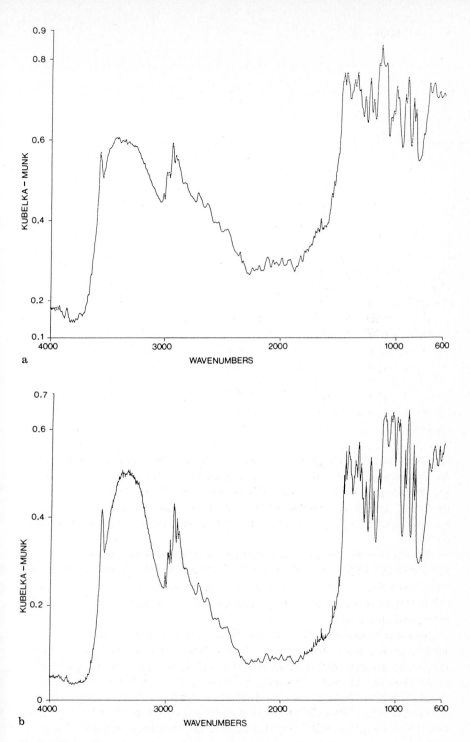

Fig. 6. Diffuse reflectance spectrum of sucrose **a**) neat, **b**) diluted to 5% in KBr

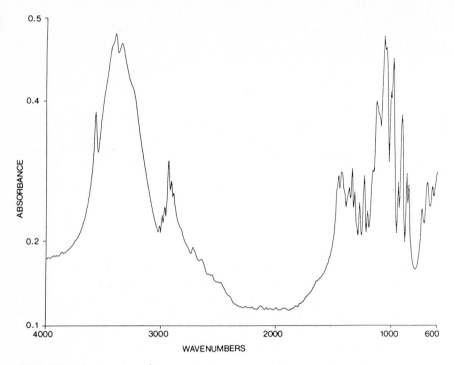

Fig. 7. KBr pellet spectrum of sucrose

the same material used for the reference (e.g. KBr). This is illustrated in (Fig. 6b) for sucrose diluted to 5% in KBr. This spectrum is considerably improved in appearance and looks more like the pellet spectrum. It is also possible to improve spectra by grinding to smaller particle size as the specular reflection decreases with particle size, so that the apparent absorbance and relative band intensities increase.

Another artifact associated with specular reflectance is a derivative shaped peak or apparent absorbance dips. This is a result of the extremely high surface reflection which certain materials exhibit at particular wavenumbers, the so-called *restrahlen* bands [4]. These effects are also extremely difficult to remove from spectra.

Sometimes quite severely distorted spectra can be produced by diffuse reflectance and cause concern to the uninitiated spectroscopist. Certainly these effects can hinder both structural assignment, library searching and quantitative analysis. Although some of these artifacts can be suppressed or reduced by dissolution in an inert matrix, the advantages of minimal sample preparation are then largely lost. However, some samples do give rise to extremely good diffuse reflectance spectra and in these cases it is an extremely powerful analytical method.

3.1 Disadvantages

In addition to the problems of specular reflectance there are some other difficulties which may be encountered when using the diffuse reflectance technique.

(a) Diffuse reflectance is mainly limited to powdered samples.
(b) In water containing samples the heating effect of the infrared beam can lead to the generation of water vapour which causes strong spectral interference.
(c) Reproducible sample cell filling for quantitative analysis is often difficult.
(d) The sometimes unusual appearance of diffuse reflectance spectra can lead to difficulty in assignment and quantification.

4   Attenuated Total Reflectance

Attenuated total reflectance (ATR) methods represent arguably the most flexible form of sample presentation and are available in numerous configurations for specific applications. However, irrespective of the particular design used, the basic components and physical principles involved are the same. The main component in an ATR accessory is the internal reflectance element. This is usually a prism of an infrared transmitting material of high refractive index. Light is focussed on one of the faces of the prism and enters the material. The angle of the prism face is cut so that the light passes into the prism undeflected but arrives at the interface at a predetermined angle. This angle θ_i is arranged such that it is greater than the so-called critical angle, θ_c where

$$\theta_c = \sin^{-1} n_2/n_1 \quad (n_1 > n_2)$$

and n_2 is the refractive index surrounding medium, n_1 the refractive index of prism material.

When $\theta_i > \theta_c$ total internal reflection occurs at the interface and so the light is trapped in the prism. After a number of such reflections the light leaves the prism. At each point of internal reflection, however, some radiation penetrates into the surrounding medium and can interact with that medium leading to attenuation of the reflected light when absorption occurs. This is the mechanism by which ATR is used to generate an infrared spectrum. The effective pathlength of such an ATR cell depends on the depth of penetration at each reflection, l, and the number of such reflections. The latter is determined by θ_i and the prism dimensions, whereas for wavelength λ we have

$$\frac{l = \lambda/n_1}{2\pi[\sin^2\theta_i - (n_2/n_1)^2]}$$

The appearance of λ and the wavelength-dependent refractive indices in this equation leads to a wavelength dependence of the penetration depth and gives

Table 2. Properties of infrared ATR prism materials

Material	Transmission	n	Comments
Germanium	5500–600	4.0	Hard and brittle, insoluble in water
ZnSe	20,000–454	2.4	Hard, easily cracked, attacked by acids and alkalis
KRS5	20,000–320	2.37	Soft, slightly soluble. Poisonous.
ZnS	17,000–833	2.2	Resistant to thermal and mechanical shock. Insoluble.
NaCl	40,00–590	1.49	Fragile, highly soluble. Cheap.

rise to the observed differences in relative intensities between ATR and transmission spectra. Typical values for l are between 0.25 and 4 μm, so that the ATR cell is loosely equivalent to a very short pathlength transmission cell. The difference between this and a transmission cell is that the whole sample need not transmit light, only the pathlength generated by the product of the penetration depth and the number of reflections. A very important criterion for good ATR spectra is that there should be good optical contact between the sample and the crystal. Thus easily malleable materials, wet materials and liquids give very good spectra. Powders, unless fine or made of soft particles, tend to be much less satisfactory. In all cases the absolute intensities in the spectrum will be dependent upon the optical contact between sample and crystal, which in general will not be reproducible in less malleable materials.

4.1   Prism Materials

Choice of prism materials is crucial. A high refractive index is required and the material must be inert and mechanically robust. For many applications resistance to water is very important. The table below illustrates the properties of some important prism materials.

4.2   ATR Accessory Geometry and Sampling

4.2.1   Solutions

Solutions may be readily sampled using a variation known as cylindrical internal reflectance (CIR) in such proprietary forms as the CIRCLE cell™ (Fig. 8). The prism is rod shaped with conical ends. This is mounted in a stainless steel "boat" so that the solution surrounds the crystal. Light is focussed into the protruding

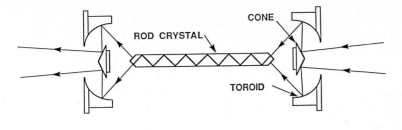

CIRCLE® OPTICS

Fig. 8. Schematic representation of CIRCLE cell™. Courtesy of Spectra-Tech Europe Ltd.

conical ends of the crystal by Cassegrain optics. Spent solutions can either be removed by tipping out and pouring new ones in, or the cell can be used "on-line" using the flow cell option. Micro-CIRCLE™ is also available for very small volumes of sample (5 ml → 120 μl) using a smaller prism. CIR with its very short effective pathlength, is extremely useful for aqueous solutions (Fig. 9).

4.2.2 Pastes and Semi-Solids

These may conveniently be studied using horizontal ATR (HATR). The prism is now a parallelogram with mirrored ends which is mounted flush with the top plate (Fig. 10). Samples can be spread over the crystal, spectra acquired and the sample wiped off; alternatively the top plate may be removed for cleaning. Some versions incorporate various pressure plates to ensure better contact between sample and prism. This is most useful for powdered samples which can sometimes give surprisingly good results. Another variation of this accessory has the top plate raised out of the spectrometer sample compartment so that it is not necessary to break the instrument purge to change samples. Such devices have become popularly known as "Skin analysers" as a consequence of their use in the

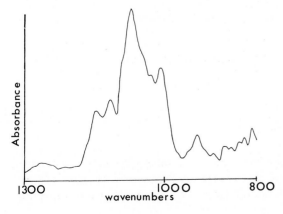

Fig. 9. Spectrum of sugars in aqueous solution acquired using CIRCLE cell™

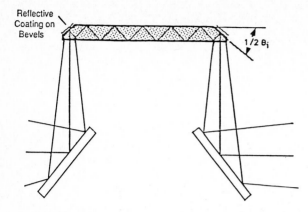

Fig. 10. Diagram showing optical configuration of horizontal ATR accessory. Courtesy of Spectra-Tech Europe Ltd.

pharmaceutical industry for the in vivo study of the behaviour of cosmetics in skin.

4.2.3   Variable Penetration Depth

The dependence of the penetration depth on θ_i leads to the possibility of depth profiling or pathlength variation by changing θ_i. This can be accomplished with such devices as the continuously variable angle ATR accessory with θ_i variable from 30–60°. This device uses a vertically mounted crystal and is sometimes known as vertical ATR.

4.3   Disadvantages

Whilst generally a very flexible technique ATR has the following serious disadvantages.

(a) It is not very successful with powders, particularly if they are hard.
(b) ATR is sensitive to contact efficiency and crystal coverage which may present problems for quantitative work.

5   Photoacoustic Spectroscopy

Photoacoustic spectroscopy (PAS) is unlike other infrared sampling methods in that it directly measures the energy absorbed by the sample. All other methods depend on the measurement of the attenuation of a light beam due to absorption, and are in essence difference methods. The photoacoustic effect depends upon the absorption of radiant energy and its conversion into acoustic energy of thermoelastic waves.

5.1   The Gas-Microphone Cell

In general, absorption causes a rise in temperature which may then be monitored by thermometry. However this process is slow and is not very accurate because of heat losses during the measurement process. A more efficient means of monitoring the heating effect is to modulate the intensity of the incident light periodically and detect the resulting temperature fluctuations. There are a number of ways of doing this, the choice for a particular application dependent upon the nature of the sample. The most universally useful of these methods is the gas-microphone cell arrangement which is designed for use with liquids and solids. The general principles of the cell are illustrated in (Fig. 11). For some

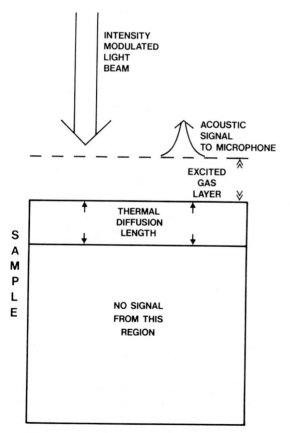

Fig. 11. Schematic representation of the photoacoustic effect

chromophores within the sample the periodic absorption of light results in periodic heating. The process of energy conversion to heat is very fast (typically 10^{-8} s) so that the absorption is effectively simultaneous with the heating process. There is thus, in sample, a source of heat which is periodic in nature, cycling between a higher temperature and the ambient temperature. During the higher temperature phase the heat diffuses towards the sample surface in what is termed a thermal wave. When it reaches the surface the heat causes an expansion of a gas layer adjacent to the surface; this expansion then drives a pressure wave through the remaining gas which may be observed as an acoustic signal by a microphone. Hence the term photoacoustic.

In normal circumstances the chromophore is distributed throughout the sample, and the mathematical problem is to solve the diffusion equation for a distributed heat source with periodic boundary conditions. This has been done by Rosencwaig [5] and co-workers in what is normally referred to as the RG theory. The solution arrived at is a general one which is hard to interpret quantitatively. In order to illustrate the basic physics of the system, Rosencwaig discussed some limiting cases in most of which the signal was linearly proportional to the optical absorption coefficient. Unfortunately this has led to an assumption by many authors that this relationship is generally true. In fact it is not so, and in general the photoacoustic gas microphone cell response is not linearly proportional to the optical absorption coefficient.

Before considering the quantitative photoacoustic response in detail, it is necessary to introduce the important concept of thermal diffusion length, usually given the symbol μ. Since the diffusive process is dissipative, the intensity of the signal decreases with distance from the heat source. Over the thermal diffusion length the intensity is attenuated by an amount $(1/e)$ of its initial value. However, this distance is also the distance moved in the time $2\pi/\omega$ where ω is the modulation frequency in radian s^{-1}. Thus when the wave has propagated a distance μ new heat is released into the system and any information taking more time to reach the surface is lost [6]. Since the microphone detects the energy arriving at the sample surface any signal originating within the sample at a depth greater than μ will not contribute to the microphone response. In effect this means that spectra are only obtained from material within a depth of μ of the sample surface.

Now μ is related to ω by the equation

$$\mu = \left(\frac{2\alpha}{\omega}\right)^{\frac{1}{2}}$$

where α is the thermal diffusivity. This equation immediately suggests the possibility of depth profiling the sample by varying ω. In practice, attempts to do this have met with relatively little success, probably because most of the signal originates much nearer to the surface than the thermal diffusion length; hence data obtained at any frequency tend to be highly weighted by signals from very near the surface. A problem arises in Fourier transform instruments because of the frequency dependence of μ. In such instruments the modulation frequency is wavenumber dependent, and hence at long wavelengths the sampling depth is greater than at short wavelengths. This tends to skew the relative intensities of the peaks in the spectrum since a greater sampling depth results in an increase in total signal intensity. However, provided sampling conditions remain constant this does not seriously impair utilization of the spectrum for quantitative analysis. Typically in Fourier transform instruments μ is in the range 20 to 200 μm.

A useful and quite general approximation to the RG expression for the photoacoustic response, H, in the gas-microphone cell is [7]

$$H = \frac{AI_0 2^{1/2} \mu\beta}{[\mu\beta^2 + (\mu\beta + 2)^2]^{1/2}}$$

Here A depends on both the sample and the instrument, but is generally wavelength-independent, and β is the optical absorption coefficient given by $\beta = 2.303\,\varepsilon C$, where ε is the molar absorption coefficient and C is the concentration. It is obvious from inspection of the above equation that the photoacoustic response is non-linear and that the frequency-dependent thermal diffusion length enters the expression in a direct way.

It is useful to consider two limiting cases. Firstly when $\mu\beta \ll 1$, then

$$H = \frac{AI_0 \mu\beta}{\sqrt{2}}$$

This is a very useful form, being linear in β, but only applies to very dilute systems or those where absorption is very weak. At the other extreme where $\mu\beta \to \infty$, we have

$$H = AI_0$$

This condition is known as *photoacoustic saturation*. Under these circumstances the signal intensity is independent of chromophore concentration and the spectrum tends to exhibit very broad lines. This condition can occur in mid-infrared spectra; it has been observed, for example, in the mid-infrared Fourier transform spectrum of starch [8], but may be eliminated with suitable dilution in an inert matrix. Photoacoustic saturation is a very useful condition for obtaining a reference background spectrum, usually carbon black. Since the photoacoustic response is proportional to the incident intensity, ratioing this spectrum to the sample spectrum eliminates the background spectrum from the lamp. The effects of the 'A' term are not totally eliminated since this depends on parameters such as sample packing in the cell. However careful experimental technique can ensure that such variations are small.

In many situations a useful approximation [9] can be made by assuming that

$$\mu\beta^2 < (\mu\beta + 2)^2$$

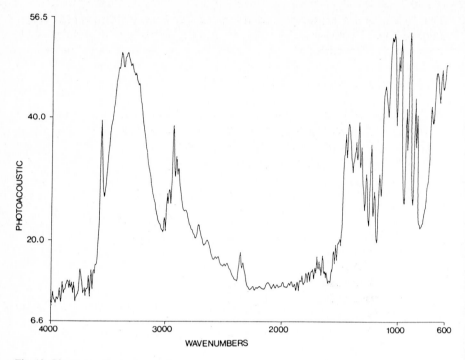

Fig. 12. Photoacoustic spectrum of sucrose

whence,

$$R = \frac{AI_0 2^{1/2} \mu\beta}{\mu\beta + 2}$$

This indicates a reciprocal relationship between photoacoustic response and intensity. It has been found to hold for a wide range of materials and wavelengths and has proved to be a generally useful quantitative means of treating data.

An alternative way to detect the heating of the gas layer adjacent to the sample surface is to observe the deflection of a light beam that is passing very close to the sample surface [10]. As the gas heats, its refractive index changes, and thus beam deflection is changed. A plot of beam deflection versus wavelength thus generates a spectrum. This technique has the great advantage that the sample need not be confined to a cell and can therefore be used on large or awkwardly shaped objects.

Photoacoustic effects in gases are much larger than those in solids. In gases the absorption is translated directly into a pressure change. With carefully designed cells very low concentrations of gas may be observed using laser excitation. There has been very little work reported on the FTIR of gases using photoacoustic detection but it has been shown that the ordinary gas microphone cell may be readily used to observe such spectra. [11].

5.2 Piezo-electric Detection

The energy coupling in the gas microphone cell is by way of a dissipative, incoherent process. Parallel with this in solids is a coherent coupling to the phonon spectrum [12]. This results in a periodic expansion of the sample which can be detected by a piezo electric transducer. In order to do this there must be strong mechanical bonding between the sample and transducer, which limits the general applicability of the method. However one advantage is that the signal comes from the whole of the sample rather than from a thermal diffusion length.

5.3 Disadvantages

(a) One of the fundamental problems of photoacoustic spectroscopy is its intrinsically low signal-to-noise ratio. This is because the efficiency of energy transfer from the light beam to the microphone or transducer is poor. In mid-infrared gas microphone cells this problem is compounded at high frequency both by the short thermal escape depth and, probably, source intensity fall off.

(b) Signal-to-noise ratios are also adversely affected by the photoacoustic band shape. Lineshapes in the gas-microphone cell will in general be broader than in the transmission of an ATR experiment; thus the relative intensity at the peak maximum will be less than in these purely optical experiments. As the absorption gets stronger, the band will tend to get broader and broader. This effect is offset to some extent by the increase in total intensity, but this is at the cost of the resolution of the band maximum.

(c) The strong photoacoustic effect in gases can give rise to some problems since if the sample has any vapour above it this may generate a spectrum which overlays the spectrum of interest. Water vapour represents a particular problem in this context since it has a very strong absorption spectrum and can often seriously distort the spectrum of the sample of interest.

6 Conclusions

The techniques described in this Chapter represent part of the armoury of modern infrared spectroscopy. It is invidious to try and decide which is the best technique. Each has its own strengths and weaknesses and for any given problem one method will be better than the rest. Comparison of figures 6, 7, 9 and 12 shows that each method gives rise to a characteristics spectral appearance. The notion that one of these, usually the transmission spectrum, is the "right" one is misguided. It is simply that most spectroscopists are more used to this type of spectrum than any other.

The judicious use of the variety of sample presentation methods now available has enormously extended the range of materials easily accessible to the infrared spectroscopist. It is to be hoped that in future, libraries of spectra will not be confined only to transmission spectra but will be extended to include spectra obtained by the other sampling methods.

Acknowledgement

The authors acknowledge the assistance of Spectra-Tech Europe Ltd., Warrington, England, with the diagrams used in this work.

7   References

1. Chantry GW (1984) Long-wave optics, vol 1, Academic, London, p 283
2. Kubelka P, Munk F (1931) Z. Tech. Phys. 12: 593
3. Kortum G (1969) Reflectance spectroscopy, Springer, Berlin Heidelberg New York
4. Chantry GW (1984) Long-wave optics, vol 1, Academic, London, p
5. Rosencwaig A (1980) Photoacoustics and photoacoustic spectroscopy, Wiley, New York
6. Carslaw HS, Jaeger JC (1960) The conduction of heat in solids, Oxford Univ. Press, Oxford, p 64
7. Poulet P, Chambron J, Unterreiner R (1980) J. Appl. Phys. 51: 1738
8. Belton PS, Saffa AM, Wilson RH (1987) Analyst 112: 1117
9. Belton PS, Tanner SF (1983) Analyst, 108: 591
10. Low MJD, Monterra C (1984) Appl. Spectrosc. 38: 807
11. Belton PS, Saffa AM, Wilson RH (1988) In: Creaser CS, Davies AMC (eds) Analytical applications of spectroscopy, Royal Soc. Chem., London, p 245
12. Vargas H, Miranda LCM, (1988) Physics Reports 161: 43

8   Bibliography

An account of these techniques is given in most modern handbooks on infrared spectroscopy see for example:
Griffiths PR, de Haseth JA (1986) Fourier transform infra red spectroscopy, Wiley, New York
Ferraro JR, Basile LJ (eds) (1978-85) Fourier transform infrared spectroscopy, vols 1–4, Academic, New York
For recent reviews of photoacoustic spectroscopy see Ref. [12] and:
Hess P, Pelzl J (eds) (1988) Photoacoustic and photothermal phenomena Springer, Berlin Heidelberg New York
Coufal H, McClelland JF (1988) J. Mol. Struct. 173: 129
Belton PS, Saffa AM, Wilson RH (1988) Quantitative analysis by Fourier transform infrared photoacoustic spectroscopy, SPIE 917
For reviews of reflectance methods see Ref. [3] and:
Harrick NJ (1967) Internal reflection spectroscopy Wiley Interscience, New York
For a review of sample preparation methods for transmission see:-
Miller RGJ, Stace BC (1979) Laboratory methods in infrared spectroscopy, 2nd edn, Heyden, London

CHAPTER 4
Electronic Absorption Spectroscopy: Theory and Practice

M.R.S. McCoustra

1 Introduction

Of the wide variety of spectroscopic techniques available to the modern spectroscopist, electronic absorption spectroscopy has perhaps the longest history, effectively tracing its origins back to the original work of Bunsen and Kirchoff on the spectroscopy of atomic species in flames. Consequently, the technique was responsible for the rapid expansion of the periodic table in the mid to late nineteenth century and, more importantly, for the development and testing of theories of atomic and molecular electronic structure in the early twentieth century. As a technique with a long history, it would be expected that electronic absorption spectroscopy would be widely used. This is true. However, we must contrast the use of electronic absorption spectroscopy with that of infrared and NMR techniques. While the latter are widely used in a qualitative sense, the former finds most use in quantitative measurements. This contrast is discussed briefly below. First, however, some basic theory needs to be considered.

2 Electronic Energy Levels and Transitions

2.1 Atoms

The theoretical description of atomic and molecular energy levels is based on the Schrödinger equation,

$$\hat{H}\Phi = E\Phi$$

where $\hat{H}$ is the Hamiltonian operator, E is the electronic energy and Φ is the wavefunction. The operator $\hat{H}$ is composed of terms describing both the kinetic energy of the electron and the potential energy, the latter including attractive terms for the electron-nucleus interaction and repulsive terms for the inter-electron potentials. When the latter is zero, as in the case of hydrogenlike (single-electron) ions, analytical solution of the Schrödinger equation to yield electronic energies and wavefunctions is possible and is well described in many textbooks of general physical chemistry. Analytical solution of the multielectron problem is, however, impossible, as a consequence of the interelectron potential terms, and approximate methods must be applied. Two families of approximation techniques are commonly used for the solution of the multielectron problem, Perturbation Theory and the Variation Method, both of which are well described in texts such as Atkins [1] and Lowe [2]. In particular, the Self-Consistent Field (SCF) method of Hartree and Fock [2,3], a method based on the Variation Theorem, has found widespread use in calculating the electronic energies and wavefunctions of atoms.

Calculations of the electronic energy levels of multi-electron atoms using the HF-SCF approach yield atomic orbitals to which the various electrons in the atom can be assigned and given energy ordering. Each orbital has a distinct set of quantum numbers which are used to characterise it. These are the principal quantum number, n, which can take values $n = 1, 2, 3, \ldots$ and largely determines the energy of the orbital, the angular momentum quantum number, l, which takes the values $l = 0, 1, 2, \ldots, (n-2), (n-1)$ and the magnetic quantum number, m_l, which is given by, $m_l = -l, -(l-1), \ldots, 0, \ldots, (l-1), l$. In addition, each electron has associated with it a spin quantum number, m_s ($= \pm \frac{1}{2}$). In normal use, atomic orbitals are labelled nl_{m_l} where the l is introduced symbolically with the letters s, p, d, f, g etc. corresponding to the numerical value of l from $0, 1, 2, 3, \ldots$; hence, labels such as 1s, $2p_{-1}$, $4f_2$ etc.

With the proviso that no two electrons should have the same set of four quantum numbers (the Pauli Exclusion Principle), we can then construct electronic configurations for the ground states of the various atoms in the periodic table. Thus, for example the ground state of carbon has the configuration, $1s^2 2s^2 2p^4$, while that of lead is

$$1s^2 2s^2 2p^6 3s^2 3p^6 3d^{10} 4s^2 4p^6 4d^{10} 4f^{14} 5s^2 5p^6 5d^{10} 6s^2 6p^2$$

Similar configurations can be constructed for all the atoms in the periodic table. Excited state configurations can then be simply generated by the promotion of one or more electrons from orbitals in the ground state to higher lying orbitals.

Such configurations, however, are of little practical use in atomic spectroscopy, as a single configuration can give rise to a vast number of atomic spectroscopic states through coupling of the various orbital and electronic angular momenta. By far the most widely applied coupling scheme is the Russell-Saunders scheme [4, 5, 6] where the individual orbital angular momenta of the various electrons in the atom, l_i, couple to give the total orbital angular momentum, L, which for two electrons is given by the Clebsch-Gordan Series,

$$L = l_1 + l_2, l_1 + l_2 - 1, \ldots, |l_1 - l_2|$$

and similarly the individual electron spins, s_i ($= |m_s|$), couple to give a total spin angular momentum, S,

$$S = s_1 + s_2, s_1 + s_2 - 1, \ldots, |s_1 - s_2|$$

from which the total angular momentum, J, is obtained from

$$J = L + S, L + S - 1, \ldots, |L - S|$$

The resulting atomic states are labelled with a term symbol of the form $^{2S+1}L_J$ appended after the configuration, for example the $1s^2 2s^2 2p^1 3d^1$ $^1P_3$ and $1s^2 2s^2 2p^1 3d^1$ $^1D_2$ electronically excited states of carbon. It should be noted that orbitals which are completely filled have no total orbital or spin angular momentum associated with them and hence need not be considered in arriving at the atomic states through the Russell-Saunders scheme, so in the above examples only the $2p^1 3d^1$ configuration is used in arriving at the spectroscopic states.

Whether an electronic transition is observed or not is determined by the selection rules. Theoretically, the basis of the selection rules is the evaluation of the transition dipole moment, R_{fi}, defined as the integral over configuration space;

$$R_{fi} = \int \Phi_f^* \hat{\mu} \Phi_i d\tau$$

where the Φ_i and Φ_f are the wavefunctions of the initial and final states linked by the transition and $\hat{\mu}$ is the dipole moment operator which is defined as

$$\hat{\mu} = -e. \sum_i r_i,$$

r_i being the position of the ith electron. Provided R_{fi} is non zero, the transition is allowed. If R_{fi} is zero, the transition is forbidden. These expressions are equally true for the simple pure electronic transitions in atoms and the more complicated transitions in molecules involving combinations of electronic, vibrational and rotational energy levels. While such calculations are useful, studies of atomic absorption and emission spectroscopy over a number a decades have given us a number of empirical selection rules to allow us to determine whether a transition is allowed or forbidden. These are listed below.

a) $\Delta L = 0, \pm 1$ except $L = 0 \leftrightarrow\!\!\!/\!\!\!\rightarrow L = 0$
b) even $\leftrightarrow\!\!\!/\!\!\!\rightarrow$ even; odd $\leftrightarrow\!\!\!/\!\!\!\rightarrow$ odd; even $\leftrightarrow$ odd

The latter is known as the Laporte Selection Rule and the even and odd refer to the arithmetic sum, $\sum l_i$, over all electrons. This rule basically forbids transitions between states arising from the same configuration.

c) $\Delta J = 0, \pm 1$ except $J = 0 \leftrightarrow\!\!\!/\!\!\!\rightarrow J = 0$
d) $\Delta S = 0$

This final rule is sometimes broken for heavy atoms where there is extensive spin-orbit coupling for each individual electron.

2.2 Molecules

In much the same manner as atoms, the energy levels of molecules can be determined by solutions of an appropriate Schrödinger equation. While it is possible to construct a Schrödinger equation exactly along the lines of an atomic system, specifying each nucleus and electron in the molecule, this approach can be seen as excessively complex. Simpler approaches are available. The earliest was the Valence Bond Method (VB–MO) of Heitler and London [7] which was rapidly superseded by the Linear Combinations of Atomic Orbitals–Molecular Orbitals (LCAO–MO) method in the early 1930s, although renewed interest has been shown in the VB–MO method in recent years.

Let us consider in some detail the LCAO–MO method. Let us assume that the wavefunction of a molecular orbital, Φ, can be expressed a linear sum of terms composed of atomic orbitals, ϕ_j, thus,

$$\Phi_i = \sum_j c_{ij} \phi_j$$

Provided that the atomic orbitals are of comparable energy, are well overlapped and have the same symmetry properties then such an assumption is quite valid. If we consider the case of a homonuclear diatomic molecule, the above can be written simply as

$$\Phi_i = c_{i1} \phi_1 + c_{i2} \phi_2$$

Substitution into the Schrödinger equation followed by a little manipulation leads to the result

$$E_i = \frac{\int \Phi_i^* \hat{H} \Phi_i d\tau}{\int \Phi_i^* \Phi_i d\tau}$$

and therefore E_i can only be calculated if the Φ_i are known. Substituting in the approximate molecular orbital wavefunction above, we have

$$\bar{E} = \frac{\int (c_{i1}^2 \phi_1^* \hat{H} \phi_1 + c_{i1} c_{i2} \phi_1^* \hat{H} \phi_2 + c_{i1} c_{i2} \phi_2^* \hat{H} \phi_1 + c_{i2}^2 \phi_2^* \hat{H} \phi_2) d\tau}{\int (c_{i1}^2 |\phi_1|^2 + 2c_{i1} c_{i2} |\phi_1^* \phi_2| + c_{i2}^2 |\phi_2|^2) d\tau}$$

where $\bar{E}$ is now an approximation to the actual energy of the molecular orbital ϕ_i. How can we simplify this cumbersome expression? If the atomic orbitals are normalised then,

$$\int \phi_i^2 d\tau = 1$$

and since the Hamiltonian operator is Hermitian then,

$$\int \phi_1^* \hat{H} \phi_2 d\tau = \int \phi_2^* \hat{H} \phi_1 d\tau = H_{12}$$

and we define the overlap integral, S, which is a measure of the degree to which the atomic orbitals overlap with each other, as

$$S = \int \phi_1^* \phi_2 d\tau$$

then the energy expression above can be re-written in the simplified form,

$$\bar{E} = \frac{c_{i1}^2 H_{11} + 2c_{i1} c_{i2} H_{12} + c_{i2}^2 H_{22}}{c_{i1}^2 + 2c_{i1} c_{i2} S + c_{i2}^2}$$

It is then possible to show using the Variation Theorem that this expression can be written in the form of two *Secular Equations*,

$$c_{i1}(H_{11} - E_i) + c_{i2}(H_{12} - E_iS) = 0$$

$$c_{i1}(H_{12} - E_iS) + c_{i2}(H_{22} - E_i) = 0$$

where E_i is now used to indicate the approximate value of the energy obtained by this variational treatment. We can then determine the energies of the resulting molecular orbitals by solution of the *Secular Determinant*,

$$\begin{vmatrix} H_{11} - E_i & H_{12} - E_iS \\ H_{12} - E_iS & H_{22} - E_i \end{vmatrix} = 0$$

where H_{12} is known as a Resonance Integral and normally symbolised by β, and H_{11} and H_{22} are Coulomb Integrals symbolised usually by α. Thus the determinant above can be re-written as

$$\begin{vmatrix} \alpha\text{-}E_i & \beta\text{-}E_iS \\ \beta\text{-}E_iS & \alpha\text{-}E_i \end{vmatrix} = 0$$

which has the solutions,

$$E_i = \frac{\alpha + \beta}{1 \pm S}$$

corresponding to symmetric and antisymmetric combinations of the two atomic orbitals. The coefficients c_{ij} can then be obtained by back substitution into the secular equations.

The development of this approach is exactly the same for larger polyatomic molecules in that we can define a secular determinant as below.

$$\begin{vmatrix} H_{11} - E_iS_{11} & H_{12} - E_iS_{12} & \cdots & H_{1n} - E_iS_{1n} \\ H_{12} - E_iS_{12} & H_{22} - E_iS_{22} & \cdots & H_{2n} - E_iS_{2n} \\ \vdots & \vdots & & \vdots \\ H_{1n} - E_iS_{1n} & H_{2n} - E_iS_{2n} & \cdots & H_{nn} - E_iS_{nn}) \end{vmatrix} = 0$$

or

$$|H_{mn} - E_iS_{mn}| = 0$$

in an abbreviated form, where the $H_{mn}(m \neq n)$ are the *Resonance Integrals*, the H_{mm} are the *Coulomb Integrals*, the S_{mn} ($= 1$ where $m = n$) are the overlap integrals and the E_i are the orbital energies.

Perhaps the simplest application to consider in polyatomic molecules is the case of a conjugated π-orbital system as originally treated by Hückel. We can then make the following assumptions.

a) Consider only the π orbitals.
b) Assume that there is no overlap between the various atomic orbitals, i.e. for $m \neq n$, $S_{mn} = 0$.

c) Assume that the coulomb integral is the same for all carbon atoms, i.e. $H_{mm} = \alpha$.

d) Assume that the resonance integral is the same for all pairs of directly bonded atoms, i.e. $H_{mn} = \beta$.

e) Assume that the resonance integral for atoms not directly bonded is zero, i.e. $H_{mn} = 0$.

Given then that the π molecular orbital wavefunctions can be described by a linear combination of only the atomic orbitals involved in the π system, we can write, for example, the secular determinant for benzene as a 6×6 determinant since only the $2p_z$ carbon atomic orbitals are involved in the π system;

$$\begin{vmatrix} x & 1 & 0 & 0 & 0 & 1 \\ 1 & x & 1 & 0 & 0 & 0 \\ 0 & 1 & x & 1 & 0 & 0 \\ 0 & 0 & 1 & x & 1 & 0 \\ 0 & 0 & 0 & 1 & x & 1 \\ 1 & 0 & 0 & 0 & 1 & x \end{vmatrix} = 0$$

where

$$x = \frac{\beta}{\alpha - E}$$

which has the solutions,

$$x = \pm 1, \pm 1, \text{ and } \pm 2$$

i.e.

$$E = \alpha \pm \beta, \alpha \pm \beta \text{ and } \alpha \pm 2\beta$$

Thus, we have six molecular orbitals, including two degenerate pairs. As β is usually a negative quantity, we can schematically represent the orbital energies in a diagram such as Fig. 1. This figure also illustrates the shapes and signs of the various orbitals.

The Hückel method is perhaps the simplest of a family of molecular orbital methods of increasing complexity. At a more sophisticated level some attempt is made to evaluate the integrals H_{mn} and S_{mn} in the secular determinant and such methods as CNDO (Complete Neglect of Differential Overlap), INDO (Incomplete Neglect of Differential Overlap) and MINDO (Modified Incomplete Neglect of Differential Overlap) allow the semi-empirical calculation of both σ and π orbital energies, with decreasing levels of approximation, in a few seconds on a minicomputer. At the other extreme, the most sophisticated ab initio methods make few assumptions and can reproduce energetic and structural parameters for small molecules to within a few percent of their spectroscopic values. However, such calculations are computer intensive and require many minutes of processor time on even the fastest supercomputer. With the rapid

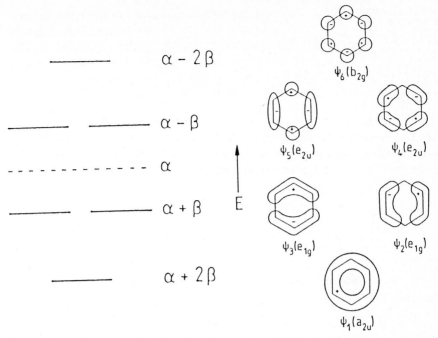

$\alpha - 2\beta$

$\psi_6(b_{2g})$

$\alpha - \beta$

$\psi_5(e_{2u})$ $\psi_4(e_{2u})$

α

$\alpha + \beta$ E

$\psi_3(e_{1g})$ $\psi_2(e_{1g})$

$\alpha + 2\beta$

$\psi_1(a_{2u})$

Fig. 1. Energies and symmetries of the Hückel molecular orbitals of benzene. Reprinted from [4] by permission of John Wiley & Sons, Ltd.

spread of small, powerful, graphics-orientated workstations, we have seen in the last few years an increasing interest in molecular modelling and molecular graphics. It is perhaps inevitable that the facility to calculate both positions and intensities of electronic transitions will become part of many sophisticated molecular graphics systems.

In much the same way as with atoms, molecular orbitals can be filled with electrons to yield electronic configurations, the various orbitals being labelled according to the symmetry of the orbital with respect to the overall symmetry of the molecule. Thus, for example in the case of N_2 which belongs to the $D_{\infty h}$ group, the ground state configuration can be written as,

$$N_2(\sigma_g 1s)^2(\sigma_u^* 1s)^2(\sigma_g 2s)^2(\sigma_u^* 2s)^2(\pi_u 2p)^4(\sigma_g 2p)^2$$

where the 1s, 2s and 2p indicate the nature of the original atomic orbital making up the molecular orbitals, while the ground state of benzene can be written as,

$$C_6H_6\ldots(1a_{2u})^2(1e_{1g})^4$$

Excited state configurations can simply be generated by promoting one or more electrons to higher lying orbitals.

As was true with atoms, a single configuration can give rise to many spectroscopic states and is therefore of little use other than an indicator as to which electron promotions occur upon electronic excitation. We can define a molecular state symbol in a similar manner to the atomic state symbols, $^{2s+1}\Gamma(\Phi_e^o)_\Omega$ where $\Gamma(\Phi_e^o)$ is the symmetry species of the orbital part of the molecular orbital wavefunction, defined as the product of the symmetry species of all the occupied molecular orbitals, i.e.

$$\Gamma(\Phi_e^o) = \prod_i \Gamma(\Phi_i)$$

The product of a set of filled orbitals is the totally symmetric species within the particular point group concerned; so, for the ground state of N_2, $\Gamma(\Phi_e^o)$ is Σ_g^+, while for C_6H_6, it is A_{1g}. For excited states, we have, for example, for the N_2 excited configuration

$$\ldots(\pi_u 2p)^3(\sigma_g 2p)^2(\pi_g^* 2p)^1$$

$$\Gamma(\Phi_e^o) = \pi_u \times \pi_g$$

$$= \Sigma_g^+ + \Sigma_u^- + \Delta_u$$

when the appropriate symmetry transformations are considered, while for the first excited state of C_6H_6

$$(1a_{2u})^2(1e_{1g})^3(1e_{2u})^1$$

$$\Gamma(\Phi_e^o) = e_{1g} \times e_{2u}$$

$$= B_{1u} + B_{2u} + E_{1u}$$

The value of the net spin angular momentum S is defined in terms of the Clebsch-Gordan series,

$$S = s_1 + s_2, s_1 + s_2 - 1, \ldots, |s_1 - s_2|$$

and naturally, for most closed shell molecules i.e. those in which all molecular orbitals are filled, this gives rise to two possible spin configurations, the S = 0 or Singlet state and the S = 1 or Triplet state. With simple open shell species, the doublet ($S = \frac{1}{2}$) and quartet ($S = \frac{3}{2}$) states are analogous. The value of Ω is defined as

$$\Omega = |\Lambda + \Sigma|$$

Table 1. The correlation of Λ with $\Gamma(\Phi_e^o)$ for diatomic molecules

$\Gamma(\Phi_e^o)$	Λ
Σ	0
Π	1
Δ	2
Φ	3
$\vdots$	$\vdots$

where $\sum$ is the projection of S on the molecular symmetry axis, simply given by

$$\sum = S, S - 1, \ldots, 0, \ldots, -(S - 1), -S$$

and Λ is the projection of the total orbital angular momentum on the same axis. This is readily related to $\Gamma(\Omega_e^o)$; for example in diatomic species we have the assignments given in Table 1. Similar relationships can be drawn for polyatomic species, but are normally appended to the state symbol only where systems exhibit large splittings due to spin-orbit or other angular momentum couplings.

Again, the electronic transitions are only allowed if the transition dipole moment is non-zero. However, with molecules we have the added complication of vibrational and rotational motions to be considered which make it difficult for us to establish rigorous rules such as exist for atoms, except for the pure electronic transitions of diatomic molecules, where the following apply:

a) $\Delta\Lambda = 0, \pm 1$
b) $\Delta S = 0$
c) $\Delta\Sigma = 0; \Delta\Omega = 0, \pm 1$
d) $+ \leftrightarrow +; - \leftrightarrow -; + \leftrightarrow\!\!\!/ -$
e) $g \leftrightarrow u; u \leftrightarrow\!\!\!/ u; g \leftrightarrow\!\!\!/ g$

For polyatomic molecules and for vibronic (electronic-vibrational) and rovibronic (electronic-rovibrational) transitions only full calculations of the transition dipole moment can determine whether a transition will be observed or not.

As in the case of atoms, the ΔS selection rule is known to break down in instances of molecules containing heavy atoms and singlet-triplet absorption spectra can consequently be observed. Such spectra are, of course, extremely weak but have been observed using intense radiation sources, in particular tunable lasers.

2.3   Chromophores

While ideally, state symbols reflecting orbital symmetry should be assigned to molecular electronic configurations and transitions, this is possible in the relatively small number of situations where detailed calculations have been performed. It is possible, however, to recognize that in a complex molecule, the absorption of light of a particular frequency range can often be associated with a particular atom or group of atoms referred to as a chromophore (c.f. group vibrations in infrared spectroscopy). These chromophore transitions can then be labelled according to the type of electrons involved in the transition. For example in simple, non-conjugated carbonyl compounds, the weak (ε_{max} typically $15 - 20 \, dm^3 mol^{-1} cm^{-1}$) transition at around 300 nm is associated with the promotion of a non-bonding electron largely associated with the oxygen atom to a delocalised anti-bonding π orbital, and hence the (n, π^*) label is normally given to the transition. In a similar way, labels have been attached to a variety of

Table 2. Labelling of simple unconjugated chromophores

Chromophore	Transition Label	Approximate Wavelength in nm
σ-Bonded Electrons		
C—C and C—H	(σ,σ^*) or $\sigma^* \leftarrow \sigma$	≈ 150
Lone-pair Electrons		
—Ö—	(n,σ^*) or $\sigma^* \leftarrow n$	≈ 185
—N̈<	(n,σ^*) or $\sigma^* \leftarrow n$	≈ 195
—S̈—	(n,σ^*) or $\sigma^* \leftarrow n$	≈ 195
>C=Ö	(n,π^*) or $\pi^* \leftarrow n$	≈ 300
>C=Ö	(n,σ^*) or $\sigma^* \leftarrow n$	≈ 190
—N=Ö	(n,π^*) or $\pi^* \leftarrow n$	≈ 650
π-Bonded Electrons		
>C=C<	(π,π^*) or $\pi^* \leftarrow \pi$	≈ 190

transitions associated with simple, unconjugated chromophores as listed in Table 2. The exact position and strength of a transition depends markedly on the degree of conjugation. In general, however, the higher the degree of conjugation, the further to the red the transition will appear. It is possible, however, to predict with some degree of accuracy the positions of electronic transitions in organic molecules using a set of empirical rules established by Woodward. These are considered in some detail in standard texts on spectroscopic methods in organic chemistry, such as Williams and Fleming [8], Kemp [9] and Scott [10], and are not discussed here.

3   Instrumentation in Electronic Absorption Spectroscopy

A detailed discussion of instrumentation in electronic absorption spectroscopy would require a major article on its own. Consequently, the reader is directed to a recent review by Threfall [11] for a more detailed discussion of the current state of the art in both instrumentation and data analysis. We will consider the field more briefly here, highlighting a few important points.

3.1   Spectrographs

The resolving power, R, of a spectroscopic instrument is defined as below

$$R = \frac{\lambda}{d\lambda}$$

where λ is the operating wavelength and $d\lambda$ the limiting resolution in wavelength units. For a classical prism spectrograph, the resolving power can be demonstrated to be given by

$$R = b\frac{dn}{d\lambda}$$

where b is the edge dimension of the prism and n is the refractive index of the material from which the prism is constructed. Clearly, to achieve a high resolving power two alternatives exist. First, b can be large. This is generally impractical for the obvious reason of cost, and also for limits imposed by the absorption of the optical material from which the prism is constructed. Alternatively, the value of $dn/d\lambda$ can be maximised. It is generally the case, however, that $dn/d\lambda$ is maximised near the absorption edges of optical materials, and hence the range of wavelengths that can be resolved is limited by loss of transmitted light due to absorption in the prism. Thus, glass prisms which absorb below 360 nm exhibit their highest resolving power in the blue and violet regions of the visible spectrum, while quartz, which absorbs below 185 nm, is optimal for the 200 to 300 nm region.

An alternative to the prism spectrograph is the grating spectrograph. Diffraction of radiation from a ruled grating or from a holographic grating is governed by the diffraction law,

$$m\lambda = d(\sin i + \sin \theta)$$
$$= d \sin \theta \quad \text{for normal incidence}$$
$$= 2 \, d\sin \theta \quad \text{if } i = \theta$$

where m is the order of diffraction, d is the groove spacing, i is the angle of incidence and θ is the angle of reflection. From these expressions, it has been shown that the resolving power of a grating spectrograph is given simply by

$$R = mN$$

where N is the number of grooves illuminated. Given that the grating operates in a reflection-like mode, the problems associated with substrate absorption are significantly reduced.

In classical spectrographs, where the sample is illuminated by broadband radiation and the radiation then dispersed, photographic plates have been and are still widely used. Large spectrographs still find use in a few laboratories where high resolution spectra of gases are of interest, although they have now been largely replaced by laser-based systems (see Chapter 8), as they offer a very flexible means of recording both absorption and emission spectra with the huge multichannel advantage of a photographic plate.

3.2   UV/Visible Recording Spectrometers

For routine analytical applications, the spectrograph has been replaced by the recording spectrometer. At its simplest such a single-beam instrument consists simply of a source of tunable, monochromatic light, a sample compartment and a detector. While this is adequate for many simple colorimetric determinations, it suffers from the severe disadvantage that as the source wavelength is altered, the instrument background also changes. Thus, to record an absorption spectrum over a range of wavelengths, two measurements must be made at each wavelength, of both the sample and reference.

Dual-beam instruments, as in the infrared, offer true ratio recording and eliminate the problem described above with single beam instruments. Modern dual-beam spectrometers, equipped with holographic gratings and sensitive photomultipliers, are quite capable of recording spectra with a typical bandpass of approx. 0.1 nm with excellent S/N ratio, or of recording absorbances of the range of up to 5 or 6. Moreover, such instruments are routinely interfaced to a microcomputer data station for digital representation of spectra and off-line data analysis. One problem associated with such systems is that to achieve high resolution, slits must be used in conjunction with the monochromator to reduce the bandwidth of the exciting radiation. Consequently, for narrow band operation, extremely low levels of light have to be measurable at each wavelength. Thus, to achieve a significant S/N ratio, considerable time must be spent in scanning through the wavelengths very slowly.

In recent years, with the increasing availability of pixel-based array detectors such as photodiode arrays (PDA) or more recently charge-coupled devices (CCD) and charge-injection devices (CID), this problem has been resolved somewhat by a return to the idea of multichannel detection originally associated with the classical spectrograph. However the current generation of such devices do not offer by any means the multichannel advantage associated with a photographic plate. Thus array-based spectrometers have by no means replaced scanning spectrometers, but merely complement them. Such array-based spectrometers have found their particular niche as detectors in HPLC, where spectra, recorded at a resolution of the order of a 1 nm or so over the several 100's of nm in the UV and visible, are required every few tenths of a second as the various components in a multi-component mixture are eluted from the column.

3.3   Interferometers

In contrast to the infrared, where the traditional dispersive, scanning spectrometer has been all but replaced by the interferometer and Fourier transform techniques, the use of interferometers in the UV and visible is distinctly limited. Several top-of-the-range FTIR instruments do offer the extension of their range into the visible and UV and one specifically designed FT-UV/Vis instrument does exist. However, the routine application of interferometry in this region has been hindered by the requirement for very high mechanical tolerances in the optical benches. Moreover, the clear advantages that interferometry has over dispersive techniques in the infrared, while still valid in the UV/visible, must be tempered with the fact that detector sensitivities in this region are many orders of magnitude greater than those in the IR and hence dispersive techniques have still not reached their limits.

3.4 Tunable Lasers

Perhaps the most important recent advance in instrument in electronic absorption spectroscopy has been the widespread and growing use of tunable lasers. A detailed discussion of these systems is given in Chapter 8.

4 Applications of Electronic Absorption Spectroscopy

4.1 Qualitative Applications

In stark contrast to infrared and NMR spectra, the vast majority of electronic spectra show little structure, appearing largely as broad coarsely structured absorption bands exemplified by the spectra in Fig. 2(b). Even in the gas phase, structure only becomes apparent when high resolution measurements are made, Fig. 2(a). Such measurements are, however, generally not feasible with standard bench spectrophotometers and are mostly regarded as the province of the high resolution spectroscopist using laser-based techniques. This lack of distinctive structure in electronic absorption spectroscopy is a severe drawback

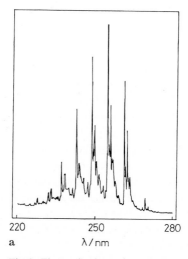

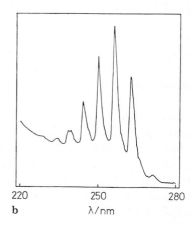

220	250	280
a	λ/nm	

220	250	280
b	λ/nm	

Fig. 2. Electronic absorption spectrum of benzene recorded in (**a**) the gas phase and (**b**) ethanol solution at a resolution of around 2Å on a Pye-Unicam SP.8100 UV Recording Spectrophotometer.

when being used in a qualitative manner, in contrast to the highly structured nature of IR and NMR spectra. This naturally means that UV/vis spectroscopy is largely restricted to providing supporting evidence for an identification based on IR and/or NMR evidence. This is particularly true when an analysis of the UV/vis spectrum is performed using the Woodward rules [8–10].

One of the significant areas in which qualitative application of electronic absorption spectroscopy is currently used is in simple colour comparisons. Numerous simple instruments allow the qualitative and semi-quantitative comparison of the colour of various samples against colour standards. These are naturally of considerable importance in the area of dye chemistry where the hue and intensity of a newly synthesised dye must be compared with previously existing materials. Naturally, this is one area where the use of theoretical calculations of electronic spectra has significant application. By simulating the electronic absorption spectrum of a proposed dye, the colour properties of the dye can be predicted reasonably well without actually synthesising it. The structure of the molecule can then be modified until the desired colour properties are obtained. Only then would the synthetic chemist take over.

4.2 Quantitative Applications

While it may appear that electronic absorption spectroscopy is rather limited in its application, it is only when one considers the quantitative application of the technique that its real power becomes apparent. Quantitative electronic absorption spectroscopy is based on the Beer-Lambert Law, which links the absorbance, A, of a substance at a particular wavelength to the concentration, c, and a molecular constant, the molar absorption coefficient, ε,

$$A = \log_{10} I_0/I$$

$$= \varepsilon cl$$

where l is the pathlength of the sample, I_0 is the intensity of the incident radiation and I is the transmitted intensity. Modern spectrophotometers are readily capable of determining absorbances of the order of 0.001 which, for a moderately strong absorber with an ε of around $100 \, m^2 \, mol^{-1}$, would correspond to a concentration of around $10^{-6} \, mol \, dm^{-3}$. With strong absorbers, such as polycyclic aromatic hydrocarbons, concentrations 2 or 3 orders of magnitude lower can be measured. With this approach, measurements on the ppm, and for certain species in the ppb, region are quite possible both for organic species and for inorganic species. The analysis of the latter through complexation to yield highly coloured species is well described in a variety of textbooks, including the classic text by Vogel [12].

The use of the Beer-Lambert law can be extended to multi-component systems as the absorbances of the various components in such a system are, at a particular wavelength, additive. Thus for a system with i components, the absorbance is given by

$$A(\lambda) = 1 \sum_i \varepsilon(\lambda)_i c_i$$

where c_i is the concentration of the ith component. By measuring the absorbance of the mixture at i independent wavelengths, it is then possible to obtain a set of i equations which can be solved to yield the concentrations of the various components when the extinction coefficients of these components are known. In recent years, this approach to multi-component analysis has received a major boost by the growing application of microcomputers in spectroscopy. Packages that will perform such analyses are available from many manufacturers of spectrophotometers both for the dedicated microcomputer sub-systems in the instruments and for stand-alone laboratory microcomputers.

Clearly, quantitative electronic absorption spectroscopy surpasses both IR and NMR techniques in terms of sensitivity. The ease of use and the relative ease with which samples can be prepared mean that quantitative UV/vis spectroscopy is probably the first choice analytical method for many simple organic and inorganic species at concentrations ranging from a few 100's of ppb up.

5   References

1. Atkins PW (1986) Physical chemistry, Oxford University Press, Oxford
2. Lowe JP (1978) Quantum chemistry, Academic, New York
3. Moore WJ (1972) Physical chemistry, Longman, London
4. Hollas JM (1987) Modern spectroscopy Wiley, London
5. Sutton(1968) Electronic spectra of transition metal complexes, McGraw-Hill, London
6. Herzberg (1944) Atomic spectra and atomic structure Dover, New York
7. Heitler W, London F (1927) Z. Physik 44: 455
8. Williams DH, Fleming I (1973) Spectroscopic methods in organic chemistry, McGraw-Hill, London
9. Kemp W (1975) Organic spectroscopy, Macmillan, London
10. Scott AI (1964) Interpretation of ultraviolet spectra of natural products, Pergamon Press, Oxford
11. Threfall TL (1988) European Spectroscopy News 88: 8
12. Vogel's Textbook of Quantitative Inorganic Analysis, revised by Bassett J, Denney RC, Jeffry GH, Mendham J (1978) Longman, London

CHAPTER 5
Luminescence Spectroscopy

C.S. Creaser and J.R. Sodeau

1 Introduction

Luminescence spectroscopy provides one of the most sensitive and selective methods of analysis for many inorganic and organic compounds. The techniques for measuring fluorescence, phosphorescence, chemiluminescence and bioluminescence spectra have become highly developed since the mid-1950s. Measurements of the wavelength distributions of luminescence provide important information on the nature and energy of the emitting species. Measurements of the variation of luminescence intensity as a function of time can assist not only the identification of excited electronic states but also the determination of the efficiency with which the luminescence process occurs. The two main aims of this chapter are to provide:

a) a basic vocabulary of the terms used by photochemists to explain the detailed molecular events behind the observation of luminescence in polyatomic organic compounds;
b) a description of the experimental methods used in the measurement of luminescence spectra and some analytical applications of the technique.

2 Fundamentals of Photophysics

2.1 Photophysical Decay Processes [1,2,3]

Both the efficiency with which a particular compound luminesces and the lifetime of the associated, emitting electronic state depend upon a complex interplay between radiative and non-radiative deactivation pathways. The dynamics involved are normally considered from a statistical viewpoint, in which the efficiencies of the competing unimolecular physical decay processes are related to events occurring after the absorption of one photon. In other words the quantum yield of a particular photophysical process cannot exceed unity. For a polyatomic molecule the most important ways in which excited electronic states decay are as follows:

Radiative Processes
a) Fluorescence
b) Phosphorescence

Non-radiative Processes
c) Internal Conversion (IC)
d) Intersystem Crossing (ISC)

Information regarding each of the photoprocesses a–d is best summarised by means of diagrams. Three types of summary are possible, the one that carries the most complete description being the most difficult to visualise.

2.1.1 One-dimensional Representation of Photophysics

Here the only variable is energy—although information about spin multiplicity is included. It is generally termed a modified Jablonski Diagram, the basic features of which are shown in Fig. 1. The representation divides the electronic states into a singlet manifold S_n, (where n = 0,1,2...) and a triplet manifold T_n, (where n = 1,2....). It is important to note that the energy of the T-state with a particular value of n is always lower than the corresponding S-state, for reasons which will be discussed later. The second important feature is that radiationless transitions occur between isoenergetic rovibronic states, whereas radiative processes clearly involve energy loss in the form of a photon. Vibrational relaxation is also indicated on the diagram and is, of course, extremely efficient in solution where the energy is transformed into heat after solvent-solute collision.

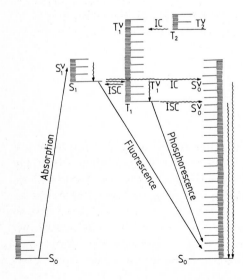

Fig. 1. One-dimensional representation of photophysics

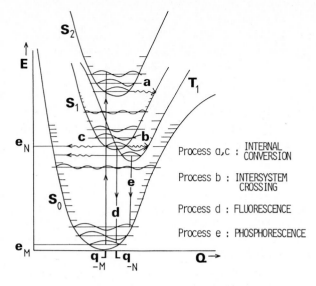

Process a,c : INTERNAL
             CONVERSION

Process b : INTERSYSTEM
            CROSSING

Process d : FLUORESCENCE

Process e : PHOSPHORESCENCE

Fig. 2. Two-dimensional representation of photophysics

2.1.2   Two-dimensional Representation of Photophysics.

This type of summary is shown in Fig. 2 and represents a plot of energy vs nuclear configuration coordinate. The diagram is much more informative than Fig. 1 as it shows that excited electronic configurations are essentially novel chemical species which need not have the same geometry (in terms of bond length or bond angle) as their non-excited counterparts. Thus the S_1 state for an excited carbonyl compound, such as formaldehyde, would show a considerable coordinate displacement from the S_0 state, whereas the corresponding change is much less pronounced for an essentially rigid, extended aromatic system such as anthracene. In the first case a considerable Stokes shift of fluorescence vs absorption is to be expected for the O—O bond whereas a mirror image relationship between fluorescence and absorption is observed for anthracene (Fig. 8).

2.1.3   Three-dimensional Representation of Photophysics.

The third variable in this representation is time. Figure 3 shows a schematic for the photophysics of acetone. Electronic energy is absorbed instantaneously in steps ① → ② from $S_0 \to S_1^\dagger$. Vibrational relaxation to give S_1° occurs in approx 10^{-12} s. Fluorescence (③ → ④) is a fast spin-allowed photoprocess and is complete in 1–2 ns, but intersystem crossing $S_1^\circ \rightsquigarrow T_1^\dagger$ takes 10^{-6} s. Phosphorescence being spin-forbidden is a relatively long-lived process (⑦ → ⑧) and the $^3n\pi^*$ triplet state of acetone has a lifetime $\tau_T \approx 1$ ms. Aromatic compounds

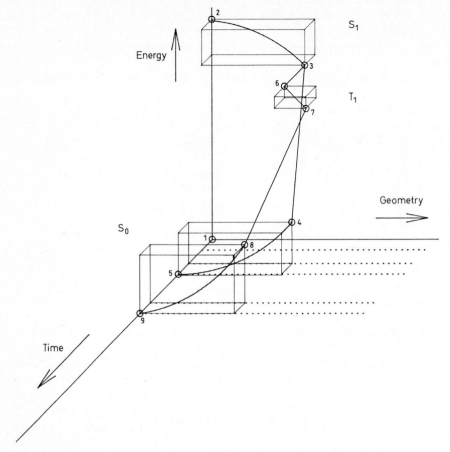

Fig. 3. Three-dimensional representation of photophysics (acetone)

which have lowest $^3\pi\pi^*$ triplet states typically have lifetimes of the order $\tau_T \approx 1$–10 s, for reasons that will be discussed below. The final geometric reorientation of $S_0^\dagger$, either ④ → ⑤ or ⑧ → ⑨ takes another 10^{-12} s. Figure 3 also shows that the geometric change from $S_0 \rightarrow S_1$ is much greater than for $S_1 \rightarrow T_1$, and that the fluorescence process is accompanied by a greater energy change than is found for phosphorescence: $\lambda_{max} = 405$ nm and 440 nm for $S_1 \rightarrow S_0^\dagger$ and $T_1 \rightarrow S_0^\dagger$ respectively.

From the above discussion it is clear that a 3-D photophysical profile for an aromatic organic compound would look very different from its carbonyl counterpart. The underlying reasons for this orginate from the contrasting electronic structures as well as from differences in molecular rigidity. A specific example can be discussed with reference to the 10^3 difference in triplet lifetime observed for $^3n\pi^*$ vs $^3\pi\pi^*$ lowest excited states. The physical principle on which this result rests is that for symmetric space wavefunctions (singlets) the electrons tend to crowd together whereas for antisymmetric space wavefunctions (triplets) the electrons tend to avoid each other. Thus, electrostatic repulsion forces are more pronounced in the former states leading to their relative destabilisation from the corresponding triplets. The magnitude of this splitting depends purely on Coulombic repulsion; it has nothing to do with magnetic effects.

For π-orbital $\rightarrow \pi^*$-orbital excitation the two available electrons occupy the same region in space. For n-orbital $\rightarrow \pi^*$ orbital excitation the two available electrons are orthogonally disposed in space. Hence electrostatic repulsion terms are much more significant in $^1\pi\pi^*-^3\pi\pi^*$ splitting than they are in $^1n\pi-^3n\pi^*$ splitting. The observed effect is that the singlet-triplet energy gap is much greater for aromatic compounds than it is for carbonyl compounds e.g. $11,500\,\text{cm}^{-1}$ for anthracene vs $2000\,\text{cm}^{-1}$ for acetone. Under these conditions, the $^3n\pi^*$ lowest triplet state of CH_3COCH_3 takes on singlet character from the $^1n\pi^*$ singlet state (even though this spin-orbit perturbation is formally forbidden); by contrast the anthracene $^3\pi\pi^*$ remains a 'pure' aromatic triplet state. The formally spin-forbidden transition $T_1 \rightarrow S_0$ (phosphorescence) is therefore slower for molecules like anthracene compared to molecules like acetone.

In the discussion so far it is clear that for polyatomic organic molecules fluorescence and the singlet are coupled, as are phosphorescence and the triplet state. Their exact relationship and definitions are described in Sects. 2.2 and 2.3 below.

2.2   Fluorescence and the Singlet State [4]

Fluorescence is a radiative photoprocess, which occurs between two energy levels of the same multiplicity, i.e. $\Delta S = 0$. From the definition it is clear that fluorescence does not necessarily involve singlet state intermediacy but for polyatomic organic compounds it is found that the process almost always takes place from the lowest vibrational level of the lowest excited singlet state and is represented as $S_1{}^\circ \rightarrow S_0$. This empirical discovery is known as Kasha's Rule and its basis depends upon the fact that radiationless internal conversion processes within the upper singlet state manifold (e.g. $S_2 \rightsquigarrow S_1$) are fast in comparison with emission (e.g. $S_2 \rightarrow S_0$). From this viewpoint, the principles which control non-radiative processes must first be explained before the reasons why compounds fluoresce efficiently or inefficiently can be understood.

Internal conversion is a non-radiative photoprocess which occurs between two isoenergetic levels of the same multiplicity, i.e. $\Delta S = 0$; intersystem crossing is the equivalent non-radiative photoprocess for states of different multiplicity i.e. $\Delta S \neq 0$. The latter is therefore formally spin-forbidden. The rate coefficient for either type of radiationless transition is theoretically proportional to:

(a) the density of the energy levels in the final state which are isoenergetic with the initial state;

(b) The square of the interaction energy between the two states.

Because the difference in energy between different *excited* states is small, e.g. $S_2 \leftrightarrow S_1$, $S_1 \leftrightarrow T_1$ (2000–12000 cm^{-1}), there is a good chance that a high vibrational level of a given final state is in correspondence with the lowest vibrational level of the initial state. Energy transfer can therefore occur quickly, leading to the formation of the lower energy excited state. However the difference in energy between S_1 and S_0 is relatively large (e.g. 30,000 cm^{-1} for acetone O—O band, in absorption) and the chance of a transition to a very high vibrational level of the ground state is small. Hence internal conversion to ground state S_0 is usually a relatively slow process and other methods of deactivation are more likely. In general the time scale for $S_2 \rightsquigarrow S_1$ is approx. 10^{-12} s whereas that for $S_1 \rightsquigarrow S_0$ internal conversion is approx. 10^{-6} s.

From the above discussion it is clear that intersystem crossing rate coefficients for carbonyl compounds will be much greater than for aromatic compounds since the $S_1 \leftrightarrow T_1$ energy gap is much smaller for the former molecules (see Sect. 2.1). In general this is true, although it should be noted that intersystem crossing can also occur between $S_1 \leftrightarrow T_n$ isoenergetic levels. In general, $1n\pi^* \leftrightarrow {}^3\pi\pi^*$ and ${}^1\pi\pi^* \leftrightarrow {}^3n\pi^*$ processes are fast whereas ${}^1n\pi^* \leftrightarrow {}^3n\pi^*$ and ${}^1\pi\pi^* \leftrightarrow {}^3\pi\pi^*$ processes are slow due to spin-orbit coupling mechanisms.

In terms of fluorescence, the preliminary expectation would be that ${}^1\pi\pi^*$ aromatic compounds fluoresce efficiently whilst ${}^1n\pi^*$-carbonyl compounds fluoresce weakly or not at all. This is indeed true for the gas phase but a more complete understanding of this observation requires consideration of other factors such as intrinsic lifetime (τ_0) and geometrical constraints such as molecular rigidity. The intrinsic lifetime, τ_0, of an excited state is the time taken for an emission intensity to reach ($1/e$) of its initial intensity assuming that no other deactivation process occurs. The probability per unit time of emitting a photon spontaneously is related to the intensity of absorption. If a few approximations are made a simple approximate relationship between τ_0 and the molar absorption coefficient ε_{max} at band maximum can be derived. It takes the form:

$$\tau_0 \approx \frac{10^{-5}}{\varepsilon_{max}} \tag{1}$$

where τ_0 is measured in seconds and ε_{max} is in units, m^2 mol^{-1}.

Due to the relative spatial overlap of the orbitals involved, n-π^* transitions are generally weak with $\varepsilon_{max} \sim 0.1$–40 m^2 mol^{-1} whereas π-π^* transitions are generally strong with $\varepsilon_{max} \sim 500$–20,000 m^2 mol^{-1}. Therefore the intrinsic singlet lifetime of a carbonyl compound is long relative to an aromatic compound. Consequently although internal conversion from $S_1 \rightsquigarrow S_0$ is slow enough to allow emission from ${}^1\pi\pi^*$ states, deactivation of ${}^1n\pi^*$ states to give ${}^3n\pi^*$ states is possible by the formally spin-orbit forbidden intersystem-crossing process. In fact the timescale for $S_1 \rightsquigarrow T_n$ (${}^1n\pi^*$–${}^3n\pi^*$ or ${}^1\pi\pi^*$–${}^3\pi\pi^*$) is approx. 10^{-6} s whereas $S_1 \rightsquigarrow T_n$ (${}^1n\pi^*$–${}^3\pi\pi^*$ or ${}^1\pi\pi^*$–${}^3n\pi^*$) is approx. 10^{-8} s.

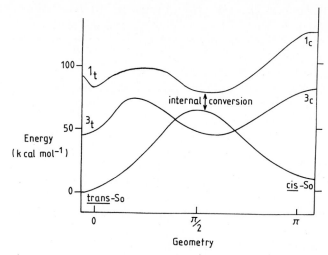

Fig. 4. Energy profiles of S_0, S_1 and T_1 states of cis- and trans- stilbene

Rigid molecules generally exhibit a higher quantum yield of fluorescence, Φ_F, than non-rigid ones; the behaviour of cis- and trans-stilbene provides a good example of how molecular rigidity can more subtly affect the fluorescence process. Figure 4 shows proposed energy profiles of the S_0, S_1 and T_1 states of stilbene as a function of the angle of twist (θ) about the double bond [5]. In solution at room temperature trans-stilbene has a fluorescence quantum yield of 0.08 which increases to approx. 1.0 at 193 K. By contrast cis-stilbene is non-fluorescent except in highly viscous solvents. These results can be explained by the fact that the activation energy (approx. $10 \, \mathrm{kJ \, mol^{-1}}$) for $^1\pi\pi^*$ trans-stilbene to achieve the twisted $\pi/2$ geometry cannot be obtained at the low temperature: the planar configuration can only fluoresce. The cis-isomer has no activation barrier to rotation and internal conversion $S_1 \rightsquigarrow S_0$ becomes an effective process from the twisted geometry: viscous solvents do however reduce the efficiency of the rotamerisation leading to the observation of fluorescence from the planar cis-stilbene configuration.

2.3 Phosphorescence and the Triplet State [6].

Phosphorescence is a radiative photoprocess which occurs between two energy levels of different multiplicity i.e. $\Delta S \neq 0$. For organic compounds the transition involved is always $T_1 \rightarrow S_0$. The triplet state is populated almost exclusively via intersystem crossing in the following sequence: $S_1 \rightsquigarrow T_n \rightsquigarrow T_1$. In its most simple consideration T_1 can be regarded as a molecule with two parallel unpaired electron spins i.e. a paramagnetic species. In fact the time-dependence of the phosphorescence emission decay process and the paramagnetic susceptibility decay profile are identical for systems such as triphenylene in boric acid "glass" at room temperature. Results of this type first provided strong evidence for the actual existence of triplet states.

From the discussion in Sect. 2.1 and 2.2 it should be clear that: (a) population of T_1 via intersystem crossing is important for carbonyl compounds but is a less efficient depletion mechanism for the S_1 state of aromatic compounds. (b) $^3\pi\pi^*$ states have triplet lifetimes, τ_T, in the order of seconds whereas $^3n\pi^*$ states have triplet lifetimes in the order of milliseconds. The relative length of τ_T or τ_P ($1-10^3$ ms) compared to the singlet lifetime, τ_S or τ_F (ns), means that the T_1 state is easily quenched and that phophorescence spectra must normally be obtained in rigid media where collisional deactivation with solvents or molecular oxygen is prevented. It also leads to the following generalised statement: "$^3\pi\pi^*$ states phosphoresce strongly in the solid but never in the gas phase, whereas $^3n\pi^*$ states may phosphoresce in either phase".

Population of the triplet states can be modified by inducing spin-orbit coupling perturbation mechanisms. The incorporation of 'heavy atoms' such as iodine in the solvent is a well known method by which this is achieved. Essentially, the spin-orbit coupling effect, in which the magnetic field produced by electron spin interacts with the field associated with orbital angular momentum, is enhanced at large atomic number, Z. Under these conditions 'pure' spin states do not exist and the rigorous selection rule $\Delta S = 0$ is inappropriate—only total angular momentum must be conserved. For molecules such as naphthalene the following observations are apparent on moving from an ethanol/methanol solvent to a system containing propyl iodide:

a) Decrease in quantum yield of fluorescence.
b) Increase in quantum yield of phosphorescence.
c) Decrease in triplet state lifetime.

The final point is to be expected as the incorporation of heavy atoms into the solvent increases the rate of all singlet-triplet processes, including $T_1 \leadsto S_0$ intersystem crossing.

Throughout the discussion so far it has been apparent that two main experimental parameters are required in photochemical systems to characterise electronic states and the efficiency with which they undergo a particular luminescence process. The two measurements are quantum yield and lifetime; their definitions are discussed in the next section.

2.4   Quantum Yields and Lifetimes

The quantum yield for a primary photoprocess represents the efficiency of photon usage for that particular deactivation pathway. Therefore:

$$\Phi_F = \frac{\text{number of fluorescence quanta emitted}}{\text{number of quanta absorbed into a singlet excited state}}$$

The corresponding equation for phosphorescence is less obvious because of the spin-forbidden nature of direct $S_0 \to T_1$ absorption. Nonetheless the definition still relates the number of quanta emitted to the number of quanta absorbed:

$$\Phi_P = \frac{\text{number of phosphorescence quanta emitted}}{\text{number of quanta absorbed into a singlet excited state}}$$

It is an appropriate description because a phosphorescent molecule is initially excited into a singlet state. However the definition does mean that Φ_P is a composite quantity, being the product of the triplet quantum yield, Φ_T, and the phosphorescence quantum efficiency, θ_P. The first component is the fraction of molecules ending up in the T_1 state and the second component is the fraction of T_1 molecules which emit a photon.

The number of quanta absorbed, which appears in the denominators of the expressions for Φ_F and Φ_P, are effectively the light intensity/density values taken from the Beer-Lambert law (see Chapter 4). These values can be expressed as the average number of photons absorbed by the photolyte per unit volume and unit time, I_a. Therefore the quantum yield equations can be reformulated in terms of rate ratios:

$$\Phi_{\text{Emission}} = \frac{\text{rate of emission}}{I_a} \tag{2}$$

This leads to an important relationship which links quantum yields to excited state lifetimes. Consider the simple photochemical scheme:

$$A^{S_0} + h\nu \to A^{S_1} \qquad \text{Rate} = I_a$$
$$A^{S_1} \to A^{S_0} + h\nu_F \qquad \text{Rate} = k_F[A^{S_1}]$$
$$A^{S_1} \to A^{T_1} \qquad \text{Rate} = k_{ISC}[A^{S_1}]$$
$$A^{T_1} \to A^{S_0} \qquad \text{Rate} = k'_{ISC}[A^{T_1}]$$
$$A^{T_1} \to A^{S_0} + h\nu_P \qquad \text{Rate} = k_P[A^{T_1}]$$

$$\Phi_F = \frac{k_F[A^{S_1}]}{I_a} \tag{3}$$

Application of the steady state approximation to the transient species, A^{S_1}, gives:

$$\frac{d[A^{S_1}]}{dT} = I_a - (k_F + k_{ISC})[A^{S_1}] = 0 \tag{4}$$

Therefore:

$$[A^{S_1}] = \frac{I_a}{k_F + k_{ISC}} \tag{5}$$

and hence by simple algebraic substitution into Eq. (3):

$$\Phi_F = \frac{k_F}{k_F + k_{ISC}} \tag{6}$$

i.e. the quantum yield is the ratio of the rate coefficient for fluorescence to the sum of the rate coefficients for A^{S_1} depletion. The expression is, in fact, also the ratio between τ_F, the measured fluorescence lifetime and τ_0, the intrinsic lifetime of the S_1 electronic state, because:

$$\tau_0 = \frac{1}{k_F} \quad \text{and} \tag{7}$$

$$\tau_F = \frac{1}{k_F + k_{ISC}} \tag{8}$$

This definition of τ_0 was implicitly described in Sect. 2.2, where it was assumed that emission was the only deactivation process for the S_1 state. It has a useful theoretical function in its link with the absorption coefficient of the electronic transition. The experimentally determined parameter τ_F represents the more realistic case when the excited singlet state is also depleted by non-radiative photoprocesses. In fact the scheme given above is generally appropriate for organic compounds as internal conversion, $S_1 \rightsquigarrow S_0$, is a very inefficient process. However, if internal conversion is significant, then the expression for τ_F is modified by inclusion of k_{IC} in the denominator (see Sect. 2.5.1).

An overall estimate of Φ_F for a particular compound may therefore be calculated as follows:

$$\Phi_F \approx 10^5 \tau_F \cdot \varepsilon_{max} \tag{9}$$

with each quantity on the right expressed in SI units. The measurement of absolute quantum yields is very difficult and involved, but values for Φ_F can readily be determined by computer-based integration methods relative to standards with a known Φ_F such as quinine sulphate in 0.5 M sulphuric acid. The key equation for the calculation of a quantum yield relative to a standard reference is as follows:

$$\Phi_F = \Phi_{REF} \cdot \frac{A_{REF}}{A_F} \cdot \frac{E_F}{E_{REF}} \cdot \left[\frac{\eta_F}{\eta_{REF}}\right]^2 \tag{10}$$

where, A values refer to the absorbances, E values corrected emission areas and η values refractive indices.

One final point of note regarding Φ_F is that for organic compounds the quantity is independent of the excitation wavelength. This is so because of Kasha's rule (Sect. 2.2) and the large rate coefficient of vibrational relaxation, i.e.

Table 1. Various photophysical parameters of a selection of organic compounds in solution

Compound	Absorption		Fluorescence			Phosphorescence[a]			Non-Radiative
	λ_{max} (nm)	ε_{max} (m² mol⁻¹)	λ_{max} (nm)	ΦF	τF (s)	λ_{max} (nm)	ΦP	τP (s)	ΦT
Acetaldehyde	292	1	406	0.84×10^{-3}	2.2×10^{-9}	430	0.0034	ca. 10^{-5}	1.0
Acetone	275	1.4	407	1.17×10^{-3}	1.5×10^{-9}	445	0.04	0.6×10^{-3}	1.0
Benzophenone	331	16.7	390	4×10^{-6}	5×10^{-12}	450	0.74	5.2×10^{-3}	1.0
	252	1860							
Naphthalene	311	25	325	0.19	100×10^{-9}	510	0.05	2.4	0.82
	275	553							
	221	1060							

[a]Quantum yield values measured in 'glasses' at 77 K

fluorescence originates from the lowest vibrational level of the lowest excited singlet state. From the Beer-Lambert law

$$I_F = I_{abs} \Phi_F \tag{11}$$

$$= I_0 \Phi_F (1 - e^{\varepsilon cl}) \tag{12}$$

where I_F is the intensity of fluorescence, c is the concentration and l the path length. At very low concentrations where the absorbance, $\varepsilon cl << 1$,

$$I_F = I_0 \Phi_F \varepsilon cl \tag{13}$$

As Φ_F is constant as a function of wavelength it follows that I_F is proportional to ε and c at a particular wavelength if the incident intensity, I_0, is the same at all wavelengths. The consequences of this are that: (a) fluorescence intensity is linearly related to the concentration of analyte; (b) the fluorescence excitation spectrum (see Sect. 4) is the same as the absorption spectrum in sufficiently dilute solutions.

To complete the discussion on quantum yields and lifetimes, Table 1 provides a summary of various photophysical parameters for a selection of representative organic compounds in solution (unless otherwise stated). It quantifies many of the issues raised in Sect. 2.1–2.4.

2.5   Luminescence quenching [7]

Fluorescence quantum yields approach a maximum value when measured in very dilute solution and when all 'impurities' (including dissolved oxygen) are rigorously removed. If the solute concentrations are increased, or if other compounds are present, Φ_F can be reduced and the fluorescence is said to be quenched.

Phosphorescence quenching is even more marked in fluid solution due to the relatively long lifetime of the triplet state. However there are several examples in the literature of room temperature solution phosphorescence under de-aerated conditions ($< 10^{-8}$ M[O_2]) e.g. acetone, benzophenone.

There is one other circumstance in which luminescence quenching is apparently observed: this is under experimental conditions which introduce the inner filter effect. The phenomenon is discussed in detail in Sect. 4. At this stage it is sufficient to mention that the effect arises generally from light absorption by the solvent or high concentrations of solute.

The kinetics of luminescence quenching are of two main types: (a) dynamic quenching; (b) static quenching. Both processes can be distinguished from inner filter effects because changes in 'true quencher' concentration alter the quantum yield of a fluorescent material whereas changes in an inner filter 'quencher' will not.

2.5.1 Dynamic Quenching

Under these circumstances there is a relative motion of probe and quencher species towards each other. The kinetics are analysed in terms of two competing processes in solution, namely a unimolecular singlet state decay route with experimentally observed lifetime τ_F, and a bimolecular quenching encounter with rate coefficient k_Q.

$$A^{S_1} \rightarrow A^{S_0} + h\nu_F \qquad \text{Rate} = k_F[A^{S_1}]$$
$$A^{S_1} \rightarrow A^{S_0} \qquad \text{Rate} = k_{IC}[A^{S_1}] \qquad \tau_F = 1/(k_F + k_{IC} + k_{ISC})$$
$$A^{S_1} \rightarrow A^{T_1} \qquad \text{Rate} = k_{ISC}[A^{S_1}]$$
$$A^{S_1} + Q \rightarrow A^{S_0} + Q \qquad \text{Rate} = k_Q[A^{S_1}][Q]$$

Clearly the bimolecular process becomes increasingly important for A^{S_1} deactivation as the concentration of Q (or A^{S_0}, if $Q = A^{S_0}$) is increased.

It is relatively simple to show from the kinetic scheme above that the ratio of Φ_F or τ_F values in the absence (Φ_F° or τ_F°) to presence (Φ_F^Q or τ_F^Q) of quencher is given by the Stern-Volmer relationship:

$$\frac{\Phi_F^\circ}{\Phi_F^Q} = 1 + k_Q\tau_F[Q] \tag{14}$$

or

$$\frac{\tau_F^Q}{\tau_F^\circ} = 1 + k_Q\tau_F[Q] \tag{15}$$

where the product $k_Q\tau_F$ is the Stern-Volmer quenching constant (k_{sv}) for the system. It was shown in Eq. (13) that I_F is proportional to Φ_F, and therefore graphical plots of I_F°/I_F^Q vs [Q] will provide a value for the quenching rate coefficient, k_Q, if the singlet state lifetime has been measured. There are countless examples of this kinetic behaviour in the literature, e.g. naphthalene/butan-2, 3-dione in cyclohexane; Ru^{2+}/(methyl viologen)$^{2+}$ in water. In solution, the rate of the quenching process is determined more by the rate of diffusion of the quenching and emitting molecules than by the rate of collision. An approximate expression for the diffusion-limited rate coefficient, k_D, is given by the Smoluchowski equation:

$$k_D \approx \frac{8RT}{3\eta} \tag{16}$$

where η is the viscosity of the solvent ($N\,s\,m^{-2}$). Self-quenching ($Q = A^{S_0}$) occurs for many compounds. This is another reason why Φ_F values should generally be determined in dilute solutions (approx. 10^{-5}–10^{-6} M concentrations).

The mechanism of the observed quenching generally depends upon the nature of the quencher. It is not appropriate to discuss the detailed molecular events in this review but it should be mentioned that the possibilities include

(a) enhancement of $S_1 \leadsto T_1$ intersystem crossing; (b) the formation of excited state collision dimers or complexes (excimers and exciplexes, respectively); (c) energy transfer (to produce an excited state of the partner molecule), which can act at short-range or over a long-range.

2.5.2 Static Quenching

If added solute forms a non-fluorescent complex with the ground state probe the quenching process is identified as static. No molecular collisions are involved and hence the process is viscosity-independent. In contrast to the dynamic quenching case the drop in fluorescence intensity that is measured with quencher addition occurs without any change in lifetime, τ_F. This is because fluorescence is only observed in encounter complexes which are devoid of quenchers.

There are cases, however, in which the quencher is not fully associated with the fluorescence probe e.g. quencher partitioning in micellar and aqueous phases. According to Perrin's sphere-of-action model there is a probability that an excited probe will lie within the 'active sphere' of volume $V = (4/3 \pi R^3)$. Under these conditions, the Φ_F values decrease exponentially with quencher concentration;

$$\frac{\Phi_F^\circ}{\Phi_F^Q} = \exp(V L [Q]) \qquad (17)$$

where L is Avogadro's number.

As a general rule if a change in the solvent produces alterations in both the absorption and fluorescence spectra, static quenching is indicated, i.e. the solvent combines in some way with the solute in the ground state. However if a solvent change only produces a shift in the fluorescence spectrum then interaction between the solute and the solvent in the excited state is indicated, i.e. dynamic quenching. Hydrogen-bonding and dielectric effects are often invoked to explain the mechanism of the shift in this latter case.

2.6   Chemiluminescence

In this process, light is emitted as the result of an energy release from an exothermic chemical reaction. The initial excitation is not thermal and, of course, is not initiated by light absorption. A detailed understanding of the excitation mechanism is, in fact, restricted to only a few systems in solution, e.g. luminol oxidation in alkaline mixtures of $H_2O_2/Fe(CN)_6^{3-}$. However the technique is of great value in analytical chemistry because very low levels of emitted light may be detected. For example a standard biochemical assay for ATP uses firefly luciferin and luciferase. The quantum yield for emission in this natural system is remarkably efficient (approx. unity). A wide-ranging review of the phenomenon is to be found in Ref. [8].

3 Luminescence Instrumentation

3.1 Introduction

Most luminescence spectrometers utilise the 90° optical system shown in Fig. 5(a). In this arrangement the light from the source is passed through an excitation wavelength selector and is then focused onto the sample. The fluorescence or phosphorescence emission is propagated in all directions and the portion emitted at right-angles to the incident beam passes through a second wavelength selector to the detector. The collection of luminescence at right-angles to the incident radiation is preferred to other configurations, because the detector directly views the emitted radiation (together with a small amount of scattered or reflected light) and not the much stronger source radiation. In chemiluminescence and bioluminescence instruments only the sample cell, emission wavelength selector and detector are required [8]. Solid samples, and those deposited on solid surfaces such as filter paper or thin layer chromatographic plates, are usually analysed using a frontal arrangement such as that shown in Fig. 5(b). Emitted radiation is usually viewed at an angle of approximately 37° to the incident light in order to avoid radiation directly reflected from the surface (specular reflection). Frontal illumination is also used for highly absorbing or opaque solution samples.

Luminescence instruments may be divided into two types; those that measure emission at fixed excitation and emission wavelengths, and those where the excitation or emission spectrum may be scanned. The first group of instruments are called fluorimeters or phosphorimeters, whilst the second group are classed as luminescence spectrometers, spectrofluorimeters or spectrophosphorimeters. A schematic diagram of the optical system of a commercial luminescence spectrometer is shown in Fig. 6. In many spectrometers, a small portion of the incident radiation is diverted away from the main path towards a reference

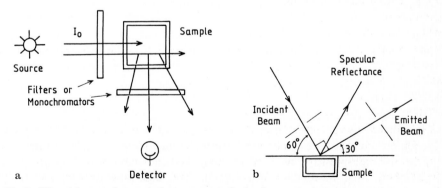

Fig. 5. 90° and front-surface optical arrangements for luminescence measurement

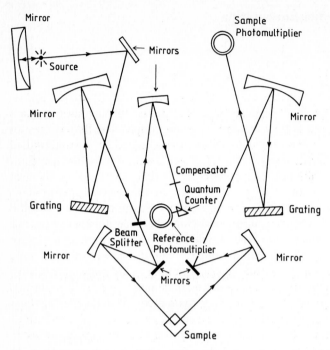

Fig. 6. Schematic diagram of the optical configuration of a commercial luminescence spectrometer (reproduced by permission of Perkin Elmer Limited)

detector. Before reaching the detector, the beam is intercepted by an attenuator, which reduces its intensity to approximately that of the fluorescence emission, or a quantum counter (see Sect 4.1). The purpose of this optical arrangement is to compensate for fluctuations in the output of the source and to correct for optical discrimination.

3.2   Components of Luminescence Instruments

The optical components of luminescence instruments are similar to those of photometers and spectrophotometers used in UV/visible absorption spectroscopy. The commonly used components are summarised in this section.

3.2.1   Sources

Light sources used in luminescence spectroscopy include deuterium, mercury and xenon arc sources, tungsten filament lamps, lasers and light-emitting diodes. Of these, the high-pressure mercury arc lamp and the xenon arc lamp are most commonly used in modern instruments. The high-pressure mercury lamp produces a series of discrete emission lines and is normally employed in fixed-

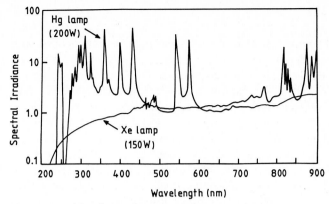

Fig. 7. Spectral irradiance of mercury and xenon arc lamps

wavelength fluorimeters or phosphorimeters. High sensitivity is achieved when the absorption band of a luminescent compound corresponds with one of the mercury emission lines. The xenon lamp provides a continuum output over the ultraviolet and visible region and is therefore used in scanning spectrometers. Xenon arc lamps may be operated in a continuous or pulsed (flash) mode. The typical output of the high-pressure mercury and xenon arc lamps are shown in Fig. 7.

Incandescent sources, such as tungsten filament lamps, have low intensity below 450 nm and are only useful in the visible region. Light-emitting diodes are similarly limited to the visible region. The use of lasers is increasing, but has so far been largely confined to research applications. The principles of laser-induced atomic and molecular fluorescence are described in detail in Section 5 of Chap. 8 (page 208).

3.2.2 Wavelength Selection

The normal method of wavelength selection is by the use of a filter or a grating monochromator. A combination of narrow bandpass transmission filters or long and short pass cut-off filters may be used to isolate a fixed range of wavelengths for the excitation and emission light. The emission filter must be chosen so that the Rayleigh-Tyndall scattering is excluded, allowing only fluorescence or phosphorescence emission to reach the detector. Both interference and absorption filters are used in fixed-wavelength instruments, although transmission characteristics are often poor in the ultraviolet region. Many spectrometers, such as the commercial instrument shown in Fig. 6, employ excitation and emission monochromators, which may be scanned over a range of wavelengths or operated at fixed wavelength. Spectral dispersion is achieved by a diffraction grating with typical bandwidths of 1-25 nm set by slits.

3.2.3 Detectors

The low intensity of luminescence emission requires high amplification and photomultipliers are generally used in luminescence instruments. The fast response of photomultipliers (10^{-8} s) makes them suitable for time-resolved measurements. A variety of multichannel detection devices has also been used for luminescence measurement, including the silicon intensified target (SIT) vidicon and array detectors. These devices detect emission intensity and spectral information over a range of wavelengths simultaneously, and are useful for the measurement of transient species such as those observed in kinetic experiments and in chromatographic detection (Sect. 4.5). However, spectral information is usually obtained at the expense of a lower sensitivity and dynamic range than fixed-wavelength detection employing a photomultiplier [9].

3.2.4 Sampling Methods

A variety of sampling methods are utilised for luminescence analysis. Solution samples are normally placed in glass (> 300 nm) or silica cuvettes. Rectangular and cylindrical cells are both used, but care must be taken in cell design to limit the scattered radiation reaching the detector. Flow cells with internal volumes down to 1μl or less are used to monitor the time-dependent concentration of luminescent species in flowing streams. Sampling methods for time-resolved phosphorescence measurements are discussed in Sect. 4.2.

4 Luminescence Measurement and Analytical Applications

4.1 Luminescence Measurement

4.1.1 Excitation and Emission Spectra

Quantitative measurement in luminescence spectroscopy is normally made at a fixed excitation/emission wavelength pair characteristic of the species under investigation. A simple filter instrument is adequate for this type of assay, although monochromators usually allow a more precise optimisation of excitation and emission wavelength. Direct measurements of fluorophores have been made in many areas including medicine, pharmacology, foods, environmental studies and forensic science. Some typical applications are listed in Table 2. In many cases luminescence procedures are possible where absorption methods have insufficient selectivity or sensitivity[10].

Qualitative and quantitative information is obtained if the emission or excitation spectrum is measured using a scanning spectrometer. An emission spectrum is recorded by scanning the emission wavelength at a fixed excitation

Table 2. Some applications of luminescence spectroscopy

Area	Typical analytes
Clinical and pharmaceutical	Salicylates, quinine, quinidine, barbiturates, steroids, catecholamines, tetracyclines, porphyrins (e.g. coproporphyrin, uroporphyrin), amphetamines, enzymes
Food	Vitamins (e.g. riboflavin), amino acids (e.g. phenylalanine, tyrosine, tryptophan), proteins, aflatoxins.
Environment and occupational hygiene	Polycyclic aromatic hydrocarbons, phenols, oils, fluorescent dye tracers (e.g. fluorescein, rhodamines), optical brighteners
Forensic	Petroleum residues, illicit drugs (e.g. LSD)
Natural products	Chlorophyll, pheophytin, humic acid, ergot alkaloids, flavonoids.

wavelength, and an excitation spectrum is obtained by a scan of excitation wavelength at a fixed emission wavelength of the sample. The excitation and fluorescence emission spectra of anthracene are shown in Fig. 8. The spectra overlap at the wavelength for the 0-0 transition (corresponding to the transition between the lowest vibrational levels of the ground state and the first excited state) (Sect. 2) due to solvent effects. A consequence of the fluorescence process is that the emission spectrum profile of a single compound is independent of excitation wavelength as long as the incident radiation falls within the absorption band (Kasha's rule). Hence all absorption processes have sufficient energy to yield all emission wavelengths and only the intensity of the emission is affected by changing the excitation wavelength.

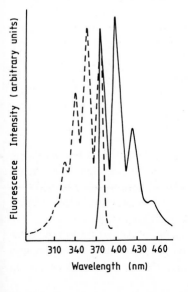

Fig. 8. Excitation (dashed line; λ_{em} 404 nm) and emission (solid line; λ_{ex} 252 nm) spectra of anthracene

4.1.2 Corrected Spectra

Emission and excitation spectra inevitably reflect some characteristics of the light source and the optical components of the spectrometer. If no attempt is made to control these instrumental parameters, then the resulting uncorrected spectra will be a composite of the true spectrum, the distribution of source intensity with wavelength, the efficiency of the monochromators and other optical components, and the detector response. It is therefore necessary to 'correct' the spectrum if data from different instruments is to be compared [11].

Excitation spectra may be corrected by diverting part of the incident radiation, using a beam-splitter or chopper, to a reference detector via a quantum counter. A quantum counter is a reference cell filled with a fluorescent compound, such as the dye rhodamine B in ethylene glycol, which absorbs incident radiation in the region 200–600 nm. Since the fluorescence quantum yield is essentially independent of the excitation wavelength, the output of the quantum counter at the emission maximum is proportional to the photon flux of the source and may be used to directly correct the excitation spectrum. A typical arrangement of the quantum counter and reference photomultiplier for a commercial instrument are shown in Fig. 6. The corrected excitation spectrum of a compound is equivalent to its absorption spectrum for reasons discussed in Sect. 2.4.

Correction of emission spectra is less straightforward, since it requires calibration of the emission optics and the detection system. One way to do this is to record the spectrum of a reference compound, for example quinine sulphate in 0.5 M sulphuric acid, and 'correct' the spectrum to match the true spectrum. If this correction is applied to subsequent sample spectra, then corrected emission spectra are obtained. An alternative procedure is to place a magnesium oxide scatterer in the sample compartment and scan the excitation and emission monochromators at the same wavelength. If the excitation source is corrected by means of a quantum counter, the resulting intensity distribution may be used to correct emission spectra. The identity of an unknown sample can then be determined by comparison with a library of corrected spectra. Corrected emission spectra are also required when quantum yields are being measured (Sect. 2.4).

4.1.3 Synchronous Luminescence Spectroscopy

In synchronous luminescence spectroscopy both excitation and emission mono-chromators are scanned simultaneously, with a fixed wavelength ($\Delta\lambda$) or frequency difference (Δv) between them [12]. The technique has the effect of enhancing resolution by narrowing the bandwidth of the emission peak. This is demonstrated for a fixed wavelength synchronous scan by reference to Fig. 9. At the limiting condition shown in Fig. 9(a), the sample is excited by the incident radiation, which falls within the absorption band, but luminescence is not

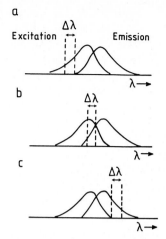

Fig. 9. Constant wavelength synchronous luminescence spectroscopy

detected because the $\Delta\lambda$ selected is too small for emission to be observed. If both monochromators are synchronously scanned, luminescence emission is observed (Fig. 9(b)) until the situation shown in Fig. 9(c) is reached, where the wavelength of the incident radiation first falls outside the absorption band. This procedure provides an increased separation for complex mixtures and may be used to discriminate between samples which give similar conventional luminescence spectra. Synchronous scanning has been successfully applied to forensic analysis [12], the identification of the source of oil spills [13], and the characterisation of other complex mixtures [14].

4.1.4 Total Luminescence Spectroscopy

The total luminescence profile of a sample may be obtained by scanning the emission spectrum at regular excitation intervals (typically 5 nm) across the entire absorption range. The resulting emission/excitation/intensity matrix is a three-dimensional spectrum giving a complete description of the fluorescence or phosphorescence emission. The data is presented as a stack-plot, resembling a panoramic view of a range of mountains, or as a contour diagram with lines connecting points of equivalent intensity, in the same way in which points of the same altitude are indicated by the contour lines on a map. A typical total fluorescence profile for a fuel oil is shown in Fig. 10. Conventional excitation and emission spectra are 'slices' taken parallel to the corresponding axes, and a fixed wavelength difference ($\Delta\lambda$) synchronous scan is a 'slice' taken at 45° to each axis. Once the total luminescence profile has been collected and stored, these and other scans, such as variable-angle 'slices' taken at angles other than 45° and fixed frequency difference ($\Delta\nu$) synchronous scans, may be generated from the data [14, 15].

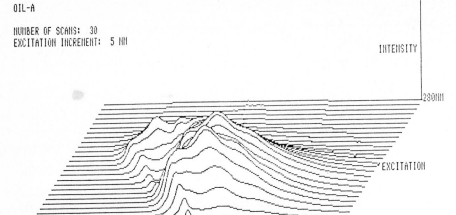

Fig. 10. Total fluorescence emission of a marine diesel oil

4.2   Time-resolved Luminescence Measurement

4.2.1   Lifetimes

The preceding discussion applies to luminescence measurements recorded on a fixed instrumental timescale. However, the different lifetimes associated with fluorescence and phosphorescence emission also allow them to be distinguished from each other by time-resolved luminescence measurement [1–3, 16]. Fluorescence is generally a rapid process which takes place within 10^{-9} to 10^{-6} s, while phosphorescence lifetimes are in the range 10^{-4} to 10^{2} s. The relationship between intensity and time during a time-resolved luminescence measurement is shown schematically in Fig. 11. If the sample is excited by a pulse of incident radiation which is short compared to the lifetime of the luminescence emission, the emission intensity will rise to a maximum (I_0) during the pulse and then (theoretically) decay exponentially. However, because of their different lifetimes, the fluorescence emission will decay much faster than the phosphorescence emission. The luminescence intensity (I_t) at time t is described by the first order rate equation:

$$dI_t/dt = -kt \tag{18}$$

where k is the rate coefficient for fluorescence or phosphorescence. Integrating this expression gives,

$$I_t = I_0 \exp(-t/\tau) \tag{19}$$

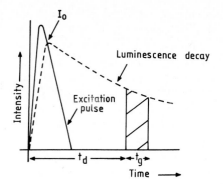

Fig. 11. Variation of intensity with time during time-resolved luminescence measurement

where I_0 is the intensity at $t = 0$ and τ is the luminescence lifetime. Taking the logarithms of both sides gives,

$$\ln I_t = -t/\tau + \ln I_0$$

and therefore

$$\log I_t = -0.434t/\tau + \log I_0 \qquad (20)$$

The lifetime for luminescence decay can therefore be determined from the slope of the line obtained by plotting $\log I_t$ against t. Since the luminescence lifetime is characteristic of the compound under investigation and the type of luminescence emission, the lifetime provides useful information about the excited state of the molecule.

The most widely adopted technique for the measurement of fluorescence lifetimes is called 'time-correlated single-photon counting'. A deuterium flash-lamp or a laser produces weak, repetitive pulses of light with picosecond to nanosecond duration, separated by orders of nanoseconds. The single-photon counting technique depends on the fact that certain photomultipliers can give a measurable output after a single photon strikes the photocathode. This allows the delay time between the excitation pulse and the arrival of individual photons to be determined. Counting the number of photons that arrive within a given time interval allows a fluorescence decay curve to be built up from which the fluorescence lifetime can be determined. Since very few molecules phosphoresce in solution at room temperature, because of the increased probability of non-radiative deactivation of the long-lived triplet state, fluorescence lifetimes can be measured by this method without interference.

4.2.2 Phosphorescence Measurement

Phosphorescence measurements are traditionally carried out by cooling the sample solution to low temperature using a liquid nitrogen cryostat. This has the effect of enhancing the phosphorescence emission by reducing collisional deactivation of the triplet state. A common solvent used for low temperature phosphorescence is EPA, a mixture of diethyl ether, isopentane and ethanol in the ratio 5:5:2 respectively, which forms a rigid, clear glass at 77 K. Other solvents such as hexane and heptane may also be used if of sufficiently high purity. Unfortunately, the smaller number of molecules which phosphoresce, together with the difficulty of making measurements at liquid nitrogen temperature, has restricted the development of phosphorescence analysis. However, several room temperature phosphorescence (RTP) techniques have recently been developed, which dispense with the requirement for a liquid nitrogen cryostat.

RTP can be observed by encapsulating the sample in a polymer matrix such as polymethylmethacrylate, which acts in the same way as the low temperature solvent glass, although the technique is limited by the difficulty of encapsulating samples without the incorporation of quenching impurities such as oxygen. Enhanced phosphorescence emission is also often observed if the sample is deposited onto a solid substrate. Several supports have been found to be suitable, including filter paper, silica gel, sodium acetate and polyacrylic acid-sodium halide mixtures, although the optimum conditions for emission have generally been determined empirically [16]. An alternative approach is to use a surfactant that forms micelles in aqueous suspension. These micelles are associations of surfactant molecules which have a hydrophilic outer surface and a hydrophobic core capable of incorporating a non-polar sample molecule. The micelle protects the sample from the surrounding aqueous environment allowing phosphorescence to be detected [17]. Cyclodextrins, which have a barrel-like hydrophobic cavity, produce a similar effect by incorporating a non-polar sample molecule [18].

Molecules that phosphoresce always fluoresce as well, so phosphorescence emission is usually measured by time-resolved spectroscopy [16]. In a typical experiment, a delay time (t_d) is allowed to elapse after the excitation pulse, to allow the fluorescence emission to decay to an undetectable level, before the phosphorescence emission is observed for a gate time, t_g (Fig. 11). Since the delay time, gate time and emission wavelength can all be varied independently, the phosphorescence spectrum may be measured by scanning the emission wavelength at fixed values of t_d and t_g, whilst the lifetime may be determined by incrementing the delay time at a fixed t_g and emission wavelength. Gating is generally carried out either electronically, using a pulsed source and a gated photomultiplier, or mechanically using a continuous source in conjunction with a rotating can or rotating disc phosphoroscope.

A simplified schematic diagram of a rotating can phosphoroscope is shown in Fig. 12. The device consists of a sample tube located in the centre of a cylindrical can, with a slot cut in its side, which is rotated about its vertical axis by a motor. If the slot is aligned with the incident radiation, then sample luminescence is excited but does not reach the detector because the emission is blocked by the can. As the can rotates, the slot comes into alignment with the emission monochromator, by which time the fluorescence has decayed to a negligible level and incident

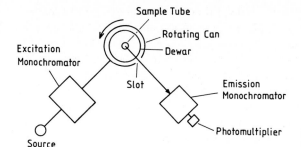

Fig. 12. Schematic diagram of a rotating-can phosphoroscope

radiation is blocked by the can, so that only phosphorescence emission reaches the detector; t_d and t_g are varied by altering the width of the slot and the speed of rotation of the can.

4.3   Factors Affecting Luminescence Emission

A number of factors such as temperature, pH, solvent, sample concentration and photodecomposition affect the intensity and wavelength of luminescence emission. Quenching processes discussed in Sec. 2.5 also affect intensity. These processes, involving the interaction of the excited state fluorophore with its surroundings, usually lead to a reduction in the quantum yield by influencing one or more of the excited state rate coefficients. The more important environmental effects are discussed in this section.

4.3.1   Temperature and Solvent Effects

Changes in temperature affect the number of collisions between the excited state molecule and surrounding solvent molecules. Increasing the temperature of a sample solution results in an increased collision rate and a consequent reduction in luminescence intensity for most molecules. However, the reverse is true for some delayed fluorescence processes in which $T_1 \rightarrow S_1$ repopulation is enhanced at higher temperatures. Normally, unthermostatted room temperature measurement of solution samples is satisfactory provided both samples and standards are allowed to equilibrate at the same temperature prior to analysis. However, the sample should be placed in a thermostatically controlled sample holder if kinetic measurements are to be made or where elevated temperatures are used. Alternatively a temperature correction curve may be used to compensate for temperature variations, such as may be encountered when making measurements in the field [19].

Solvent viscosity also influences the luminescense quantum yield. A higher viscosity reduces the number of collisions an excited molecule undergoes and the net effect is an increase in intensity. Other interactions between the solvent and the excited state or the ground state of the molecule may have a considerable effect on luminescence in solution. For example, a shift in the emission maximum to longer wavelength with increasing solvent polarity is often observed for polar molecules as a result of the greater stabilisation of the excited state.

4.3.2   pH

The luminescence intensity of some ionisable molecules, where the quantum yields of the dissociated and undissociated states are different, is highly dependent on pH. For example, the anilinium ion $(C_6H_5NH_3^+)$ fluoresces very weakly while aniline $(C_6H_5NH_2)$ is strongly fluorescent. The emission wavelength and the spectral bandwidth may also be affected by the degree of ionisation of the sample molecule and hence the pH. The emission of ionisable molecules can be enhanced by controlling the pH to favour the form with the highest quantum efficiency. However, if dissociation is likely to occur, the pH of samples and standards should always be adjusted to the same value, or the solutions buffered, prior to measurement.

4.3.3   Inner-Filter and Self-Absorption Effects

Figure 13 shows the typical calibration curves for right angle and frontal illumination. The linear relationship between the fluorescence intensity and concentration, Eq. (13), holds only for dilute solutions with absorbances below approximately 0.05. Deviations from linearity occur at high concentration and are attributable to inner-filter and self-absorption phenomena. Inner-filter effects arise in 90° optical sampling systems where the incident radiation is strongly absorbed at the front of the cuvette. The result is that less radiation is available to illuminate the part of the sample furthest from the source. Since most of the length of the cuvette is observed by the detector, the attenuating or inner-filter effect of a highly absorbing sample reduces the luminescence reaching the detector. A frontal illumination system, such as that shown in Fig. 5(b), is less susceptible to inner-filter effects and may be used for highly absorbing or opaque samples.

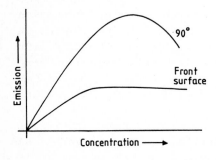

Fig. 13. Typical calibration curves for 90° and frontal illumination

Self-absorption occurs when emission and absorption bands overlap and part of the emitted radiation is re-absorbed. For single components, a small overlap is often observed in the region of the 0-0 transition, as a result of solvent effects, which can lead to self-absorption of fluorescence emission. However, self-absorption is most pronounced in samples containing complex mixtures of luminescent compounds, where there is often a considerable overlap between the emission band of one compound and the absorption band of another. The effect of self-absorption is demonstrated for a crude oil sample in Fig. 14. At a concentration of 200 ppm, the fluorescence emission profile of the oil differs substantially from the spectra recorded at 20 ppm and 0.6 ppm. This occurs at the higher concentration because short wavelength emission is re-absorbed by other components of the mixture and re-emitted at longer wavelength. The result is a distortion of the spectrum, with a reduction in the intensity of short wavelength emission and an increase in the intensity at longer wavelengths.

Inner-filter and self-absorption effects can be overcome by working within the linear range of the sample, which may be determined by preparing a calibration curve. A simple alternative is to measure the absorbance of the sample, which should be less than about 0.05 as a consequence of Eq. (13), or to dilute the sample successively until the emission profile remains unchanged.

4.3.4   Photodecomposition

Photodecomposition of sample molecules to produce less or more luminescent products is a frequent problem in fluorescence and phosphorescence. The

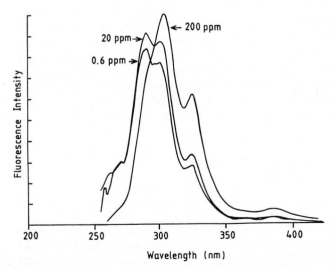

Fig. 14. Fluorescence emission of a crude oil at 200 ppm, 20 ppm and 0.6 ppm in hexane

problem becomes more acute if intense light sources and short excitation wavelengths are used. Photodecomposition is easily detected by changes in emission intensity with time. Since decomposition occurs principally at the point where the incident radiation is focused onto the sample, the existence of photodecomposition may be confirmed by blocking the incident radiation to prevent it falling on the sample for a short time while fresh sample diffuses to the focal point. If the signal rises to its original intensity when the sample is reilluminated, then the sample is photodecomposing. Photodecomposition can usually be reduced by selecting a longer excitation wavelength or reducing the source intensity.

4.4   Measurement of Non-Luminescent Compounds

Molecules which possess native luminescence may be directly measured and most fluorescence and phosphorescence methods utilise the native luminescence of the analyte (Table 2) [10]. Unfortunately, many compounds do not luminesce, or have quantum yields which are too low to be analytically useful. These compounds may often be determined by attaching a fluorescent or phosphorescent label which imparts luminescence to the product. It is even possible to improve selectivity by the choice of a derivatising reagent which reacts specifically with one particular analyte. High sensitivity is also possible if the sample is pre-concentrated during the derivatisation procedure. Some common fluorescence derivatising reagents and ligands are presented in Table 3.

Many metal cations form highly luminescent chelates with aromatic organic ligands, and fluorescence methods based on chelate formation have been developed for several elements including, Be, B, Al, Ca, Mg, Se, Zn and Sn. Transition-metal ions that are paramagnetic have increased rates of intersystem-crossing to the triplet state and show phosphorescence rather than fluorescence emission. However, transition metal complexes have closely-spaced energy levels, which enhance the probability of deactivation by internal conversion, so the emission is usually weak. Simple anions may also be determined by reaction to form a fluorescent derivative [20]. For example nitrite may be determined by reaction with 2,3-diaminonaphthalene to form 2,3-naphthotriazole:

$$\text{(naphthalene with two } NH_2 \text{ groups)} + HNO_2 \rightleftharpoons \text{(naphthotriazole)} + 2H_2O$$

Indirect fluorescence detection, where the quenching action of the analyte is used for quantitative determination, has also been widely used for non-luminescent species, and inorganic anions are readily determined by this method. However, the presence of more than one quenching species in real samples limits

Table 3. Derivatising reagents for luminescence measurement

Reagent	Typical analytes	Wavelengths (nm)	
		Ex	Em
Dansyl chloride	Phenols, amino acids	298	545
o-phthalaldehyde	Primary amines	350	470
Bromomethyl-7-methoxycoumarin	Carboxylic acids	328	380
Benzoin	B, Zn, Ge,	410	480
Alizarin garnet R	Al, F, NO_3	470	500

this approach unless some pre-separation, or chromatographic separation (Sect. 4.5) has been carried out. Fluoroimmunoassay, where an antigen tagged with a fluorescence label competes with an unlabelled antigen for binding sites on the antibody, also frequently relies on indirect detection [21].

4.5 Luminescence Detection in Chromatography

Luminescence has been widely used as a spectroscopic detector in all areas of chromatography as a consequence of its inherent advantages of sensitivity and selectivity[10,22] The application of luminescence detection to planar chromatography, high performance liquid chromatography and gas chromatography is discussed in this section.

4.5.1   Thin Layer and Paper Chromatography

Detection is carried out *in situ* after development of the chromatogram, either by observation of the native fluorescence of the separated compounds, or by indirect detection of the sample spots of non-fluorescent molecules against a fluorescent background impregnated in the coating of a thin layer chromatography (TLC) plate. Spray reagents are also used to form fluorescent derivatives for some compounds. Quantitative measurement may be made by using a scanning densitometer, although measurements are susceptible to variations in spot dispersion for samples and standards. An alternative method of quantitation is to extract the sample from the TLC plate or paper using a solvent, followed by measurement of luminescence emission compared to standards using a normal solution cuvette.

4.5.2   High performance liquid chromatography

The effluent from the HPLC column is usually passed down narrow bore tubing to a low volume flow cell with an internal volume in the range $1-50\,\mu$l. The choice of flow cell volume is a compromise between the need to maintain chromatographic integrity, whilst maximising luminescence sensitivity [22]. The normal method of detection uses a flow cell in conjunction with a conventional 90° optical system, although many other detector and cell designs have been described, including a frontal illumination arrangement with appropriate collection optics, and on-column laser-induced fluorescence detection for capillary columns [23, 24]. Fixed excitation and emission wavelengths are used in luminescence detection, since even the fastest scan speeds of most commercial spectrometers are unsuitable for recording the spectra of transient species eluting from the chromatograhic column. High sensitivity throughout a chromatographic separation is possible if the excitation and emission wavelengths are changed during the run to the optimum for eluting compounds of interest. The analysis of a fly ash sample using this technique to optimise the detection of several polycyclic aromatic hydrocarbons is shown in Fig. 15. Spectral information is either obtained by stop-flow methods, in which the flow through the cell is halted and the chromatographic effluent temporarily diverted, or by employing multichannel devices. Wide bandpass filters or large slit settings (> 10 nm) are most frequently used for HPLC detection in order to improve sensitivity, with the maximum bandpass often determined by the illuminated volume of the flow cell. [22]

Non-fluorescent compounds may be detected by pre-column or post-column derivatisation [25, 26]. Some typical derivatising reagents are listed in Table 3 [27]. Pre-column methods are normally carried out off-line prior to injection onto the chromatograph. They have the advantage that long derivatisation procedures may be employed and that excess reagent can be separated chromatographically to prevent it interfering with measurement. An example of

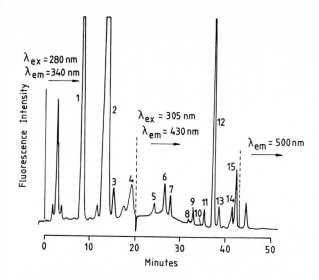

Fig. 15. HPLC chromatogram of a fly ash extract using fluorescence detection. (1) Naphthalene, (2) fluorene, (3) phenanthrene, (4) anthracene, (5) fluoranthrene, (6) pyrene, (7) benz[*a*]anthracene, (8) chrysene, (9) benzo[*e*]pyrene, (10) benzo[*b*]fluoranthrene, (11) benzo[*k*]anthrene, (12) benzo[*a*]pyrene, (13) dibenz[*a, h*]anthracene, (14) benzo[*ghi*]perylene, (15) indeno[1, 2, 3-*cd*]pyrene (reproduced by permission of Perkin Elmer Limited)

the application of pre-column derivatisation to the detection of amino acids in serum as their dansyl derivatives is shown in Fig. 16. Post-column derivatisation techniques are carried out on-line by mixing the reagent with the chromatographic eluant via a T-union located between the end of the column and the detector. A variety of reactors have been developed for post-column derivatisation to allow for mixing of reagents and to provide sufficient reaction time [27]. Room temperature phosphorescence (RTP) detection in HPLC has also been reported using post-column mixing of a surfactant to the chromatographic eluant to enhance the phosphorescence emission [17]. In contrast to pre-column techniques, the reaction does not need to go to completion in post-column derivatisation. The only requirement is that the reaction is repeatable.

Indirect fluorescence detection of non-fluorescent compounds may be carried out using pre- and post-column procedures. In a typical experiment, a fluorescent compound is added to the solvent reservoir, or to the column effluent, to produce a steady background signal at the detector. If a non-fluorescent compound enters the detector, it will displace or quench the background fluorescence and a negative peak will be detected [28]. Indirect detection has been successfully used in HPLC, ion chromatography and capillary zone electrophoresis [29, 30]. The range of techniques and applications of fluorescence detection in HPLC have been discussed in several reviews [14, 23].

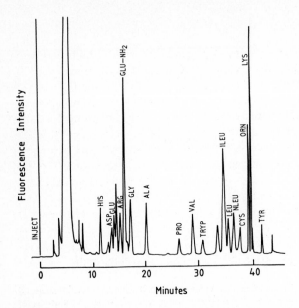

Fig. 16. HPLC chromatogram of dansylated amino acids in a dialysed serum standard using fluorescence detection (reproduced by permission of Perkin Elmer Limited)

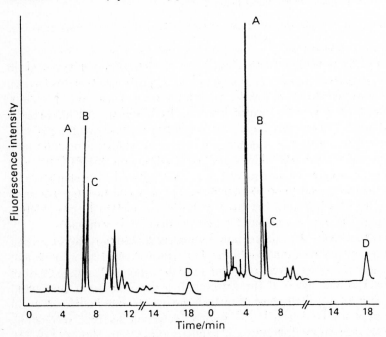

Fig. 17. Capillary gas chromatogram for (**a**) kerosene and (**b**) petrol using fluorescence detection: A, naphthalene; B, 2-methylnaphthalene; C, 1-methylnaphthalene (λ_{ex} 264 nm, λ_{em} 335 nm); and D, fluorene (λ_{ex} 252 nm, λ_{em} 308 nm). (Reproduced from Ref. [33])

4.5.3   Gas Chromatography

Fluorescence detection in gas chromatography (GC) is less well established than it is for HPLC, although the two techniques may be readily combined [31, 32]. The most common types of interface use a heated transfer line and flow cell to transfer the packed or capillary effluent to the fluorescence spectrometer, or employ fibre optics to detect vapour-phase fluorescence at the chromatograph. Vapour-phase fluorescence emission spectra for many compounds are similar to the corresponding solution spectra, although the emission intensity is usually weaker. Good quantitative measurement has been demonstrated for the determination of polycyclic aromatic hydrocarbons in light petroleum fractions with sub-nanogram detection limits in favourable cases (Fig 17) [33]. Two other methods of interfacing fluorescence with GC, matrix-isolation on a cryostatically cooled surface and laser-induced fluorescence of rotationally cooled molecules in a supersonic jet, both give spectra showing fine structure [34, 35]. Indirect fluorescence detection has also been demonstrated for the determination of non-fluorescent compounds.

5   References

1. Wayne RP (1988) Principles and applications of photochemistry, Oxford Science Publications, Oxford
2. Barltrop JA, Coyle JD (1975) Excited states in organic chemistry, Wiley, New York
3. Becker RS (1969) Theory and interpretation of fluorescence and phosphorescence, Wiley, New York
4. Birks JB (1970) Photophysics of aromatic molecules, Wiley, New York
5. Borrell P, Greenwood HH Proc. Roy. Soc. A. 298:453 (1967).
6. McGlynn SP, Azumi T, Kinoshita M (1969) Molecular spectroscopy of the triplet state, Prentice-Hall, New York
7. Kalyanasundaram K (1981) Photochemistry in microheterogeneous systems, Academic Press, New York
8. De Luca MA, McElroy WD (eds) (1981) Bioluminescence and chemiluminescence, Academic Press, New York
9. Wehry EL (ed) (1981) Modern fluorescence spectroscopy, vol 4, Plenum, New York
10. Meites L (ed) (1963) Handbook of analytical chemistry, McGraw-Hill, New York
11. Miller JN (ed) (1981) Standards in fluorescence spectrometry, Chapman and Hall, London
12. Lloyd JBF (1971) J. Forensic Sci. 2:83
13. John P, Soutar I (1981) Chem. Br. 17:278
14. Christian GD, Callis JB (eds) (1986) Trace analysis: spectroscopic methods for molecules, Wiley, New York
15. Warner IM, McGown LB (1982) Crit. Rev. Anal. Chem. 13:155
16. Schulman SG (ed) (1988) Molecular luminescence spectroscopy. Methods and applications: part 2, Wiley, New York
17. Gooijer C, Baumann RA, Velthorst NH (1987) Prog. Analyt. Spectrosc., 10:573
18. Vo-Dinh T (1984) Room temperature phosphorimetry for chemical analysis, Wiley-Interscience, New York
19. Smart PL, Laidlaw IMS (1977) Water Res. 13:15
20. Gomes-Hens A, Valcárcel M (1982) Analyst 107:465
21. Weir DM (1975) Handbook of experimental immunology, Blackwells, Oxford
22. Wehry EL (ed) (1981) Modern fluorescence spectroscopy, vol 3, Plenum, New York

23. Weinberger R, Sapp E (Sept. 1984) Int. Lab. 80
24. Takeneuchi T, Ishii D (1988) J. Chromatogr. 435: 319
25. Johnson E, Abu-Shumays A, Abbot SR (1977) J. Chromatogr. 134: 107
26. Frei RW, Michel L, Santi W (1976) J. Chromatogr. 126: 665
27. Frei RW, Lawrence JF (eds) (1981) Chemical derivatization in analytical chemistry, vol 1, Plenum, New York
28. Su SY, Jurgensen A, Bolton D, Winefordner JD (1979) Anal. Lett. 51: 1444
29. Sherman JH, Danielson ND (1987) Anal. Chem. 59: 1483
30. Kuhr WG, Yeung ES (1988) Anal. Chem. 60: 1832
31. Burchfield HP, Wheeler RJ, Bernom JB (1971) Anal. Chem. 43: 1876
32. Creaser CS, Stafford A (1987) Analyst 112: 423
33. Bagheri H, Creaser CS (1988) Analyst 113: 1175
34. Contrad VB, Carter WJ, Wehry EL, Mamantov G (1983) Anal. Chem. 55: 1340
35. Small GJ, Hayes JM (1982) Anal. Chem. 54: 1202

CHAPTER 6

An Introduction to Nuclear Magnetic Resonance in Fluids

P.S. Belton

1 Basic Principles

NMR arises because of the magnetic moment associated with the non-zero spin quantum number of certain nuclei. Fortunately the occurrence of non-zero spin quantum numbers is common in the periodic table, and thus NMR can be observed in isotopes of most elements. Examples are given in Table 1. The magnetic moment, μ, associated with the quantum number, $|I|$, responds to a macroscopic magnetic field by orientating in it. $2|I| + 1$ orientations are allowed. Each orientation corresponds to one energy level. Hydrogen ($I = \frac{1}{2}$) has two energy levels, which correspond to orientations parallel (low energy) and anti-parallel to the magnetic field. Only one transition is observed therefore. Deuterium on the other hand has $I = 1$ so that three levels and two transitions occur; in practice however these two transitions may be degenerate so that only one resonance line is observed.

The energy of a transition is mainly determined by the magnetic field, B_o, hence;

$$\Delta E = \mu B_o / I$$

It is usual to express the nuclei magnetic moment in terms of the magnetogyric ratio, γ:

$$\mu = \gamma h I / 2\pi$$

$$\therefore \text{Since } \Delta E = h\omega_o / 2\pi = h\nu_0$$

$$\omega_o = \gamma B_o; \tag{1}$$

ω_o is called the Larmor frequency. Irradiation at the Larmor frequency will cause transitions between the energy levels and this is the basis of the spectroscopic method. ΔE is a very small quantity (typical values of ν_0 are in the range 1–600 MHz). This has two important consequences. The first is the insensitivity of the technique: since signal strength depends on the difference in populations of the two levels, the effect of a small ΔE is to make a small difference in populations. If N is the population of the highest level and N_o that of the lower then the

Table 1. NMR properties of selected nuclei
The field strengths for A, B and C are 2.35, 4.70 and 7.05 T respectively

Nucleus	Spin	Natural Abundance	Resonance Frequency/MHz		
			A	B	C
^{1}H	$\frac{1}{2}$	99.985	100	200	300
^{2}H	1	0.015	15.3	30.7	46.1
^{13}C	$\frac{1}{2}$	1.108	25.1	50.3	75.5
^{14}N	1	99.63	7.2	14.4	21.7
^{15}N	$-\frac{1}{2}$	0.37	10.1	20.3	30.4
^{17}O	$-\frac{5}{2}$	0.037	13.6	27.1	40.7
^{19}F	$\frac{1}{2}$	100	94.1	188.2	282.3
^{23}Na	$\frac{3}{2}$	100	26.5	52.9	79.4
^{27}Al	$\frac{5}{2}$	100	26.1	52.1	78.2
^{29}Si	$-\frac{1}{2}$	4.70	19.9	39.7	59.6
^{31}P	$\frac{1}{2}$	100	48.5	81.0	121.5
^{39}K	$\frac{3}{2}$	93.1	4.7	9.3	14.0
^{43}Ca	$-\frac{7}{2}$	0.145	6.7	13.5	20.2
^{57}Fe	$\frac{1}{2}$	2.19	3.2	6.5	9.7
^{95}Mo	$\frac{5}{2}$	15.72	6.5	13.0	19.6
^{111}Cd	$-\frac{1}{2}$	12.75	21.2	42.4	63.6
^{113}Cd	$-\frac{1}{2}$	12.26	22.1	44.3	66.6
^{119}Sn	$-\frac{1}{2}$	8.58	37.3	74.6	111.8
^{119}Hg	$\frac{1}{2}$	16.84	17.9	35.7	53.5

Boltzmann equation gives

$$N = N_o \exp(-\Delta E/kT)$$

For protons in a fairly high field of 7 Tesla ($v_0 = 300\,\text{MHz}$), $N/N_o = 0.99952$ at 300 K. Thus only about 5 in 10^4 nuclei give rise to a signal.

The second effect of low ΔE is rather more subtle but is much more important. ΔE represents the energy of one proton involved in the transition; at 300 MHz this has a value of 19.8×10^{-26} Joules. If then it is required to excite a bandwidth of 50 kHz with a radiofrequency pulse of 10 μs a power output of 100 watts will generate about 10^{17} photons per frequency unit (Hz). Because of the very large number of photons created, emission and absorption induced by the radiation field far outweighs any spontaneous processes. Thus the transitions are completely controlled by the radiation field.

In general the phase of a group of photons has an indeterminacy related to the number of photons in the group by the uncertainty principle, thus

$$\Delta\Theta\Delta n \sim h$$

Θ is the phase angle and n is the number of photons. Since n is very large Δn may be large and consequently $\Delta\Theta$ is very small, hence the phase may be very precisely determined. The radiation field therefore controls both the transitions and the phase coherence of the system, hence the state of the spin system may be completely determined by the radiation field and NMR is said to be a coherence spectroscopy. It is that makes NMR such a versatile form of spectroscopy.

It is instructive to compare NMR with infra-red spectroscopy. Here the "resonance" frequency is about 10^{13} Hz, hence excitation equivalent to that in NMR would require 10^6 to 10^7 watts. This is only achievable with powerful lasers. In practise infra-spectroscopy is not carried out coherently but incoherently, with small numbers of photons of unknown phase. As a consequence essentially only one type of infra-red experiment is possible compared to a plethora of NMR experiments.

2 Spectral Acquisition

The basis of the modern NMR experiment is to manipulate the spin system into the required state and thus observe its evolution in time. The manipulation is achieved by a series of pulses of radiofrequency energy, although in some applications the static magnetic field is also pulsed. For many purposes the interaction of the spin system of the nuclei with the radiofrequency (RF) or static magnetic fields can be described by a simple classical model. For illustrative purposes the simplest spin $\frac{1}{2}$ system is considered. The ensemble of nuclear magnetic moments that comprise the spin system may be represented by a collection of nuclear magnets aligned parallel or anti-parallel to the main magnetic field, B_0; they precess about this with a frequency ω_0. The magnets parallel to the field will be in a slight excess to those anti-parallel and summation therefore results in a net nuclear field parallel to the main field. By convention, B_0 is defined as being in the z direction. Precession at ω_0 is removed by the mathematical convenience of rotating the x and y coordinates at a frequency ω_0 whence the nuclear magnets are static. This coordinate reference frame is called the rotating frame; in this frame the net nuclear magnetisation can be described by a single static vector along z. Electromagnetic radiation consists of oscillating electric and magnetic field vectors at right angles to each other. It is the magnetic field to which the nuclear magnetic moments respond. If it is polarised and has a frequency ω_0 it may be represented as a static vector in the rotating x'y' plane. Nuclear magnets can only induce a signal in the sampling coil when they are in the x'y' plane. Thus in order to observe any signal the z magnetisation must be rotated towards the transverse, x'y', plane. Figure 1 shows how this happens. In the rotating frame the RF field is represented by a static field vector B_1 along x'. The nuclear magnetisation responds to this by precessing about it with a frequency ω_1 given by

$$\omega_1 = \gamma B_1 \tag{2}$$

After a time, t, the angle Θ through which the magnetisation has precessed

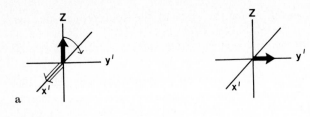

90° Pulse

180° Pulse

Fig. 1. (a) a 90° pulse with the radiation field in the x′ direction (b) a 180° pulse with the same field orientation–note for this particular experiment the final result is independent of the initial pulse direction

is

$$\Theta = \gamma B_1 t \tag{3}$$

A maximum signal is observed when $\Theta = \pi/2$ radians. A pulse of sufficient duration and strength to cause a rotation of 90° ($\pi/2$ radians) is called a "90°" pulse. If instead of a 90° pulse a 180° pulse is applied, no signal is observed but the magnetisation is completely inverted; this situation is one in which the populations of the energy levels are inverted.

Consideration of what happens after the application of 90° and 180° degree pulses illustrate the two main types of relaxation phenomena in NMR, transverse and spin lattice relaxation and some general features of the spectrum.

In situations where motion is sufficiently rapid it is possible to consider the nuclear spins as if they were isolated from each other and their interactions were only a small perturbation on their behaviour. In quantum mechanical terms this is a weakly coupled system. When motional rates are comparable to the interaction strength (expressed in frequency units) this condition no longer obtains and the system moves from the weakly coupled "motional narrowing regime" to the strongly coupled "rigid lattice regime", where for many purposes the nuclear spin system must be considered as an entirety, and fine structure often appears in the broad line spectra. The phenomena are generally seen in solids, a subject dealt with in Chapter 7. However, the motional narrowing regime is an important one in NMR and illustrates many of the basic principles of the subject. In order to simplify the discussion further we shall once again limit ourselves to the spin $\frac{1}{2}$ systems.

The resonance frequency of a nucleus at an instant in time may be written as

$$\omega_o = \gamma(B_o + B_\sigma + B_J + B(t)) \tag{4}$$

The term B_o is, in most NMR experiments, much bigger than any of the other terms by several orders of magnitude. Since in general the magnetogyric ratios of nuclei differ by large amounts, this serves to give each isotope its own resonance frequency which is separated from other isotopes by amounts greater than the remaining terms. Examples of resonance frequency are given in Table 1.

The term B_σ is the chemical shift term. It represents the way in which the nucleus is shielded from the main field by the electron cloud in which it is embedded. The shielding is very sensitive to the electronic environment and thus the chemical environment. A very large body of literature now exists which details the relationship between chemical shifts and molecular structure. The chemical shifts effect is field-dependent and is usually expressed as the dimensionless parameter σ,

where

$$B_\sigma = B_o(1 - \sigma) \tag{5}$$

Typically σ is of the order of parts per million. In general σ is a tensor quantity, that is its magnitude and sign is direction dependent. In general, unsymmetrical molecules will have different values of σ in each direction. Where there is rapid isotropic rotation, this effect is averaged to give an average value of σ but where motion is slow, anisotropic line broadening or structural effects can appear due to anistropy in σ, usually termed chemical shift (or shielding) anisotropy.

The term B_J arises from spin-spin coupling effects which are transmitted by an electron cloud. If two nuclei in a molecule have different resonance frequencies then the polarisation states of one may be communicated to the other via the electron cloud. For a spin $\frac{1}{2}$ nucleus two states are permitted, thus a coupled nucleus will experience the main magnetic field plus or minus the effect due to the neighbour. As a result two slightly different resonance frequencies will occur even though only one chemical environment exists. In general where a nucleus is coupled to n other nuclei 2nI + 1 lines will result. Spin-spin coupling of this sort is often referred to as scalar coupling or J coupling, its magnitude is independent of the magnetic field and it may occur between hetero- and homo-nuclei. The sizes and patterns of spin coupling can be related to molecular structure and geometry.

Very often, the presence of scalar coupling results in considerable complication of the spectrum and it is desirable to remove it. In the case of heteronuclear scalar coupling this can be arranged by irradiating the unobserved nuclei with a suitable radiofrequency field for the duration of the spectral acquisition. The frequency range of the field must be such that it covers the whole spectrum of the unobserved nuclei and its intensity must be greater than B_J. The effect of this field is to cause rapid transitions amongst the spin states and hence remove the coupling effect. Where homonuclear coupling is involved, such a crude approach is not possible and irradiation covering only very limited frequency ranges must be used, otherwise the entire spectrum may be distorted.

Often when decoupling fields are applied, intensity changes in the resonance lines of the observed nuclei are seen. This phenomenon arises from the nuclear Overhauser Effect (NOE). This effect is due to dipolar interactions and occurs through space rather than via chemical bonds; however since scalar coupled nuclei are often close to one another the effect is often observed in such systems. The NOE arises from the redistribution of populations of energy levels resulting from the irradiation of one or more of the spins involved. The redistribution will be different to that given by the Boltzmann equation and will depend on the ratios of the magnetogyric ratios of the two nuclei and can result in either positive or negative signal enhancements. In the case of ^{13}C (^1H) (i.e. ^{13}C observed, ^1H irradiated) the enhancement is always positive, for ^{15}N (^1H) the enhancement is negative; thus diminished, zero or negative signals may be observed. For $^1H(^1H)$ the enhancement can be positive or negative.

3 Transverse Relaxation

The term B(t) in Eq. (4) arises because of the random perturbations of the local field due to thermal motions. There are numerous mechanisms by which this process can occur and useful descriptions are given by Harris, Abragam, and Farrar and Becker. In protons for example, random fluctuations in field can result from the motion of magnetic dipoles in neighbouring protons.

Since the perturbations are a random function of time the resonance frequency of a nucleus will also be a random function of time. If it were possible to take an instantaneous snapshot of a collection of nuclei there would be a range of frequencies covered by the nuclei due to this effect. In spectroscopic terms, these random perturbations generate the linewidth of the spectrum. The linewidth is dependent upon the rates of motion in the system. Generally, rapid motions result in narrow lines and slow motions result in broad lines. It is fortunate that for many experimental systems of interest in NMR, the line widths are less than the chemical shift differences so that chemical information can be obtained.

We are now in a position to follow the processes occurring in the x'y' plane following a 90° pulse. Immediately after the pulse all the nuclear magnetic vectors will be aligned along some direction determined by the phase of the RF radiation. Let us say it is y'. The normal procedure in detecting the NMR signal is to mix it with RF at the same frequency as the radiofrequency field so that only frequency differences are observed. This amounts to observing in the rotating frame. The phase of detection is set to observe along one axis of the x'y' plane. If this axis is y' then detection is said to be in phase and the experiment effectively measures the length of the vectors projected onto y'. For a one line spectrum where the reference frequency is set to that of the centre of the line, that is it "on resonance" the signal decays exponentially with time. This decay arises because of the effect of B(t). Random fluctuations in local magnetic fields will randomly shift the resonance frequencies of individual spins; in the rotating frame a frequency shift results in a rotation to the left or right of the y' axis. Thus as time goes on the net projection on this axis will be diminished. After a long enough time has elapsed all possible angles to the axis will be occupied and the net megnetisation will be zero. The process is shown in Fig. 2a.

When chemically shifted and/or scalar coupling resonances are present, the signal intensity oscillates with a frequency which is equal to the difference between the reference frequency and the resonance frequency. The observed signal, usually called the free induction decay (FID), is an interferogram of all the frequencies present upon which is superimposed the decay process (Fig. 3a). Fourier transformation of the interferogram results in the spectrum (Fig. 3b). The use of the Fourier transform is of major importance in modern spectroscopy (see also Chapter 3). The traditional method of continuous wave acquisition of

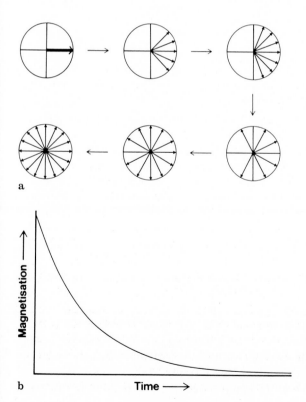

a

b   Time ⟶

Fig. 2. (a) The effects of relaxation on the distribution of vectors in the x'y' rotating plane. (b) The observed decay of magnetisation corresponding to (a)

Magnetisation ⟶

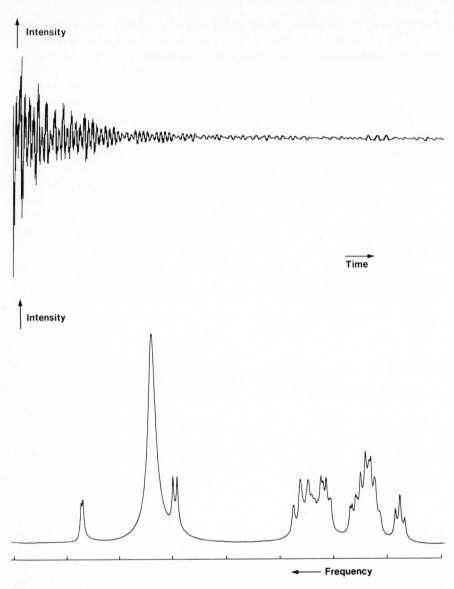

Fig. 3. (**a**) The interferogram resulting from the relaxation of a collection of chemical shifted nuclei. (**b**) The spectrum resulting from the Fourier transformation of (a)

spectra which depended on scanning sequentially from one end of the spectrum to the other was very slow and as a consequence did not allow effective signal averaging. Fourier transform methods are much faster (by as much as 3 orders of magnitude) and thus permit extensive signal averaging.

The decay of intensity following an excitation pulse is termed transverse relaxation and is characterised by one or more decay constants T_2. T_2 is related to the line width for Lorentzian lines by the equation:

$$\Delta v_{\frac{1}{2}} = \frac{1}{\pi T_2} \tag{6}$$

where $\Delta v_{\frac{1}{2}}$ is the full width at half height of the line.

The rate of transverse relaxation is a subtle indicator of the motions of the nuclei within the system. Nuclear motion resulting from molecular motion causes fluctuations of the local magnetic fields which in turn causes transverse relaxation. In general, the slower the modulation rate of local field the faster is the transverse relaxation rate. Typically for a given nucleus transverse relaxation, rates may vary over two or three orders of magnitude dependent on the molecular environment. This makes transverse relaxation a very sensitive probe of molecular dynamics and allows discrimination between different dynamic regions in heterogeneous systems.

When large molecules are involved relaxation rates may be such that extensive line broadening occurs and obscures much of the chemical shift information. This is a particular problem in NMR of proteins where proton spectra containing chemically shifted resonances can only be obtained from the smaller type of protein molecules or mobile regions of larger proteins.

4   Spin Lattice Relaxation

A 180° degree pulse results in a net magnetisation pointing along the -z axis. This corresponds to an inversion of the populations of the energy levels and is thus a non-equilibrium state. Recovery from this state to the equilibrium state is by a process called spin lattice relaxation and proceeds by way of transfer of energy between the spin system and the thermal bath formed by the matter of which it is a part. Spin lattice relaxation is an exponential process and is characterised by one or more time constants T_1, called the spin lattice relaxation time. In general any RF pulse will cause some perturbation of the populations and it is thus necessary to wait for about $5 \times T_1$ to allow the system to reach equilibrium after a pulse or sequence of pulses. Failure to do this results in a phenomenon called saturation in which subsequent signals are diminished in intensity compared to the first signal.

The time constant T_1 characterises spin lattice relaxation in the laboratory frame, since it takes place along z which is the same both in the laboratory and rotating reference frames. A second kind of spin lattice relaxation can be observed by first bringing the magnetisation into the $x'y'$ plane and then applying a continuous RF field. This behaves as a small static magnetic field in the rotating frame. The magnetisation will gradually relax to its (zero) equilibrium value in this field by a process known as spin lattice relaxation in the rotating frame which is characterised by a time constant $T_{1\rho}$.

Both T_1 and $T_{1\rho}$ are dependent on the motional state of the nuclei. T_1 is dominated by motions which have frequencies near to the Larmor frequency, ω_0. The greater the component of this frequency the faster is the relaxation. $T_{1\rho}$ on the other hand is dominated by processes at frequency ω_1 related to the strength of the radiofrequency field. Both the relaxation processes can thus exhibit maxima in their rates (minima in their relaxation times) as the motions in the system are varied, for example by temperature. In contrast T_2 shows a monotonic decrease as the motional rate is lowered.

5   Spin Labelling Methods

From Eq. (1) it is clear that all chemically equivalent nuclei in a sample will only have the same resonance frequency if the magnetic field is spatially homogeneous. In practice it is never perfectly homogeneous and much of the art of the magnet manufacturer is concerned with minimising the magnetic field inhomogeneity. A spread of B_0 values will result in increased linewidth through a spread of ω_0 values. If, however, the field inhomogeneity is applied in a controlled way then the position of a nucleus in space is labelled by its resonance frequency. Normally a linearly varying field is applied, say in the x direction of the laboratory reference frame. Hence

$$B_x = B_0 + Gx \qquad (7)$$

where B_x is the field at x and G is the field gradient dB/dx. The local resonance frequency ω_x is thus given by

$$\omega_x = \gamma(B_0 + Gx) \qquad (8)$$

Herein lies the principal of NMR imaging shown in Fig. 4. An object consisting of two cylinders is placed in the magnetic field and a field gradient applied in the x direction. The resulting spectrum consists of two lines whose frequency difference corresponds to their distance apart. A similar gradient in the y direction will give a spectrum corresponding to their positions projected onto the y plane. As can be

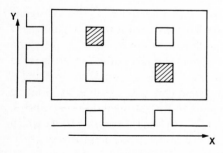

Fig. 4. The principle of magnetic field gradient imaging. The shaded areas represent the real objects, the unshaded areas are the two other possible positions that could give rise to the same spectra for the x and y gradients

seen from the diagram these two measurements are insufficient to define the original object and gradients must be applied along directions intermediate between x and y to remove ambiguities. Many sophisticated strategies now exist for the rapid construction of NMR images but all rely on relationships of the form of Eq. (8).

The most important aspect of NMR imaging is medical imaging. An advantage of the technique is that, unlike X-ray imaging, it does not use ionising radiation and it is thus intrinsically safer. Another advantage is that image contrast may be improved by exploiting differences in relaxation times between different types of tissue. Thus tissue with short transverse relaxation times may be selectively removed from the image by delaying signal aquisition until after its signal has decayed but whilst there is a residual signal from slower relaxing tissues. The effects of short relaxation times in solids usually mean that signals from bone are not observed.

In clinical practice, it is usual to observe signals only from protons. Since water is the most abundant protonacious material in the body, NMR imaging is often the imaging of water proton density, although fats can contribute to the image in certain circumstances. An interesting development is imaging of other biologically important nuclei such as phosphorus and sodium. Whilst this is far from routine it does offer the possibility of widening the range of information available to the clinician. Probably the ideal situation for diagnosis is to combine the spatial specificity inherent in imaging with the chemical information available in the NMR spectrum. Such a measurement allows non-invasive biochemical assessment of a particular tissue region. This exciting possibility is already being explored by some groups and holds out the exciting possibility of a major development in diagnosis techniques.

Spin labelling may also be used to measure diffusion and flow. The pulse sequence is shown in Fig. 5. First a 90° pulse is applied to bring the magnetisation into the x'y' plane of the rotating frame. A pulse of field gradient is then applied for a time t, following a period τ after the pulse. The response of the magnetisation vector of each spin depends on its position in space. The angle Θ_x through which

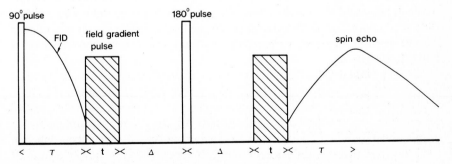

Fig. 5. The principle of the pulsed field gradient experiment for measuring diffusion and flow

it is rotated in the $x'y'$ plane is given by

$$\Theta_x = \omega_x t = \gamma B_x t$$

At some time $t + \tau + \Delta$ a $180°$ pulse is applied in a direction perpendicular to the first RF pulse. This has the effect of reversing the angle of the spin but not its frequency, -i.e. the magnetisation vector is rotated by $180°$ about the direction of the RF field.

Thus immediately before the pulse the total angle by which a spin has rotated is given by

$$\Phi_x = \omega_x t + \omega_r(t + \tau + \Delta)$$

where ω_r is the difference between the measuring frequency and the actual frequency of the spin. Immediately after the pulse the angle is now

$$-\Phi_x = -\omega_x t - \omega_r(t + \tau + \Delta)$$

After a further time Δ a second identical pulse of field gradient is applied which will rotate the magnetisation vector by $+\omega_x t$. Provided no translational motion has occurred the angle will be given by

$$-\Phi_x = -\omega_r(t + \tau)$$

Two effects have occurred here. One is the reversal of the magnetic field gradient effect, the other is the motion of the spin back towards its initial position due to the effects of the $180°$ pulse. At a time $2(t + \tau + \Delta)$, Φ_x will be zero and all the magnetisation will be recovered. Thus a new signal will appear, which in the absence of relaxation and translational motion will be as intense as the signal following the $90°$ pulse. This recovered signal is called the "spin-echo". The effects of relaxation on the spin echo intensity may be determined by experiment in the absence of field gradients, and its effects need not be considered further when considering the general principles involved. It should be noted however that relaxation can cause several practical difficulties. If translational motion has occurred, the second field gradient pulse will not exactly reverse the effects of the first since the effect of motion will be to move the nucleus to a different position in space and hence a different value of ω_x. The effect of motion is thus to attenuate the echo height by an amount related to the mean displacement. Thus a plot of echo attenuation versus the time between the field gradient pulses will allow a measurement of the mean displacement with time and hence a measurement of diffusion and flow.

An important development in the measurement of diffusion in mixed systems has been the development of pulsed field gradient Fourier transform (PFGFT) NMR. In this the half echo following the maximum is Fourier transformed to give a spectrum. In the spectrum the heights of the various resonances are proportional to their attenuation by the field gradients and hence their diffusion coefficients. In this way diffusion in multicomponent systems may be examined and the diffusion coefficients of different chemical species obtained.

6   Two-Dimensional NMR

The conventional Fourier transformation of an interferogram to a spectrum is referred to as a one-dimensional Fourier transform, because only one time variable and one intensity variable is involved. In two dimensional methods a double transform involving two intensity and two time variables is used. In order to see the rationale of dramatically increasing the computational complications of spectral acquisition one must consider the inadequacies of one-dimensional spectral acquisition. Consideration of Fig. 3b shows that whilst some regions of the spectrum have well resolved structure in them other regions contain overlapping peaks which are likely to defy simple analysis. Structure due to homo- or heteronuclear scalar coupling may be present but is not resolved. In many cases of interest, molecules contain more than one element suitable for NMR and it would be useful to determine how the spectrum of one element correlates with that of another. An important example of this is in molecules containing carbon and hydrogen where it is extremely useful to know which carbon atoms are attached to which hydrogen atoms. This can be done by detection of the heteronuclear bond scalar coupling. However, to work through a complex spectrum and sequentially irradiate each resonance for ^{1}H and observe its effect on the ^{13}C spectrum would be immensely time consuming. Similarly if it were required to measure all the possible NOE's in a spectrum large amounts of labour and instrument time would be required.

In the same way that one-dimensional transforms are much more efficient than CW methods for obtaining spectra, two-dimensional methods offer enormous savings in time. An additional advantage is that the two-dimensional display spreads out the spectral data so much that in appropriate cases much better spectral resolution may be obtained.

The general sequence for a 1 or 2D FT experiment is shown in Fig. 6. It consists of a preparation period, followed by a period of evolution of duration t_1, a mixing period which is of fixed duration and a detection period of duration t_2. In the preparation period a pulse or series of pulses is applied to the spin system of interest, in the most general case pulses will be applied to hetero and homonuclei. The object of this is to get the spin system in the correct state for the evolution process. In the two dimensional experiments the evolution period is critical since its duration t_1 will affect the FID acquired during the detection time t_2. The mixing period may be of zero or finite duration but remains fixed. Its function is to enable further manipulation of the spin system to obtain a suitable state for

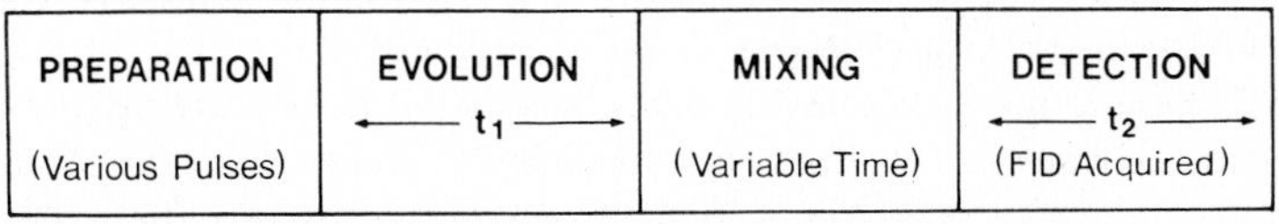

PREPARATION	EVOLUTION	MIXING	DETECTION
	←——— t_1 ———→		←——— t_2 ———→
(Various Pulses)		(Variable Time)	(FID Acquired)

Fig. 6. The general sequence for a two dimensional NMR experiment

detection during the detection period, t_2. During t_2, which is also fixed, the FID is acquired.

The essence of two dimensional experiments is that the time period t_1 is used to modulate the FID. Fourier transformation of the FID acquired during the fixed time t_2 yields a series of spectra each corresponding to a different value of t_1; a second transformation is then carried out over the period t_1 which gives the two dimensional spectrum.

In order to illustrate the process it is useful to consider the use of 2D FT's in imaging. Imagine a nucleus at some point x^*y^* in the imaging plane. If a gradient is applied in the x direction for a time t_1 its frequency will be

$$\omega_x^* = \gamma(B_o + G_x x^*)$$

At the end of the period t_1 the x gradient is switched off and the y gradient is switched on for a time t_2 during which the signal is recorded. The frequency during acquisition will be

$$\omega_y^* = \gamma(B_o + G_y y^*)$$

For convenience the signal is detected at this frequency so that the on resonance FID illustrated in Fig. 2b is observed.

If the signal were observed during the period t_1 it would appear to oscillate with a frequency

$$\omega_y^* - \omega_x^* = \gamma(G_y y^* - G_x x^*) = \Delta\omega$$

and a period $1/2\pi\Delta\omega$. The process is illustrated in Fig. 7a. When t_1 is zero a large positive signal will result. If t_1 is a quarter period of the frequency then a zero signal will result (signal 1 in the diagram). As t_1 increases to a half period an inverted signal appears (signal 2): at a three quarter period a zero signal is observed (signal 3) and, finally, at a whole period (signal 4) the large positive signal, observed when t_1 equals zero, returns. The effect is that the FID has its intensity modulated by the duration of the x gradient but its frequency is determined by its response to the y gradient. After the first Fourier transform a series of spectra as shown in Fig. 7b are obtained. These spectra correspond to positive, zero and inverted FID's. It will be seen that the intensity of the peak oscillates in a sinusoidal manner. A second transform changes this into a single peak, the position of which corresponds to the frequencies generated by the x and y gradients and hence determined by the positions of the nuclei in the x, y plane.

The presentation of data from 2D FT's can take a variety of forms. In imaging peak heights are translated into a grey or false colour scale and presented as an optical image as in a photograph. More useful for spectral information are displays as stacked plots (Fig. 8a) or contour plots (Fig. 8b).

The range of 2D experiments now available is vast. Apart from imaging there are two important classes of experiments in liquids; these are J-resolved experiments and correlation experiments. In J-resolved spectroscopy the scalar couplings are spread out along one axis of the plot whilst the other axis represents

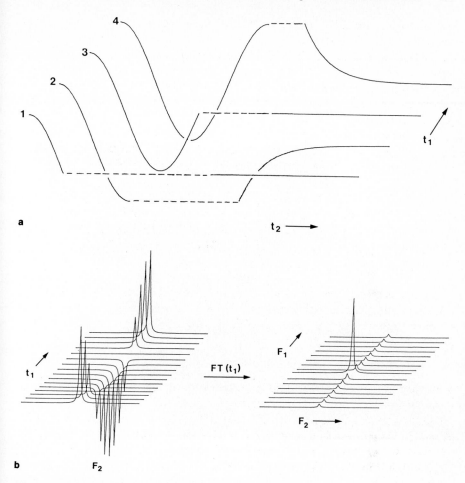

Fig. 7. (a) The changes in intensity of magnetisation as a result of changes in the evolution period t_1. the signals labelled 1, 2, 3 and 4 correspond to increasingly long t_1 values. (b) the one and two dimensional Fourier transforms corresponding to the experiment described in the text. The dimensions F_1 and F_2 correspond to the times t_1 and t_2

chemical shift. In this way the often confused and overlapping coupling patterns can be resolved. Both homonuclear (e.g. ^{1}H-^{1}H) and heteronuclear (e.g. ^{13}C-^{1}H) experiments are possible.

The second class of experiments results in a plot in which correlations between chemical shifts may be obtained. These experiments fall into two main classes. Those which observe nuclei which interact with each other by scalar interactions are called, rather unhelpfully, COSY (Correlation Spectroscopy) experiments. Those which observe nuclei interacting by Overhauser effects or by chemical exchange are called NOESY (Nuclear Overhauser Exchange Spec-a

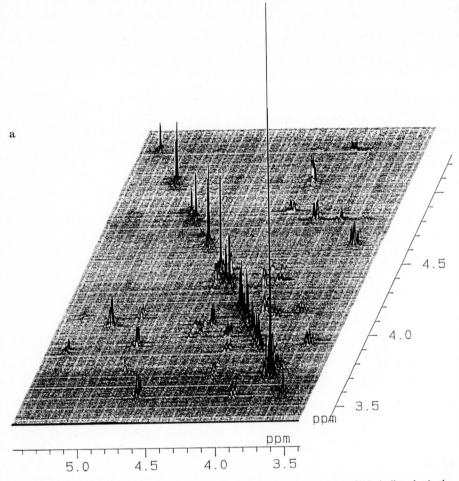

Fig. 8. (a) A stacked spectral plot from a two dimensional NMR experiment (b) A similar plot in the contour display mode. Both (a) and (b) were obtained using proton NMR and used the COSY method to determine connectivity by scalar coupling. The conventional spectrum is plotted on the diagonal and the off diagonal peaks have coordinates in F_1 and F_2 which represent coupled pairs of peaks

troscopy). COSY experiments in general give information on which nuclei are bonded to each other, which may be homo or heteronuclei. NOESY experiments on the other hand give information about the spatial proximity of nuclei or about which nuclei are in chemical exchange.

Together the COSY and NOESY experiments have made a considerable impact on NMR, particularly in the field of biological molecules. A very good account of the applications has been given by Wüthrich.

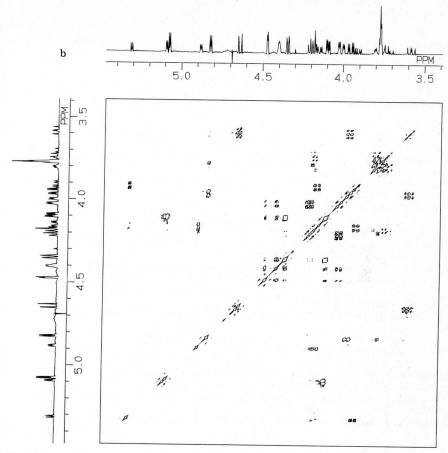

Fig. 8(b)

7 Summary

The coherence apsects of NMR spectroscopy make it an extremely versatile and powerful technique. It may be applied to a wide variety of nuclei in almost any environment. Measurement of relaxation times may be used to examine details of molecular dynamics. Chemical shifts and coupling constants can give very valuable chemical information. Further sophistications allow NMR to be used for imaging and for the measurement of diffusion and flow.

Since only radiofrequency energy and magnetic fields are used, NMR is essentially a non-invasive technique and may be used on live biological systems including human beings.

Acknowledgements

Thanks are due to B. Goodfellow for supplying Figures 6a and 6b and to I. Colquhoun for supplying Figures 7a and 7b.

8 Bibliography

There are a number of excellent textbooks on NMR. The most profound, though not the most readable is:-
Abragam A (1978) The principles of nuclear magnetism, Oxford Univ. Press, Oxford
More readable are as follows:-
Harris RK (1983) Nuclear magnetic resonance spectroscopy, Pitman, London
Farrar TC, Becker EC (1971) Pulse and Fourier transform NMR, Academic, New york
Akitt JW (1983) NMR and chemistry, Methuen, London
Chandrakumar N, Subramanian S (1987) Modern techniques in High Resolution FT NMR, Springer, Berlin Heidelberg New York
Freeman R (1988) A handbook of nuclear magnetic resonance, Longman, London
Multinuclear aspects of NMR are covered in:-
Harris RK, Mann BE (1978) NMR and the Periodic Table, Academic, London
Lambert JB, Riddell FG (1983) The multinuclear approach to NMR spectroscopy, D. Riedel, Dordrecht
For Proton and Carbon NMR see:-
Abraham R, Loftus J (1978) Proton and Carbon 13 spectroscopy: An integrated approach, Heyden, London
Levy GC (1984) Topics in ^{13}C NMR spectroscopy, vol 4 and preceding volumes, Wiley, New York
For biological applications see:-
Wüthrich K (1986) NMR of protons and nucleic acids, Wiley, New York
Gadian DG (1982) NMR and its applications to living systems, Oxford Univ. Press, oxford
Foster MA (1984) Magnetic resonance in medicine and biology, Pergammon, Oxford
Jardetsky O, Roberts GCK (1981) NMR in molecular biology, Academic, New York
For imaging see:-
Mansfield P, Morris PG (1982) NMR imaging in biomedicine, Academic, New York
Roth K (1984) NMR tomography and spectroscopy in medicine, Springer, Berlin Heidelberg New York
For two dimensional methods see:-
Bax AD (1982) Two dimensional NMR in liquids, D. Reidel, Dordrecht
Turner DL (1985) Progr. NMR Spectrosc, 17: 281
For diffusion and flow see:-
Stilbs P (1987) Prog. NMR Spectrosc, 19: 1
(This paper is particularly concerned with PFGFT methods).
Tyrrell HJV, Harris KR (1984) Diffusion in liquids, Butterworths, London
Packer KR, Rees C, Tomlinson DJ (1972) Adv. Mol. Relaxation Processes. 3: 119
For quadrupolar nuclei see:
Forsen S, Lindman B (1981) Methods of Biochemical Analysis 27: 289
(This mainly deals with ion binding but covers the principles of NMR of quadrupolar nuclei very well).
For reviews of specific applications see:-
Belton PS, Ratcliffe RG (1985) Prog. NMR Spectrosc. 17: 24
Belton PS (1984) Chan HW-S (ed) In: Biophysical methods in food research, Blackwell, Oxford
Hampson P (1984) Anal. Proc. 21: 208
Howe RF (1984) Springer Ser. Chem. Phys. 35: 39
Moore GR, Ratcliffe RG, Williams RJP (1983) Essays Biochem, 19: 143
Smith RL, Oldfield E (1984) Science 225: 280
A very useful sources of references including a list of reviews and books is in "Specialist Periodical Reports, Nuclear Magnetic Resonance", published yearly by the Royal Society of Chemistry.

CHAPTER 7
Multinuclear High-Resolution NMR in Solids

M.E.A. Cudby and D.J. Williamson

1 Introduction

The high-resolution NMR spectroscopy of isotropic liquids or solutions has provided valuable structural information for many years (see P. Belton, Chapter 6). Considerable effort has been applied to provide more and more sensitive equipment, and more and more sophisticated analytical computer software and pulse programs. However, there are materials which are not liquids and will not dissolve in suitable solvents making this information unavailable. The past fifteen years has seen the advancement of an NMR technique which has allowed a great deal of structural information to be obtained from solid materials.

The most obvious difference between the NMR spectra of liquids and solids lies in the linewidths of the observed resonances. The liquid or solution sample generally has ^1H and ^{13}C linewidths of a few tenths of a hertz, whereas the solid typically gives linewidths of tens of kilohertz. This difference is due to rapid molecular motion in liquids or solutions very greatly reducing the magnetic coupling of nuclear dipoles, whereas this coupling interaction is preserved in the solid.

An attempt will be made to outline techniques which will persuade a solid to give an NMR spectrum which resembles that produced by an isotropic liquid and a number of applications will be discussed to illustrate the usefulness of the technique. There are three major difficulties to be overcome if high resolution spectra with reasonably narrow resonances are to be obtained from a solid. The first is the coupling linewidths problem mentioned above. Second is the orientation-dependence of shielding constants in a static magnetic field, which gives rise to a distribution of chemical shifts in solids and so results in further line broadening. Third, low sensitivity and long spin-lattice relaxation times in solids, particularly in the case of ^{13}C, lead to weak signals in repetitive pulse experiments.

These broadening effects and sensitivity problems may be dealt with in the following manner.

2   Techniques Based on Carbon-13

2.1   High Power Decoupling

The major line-broadening effect in ^{13}C NMR spectroscopy is due to dipolar interactions between the dilute ^{13}C nuclei and the abundant spin ^{1}H nuclei, that is heteronuclear interactions. In this case the heteronuclear dipolar interactions can be reduced towards zero by the application of a very high power ^{1}H decoupling field. Decoupling experiments are commonplace in solution NMR studies, but in the solids experiment, much higher decoupling powers are required. This heteronuclear decoupling reduces the $^{13}C-^{1}H$ dipolar interaction towards zero leaving the small contribution of $^{13}C-^{13}C$ dipolar interactions when they occur.

2.2   Magic Angle Rotation (MAR)

A powder or amorphous sample contains molecular groupings which are oriented in all possible directions with respect to the magnetic field and this gives rise to a variety of chemical shifts producing a broad resonance. This is known as chemical shift or shielding anisotropy and is represented by the diagram known as a 'powder pattern', shown in Fig. 1.

Organic solids generally contain many nuclei in different chemical environments so that considerable overlap of the powder patterns occurs giving rise to a broad resonance. The narrow lines produced in the solution NMR experiment result from molecular motion which reduces the shielding term to the isotropic value. Some means must therefore be found to reduce the powder anisotropic shifts to the isotropic value in the solid state.

The orientation dependence of shielding for nuclei in an axially symmetric environment can be expressed as:-

$$\sigma(\theta) = \ddot{\sigma} + (3\cos^2 \theta - 1)\ \ (\sigma_{\parallel} - \sigma_{\perp})/3$$

where θ is the angle between the symmetry axis and the static magnetic field,

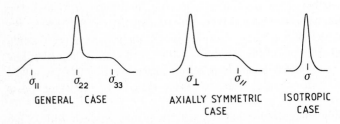

GENERAL CASE   AXIALLY SYMMETRIC   ISOTROPIC
                      CASE   CASE

Fig. 1. NMR powder patterns for solid materials

$\sigma_\parallel$ and $\sigma_\perp$ are the shielding constants at $\theta = 0$ and $90°$ respectively, and $\bar{\sigma}$ is the isotropic average of $\sigma(\theta)$. It is sufficient to indicate that if the factor $(3\cos^2 \theta - 1)$ is made equal to zero, then the anisotropic terms will vanish. This is achieved by spinning the sample at an angle of $54°44'$ to the magnetic field and this is the so-called "Magic Angle" [1]. In addition to the removal of chemical shift anisotropy, the $^{13}C-^{13}C$ homonuclear dipolar interactions are also reduced toward zero since the $(3\cos^2 \theta - 1)$ factor also appears in the mathematical description of these interactions.

The speed at which the sample must be rotated is related to the field strength of the instrument and to the extent of the chemical shift anisotropy. In general, these speeds are in the range 2–6 kHz. (Although, as stated earlier, dipolar interactions are averaged towards zero by MAR, very high speeds indeed are required to remove $^{13}C-^1H$ interactions and in practice this is best achieved by high-power decoupling).

2.3   Cross-Polarisation

As stated earlier, a major problem to be overcome is the inherently low sensitivity of the ^{13}C nucleus and its long relaxation time. The cross polarisation method was first developed by Pines, Gibby and Waugh[2] to make use of the strong polarisation of abundant spins, i.e. ^1H(protons), to enhance the inherently weak spin polarisation of dilute nuclei, i.e. ^{13}C. The experiment is perhaps most easily understood in terms of spin temperature. A low spin temperature means that there is a large polarisation of magnetic moments and it follows that the NMR signals are intense. Conversely, a high spin temperature indicates a small polarisation of magnetic moments and therefore the NMR signals are weak. Let us assume that the abundant spin system, i.e. ^1H, can be prepared in a low temperature spin state. We then have a large cold reservoir of magnetisation which could provide a means of cooling the dilute spin system, i.e. ^{13}C without significantly raising the spin temperature of the abundant species. If some method can be devised to make efficient thermal contact between the two different spin systems, it should be possible to enhance the NMR signal of the dilute spin system from the reservoir of magnetisation of the abundant spin system. Obviously the dilute spin system will transfer energy to the abundant spin system, but now the flow of energy to the surrounding lattice will be via the abundant spin relaxation process T_1. Therefore, for example, it is the shorter relaxation times of protons which will determine the repetition rate of the experiment, and not the long relaxation times of the ^{13}C nucleus.

The thermal contact required is made under the Hartmann-Hahn[3] condition in the rotating frame of reference when $\gamma_C B_{1C} = \gamma_H B_{1H}$, where γ_C and γ_H are the magnetogyric ratios for ^{13}C and ^1H respectively and B_1 is the appropriate radiofrequency field. The rotating frame of reference is a coordinate system set to rotate at the Larmor or resonance frequency about the applied field B_0. Since no

Larmor precession is present under these conditions only the radiofrequency field B_1 is active. Thus, by appropriate choice of the B_1 field strength for each nucleus the Hartmann-Hahn condition can be fulfilled. Consequently, both nuclei will have the same Larmor frequency and an exchange of energy is possible resulting in mutual spin transitions or 'flips' of these dissimilar nuclei. The sensitivity increase observed is approximately γ_H/γ_C, a value of 4.

2.4   Pulse Sequence

The double resonance situation required is illustrated by the pulse sequence shown in Fig. 2. A 90° pulse is applied in the proton channel followed by a phase change to retain the magnetisation in the spin-locked situation. The ^{13}C pulse is now applied so that the previously mentioned Hartmann-Hahn condition can be satisfied. Once cross-polarisation is complete the ^{13}C pulse is removed and the FID obtained. Following a waiting time indicated by the decay of proton magnetisation (T_1 ^1H) the whole sequence may be repeated and the ^{13}C signal accumulated. The efficiency of the cross-polarisation is affected by $T_{1\rho}$ ^1H since this is a measure of the rate at which protons lose magnetisation during the spin-locked time. It is clear then that it is quite vital to know the values of proton T_1 and $T_{1\rho}$ relaxation times in order to optimise the experimental condition. The successive application of these various techniques was first used by Schaefer and Stejkal [4]. An illustration with reference to calcium acetate hemihydrate shows how the broad dipolar coupled spectrum is transformed to that closely resembling a high resolution dipolar decoupled spectrum (Fig. 3).

2.5   Applications of the Method

The sample for examination is placed in a small container known as the rotor. The amount of material required will vary but 300 to 500 milligrams is a reasonable estimate for most systems. It is necessary for the rotor to attain high

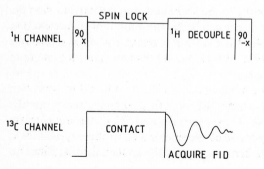

Fig. 2. Double Resonance experiment to satisfy the Hartmann-Hahn condition

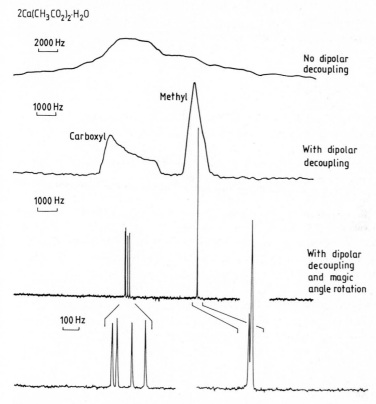

^{13}C High-resolution NMR spectra of powdered Calcium Acetate Hemihydrate obtained using different combinations of the experimental techniques.

$2Ca(CH_3CO_2)_2 \cdot H_2O$

2000 Hz

No dipolar decoupling

1000 Hz

Methyl

Carboxyl

With dipolar decoupling

1000 Hz

With dipolar decoupling and magic angle rotation

100 Hz

Fig. 3. The effect of different experimental conditions on the ^{13}C spectrum of calcium acetate hemihydrate

spinning speeds between 2 and 6 kHz, so it is important to distribute the weight of the sample evenly within the rotor. Powders are perhaps the most suitable type of sample but discs of material can be cut to size and stacked to fill the rotor, or films may be rolled and placed in the rotor so that an even weight distribution is achieved. A variety of materials has been used in the past such as machineable ceramics or polymers such as a polyoxymethylene, "DELRIN", or for preference a chlorofluorocarbon polymer known as "KEL-F" since this material is robust and does not produce an interfering signal.

As with solution-state NMR the observed resonances help to characterise the molecular structure of the material under examination. The spectra of some organic systems shown in Figs. 4–6 illustrate this particular property. However, with this technique there may be additional information within the spectrum

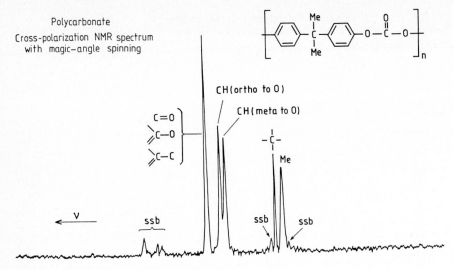

Fig. 4. ^{13}C spectrum of bisphenol-A polycarbonate

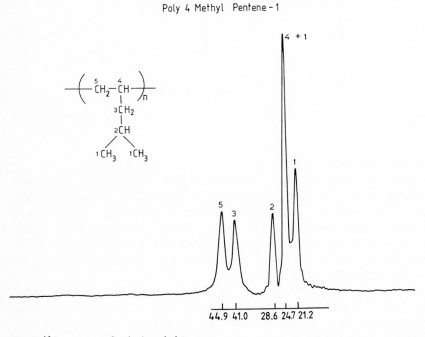

Fig. 5. ^{13}C spectrum of poly 4-methylpentene

Poly Ethyl Cellulose

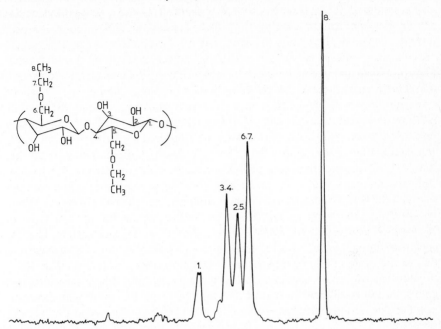

Fig. 6. ^{13}C Spectrum of polyethyl cellulose

Isotactic Polypropylene

α—crystalline form

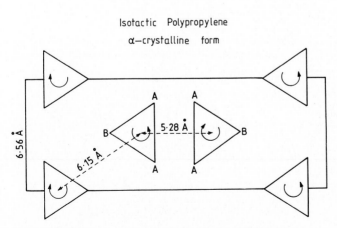

Fig. 7. Unit cell of α-crystalline form of isotactic polypropylene

which relates to the solid state condition, especially if the material being studied is able to exhibit different ordered forms. For example, isotactic polypropylene has a molecular conformation in the solid which is based on a three monomer repeat unit for each complete turn of the helix. These helical molecules can be arranged into various ordered situations which have been studied by X-ray crystallography. Two arrangements which have been closely defined are known as the α and β forms, and these structures are illustrated in Figs. 7 and 8. The ^{13}C solids spectrum Fig. 9 displays the expected three resonances due to the methyl, methine and methylene carbons [5]. If the polymer is annealed at a temperature just below the melting point to produce well ordered crystalline regions, these resonances now appear as doublets. From the X-ray studies it is recognised that the α-crystal form contains four helical molecules which are arranged in left and right handed pairs, Fig. 7. As can be seen from this figure, there is closer contact between the helical molecules in one pair than between the pairs of helices. Thus, two different environments exist in the crystalline regions for all three types of carbon, and as a result two chemical shifts are observed for each carbon. The ^{13}C spectrum of the β-form, however, only shows a single resonance for each carbon due to the more nearly equal spacing of the molecules in this form, Fig. 10.

Polypropylene can also exist in the syndiotactic form where the molecular configuration is different to that of the isotactic form. The molecular conformation of this stereoisomer has also been studied [6] and is shown in Fig. 11. The observed ^{13}C solids spectrum is a little surprising since four distinct resonances are observed from only three carbons. The explanation lies in the molecular

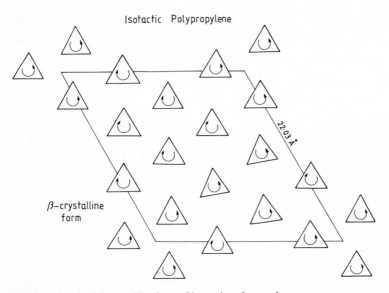

Fig. 8. Unit cell of β-crystalline form of isotactic polypropylene

Isotactic   Polypropylene

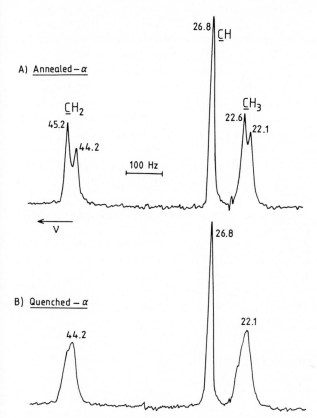

Fig. 9. ^{13}C spectrum of quenched and annealed α-crystalline form isotactic polypropylene

conformation of the solid shown in Fig. 11. The even numbered circles represent the methylene groups and clearly two environments exist for 2 plus 6 and 4 plus 8.

The study of the solid form of the steroid sodium glycocholate reveals an interesting phenomenon. The ^{13}C spectrum of the 'as made' product is shown in Fig. 12 and appears to have rather broad lines. It is observed that heating of the sample below its melting point during the high power decoupling cycle changes the physical state of the sample. The spectrum obtained from this heated sample (also shown in Fig. 12) displays much sharper resonances and, more significantly, some very different chemical shifts. Thermal analysis produces the DSC (differential scanning calorimetry) trace shown in Fig. 13 exhibiting a transition near to 100 °C. This is believed to be due to water which has been included in the structure being released. It would appear therefore that water in the structure of

Isotactic Polypropylene β—form

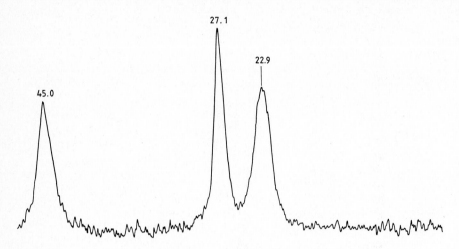

Fig. 10. ^{13}C spectrum of β-crystalline isotactic polypropylene

Syndiotactic Polypropylene

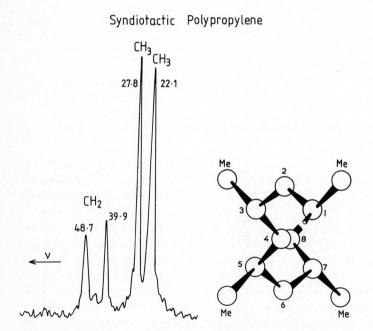

Fig. 11. ^{13}C spectrum and conformation diagram of syndiotactic polypropylene

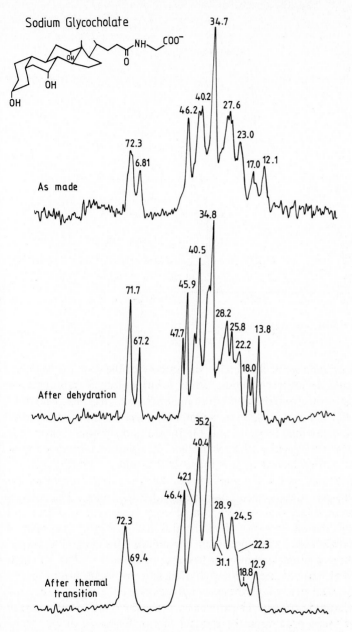

Fig. 12. ^{13}C spectra of various polymorphic forms of sodium glycocholate

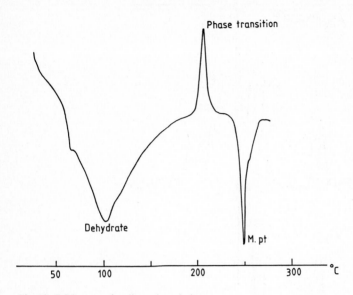

DSC   Na   Glycocholate

Phase transition

Dehydrate

M. pt

50   100   200   300   °C

Fig. 13. DSC trace of sodium glycocholate

the solid provides a number of chemical environments for the sodium glycocholate molecule, hence the different chemical shifts. A further transition is observed near 200 °C and the spectrum obtained from material which has been heated to just above 200 ° C, but not melted, shows yet more changes in both chemical shift and width of resonance lines (Figs. 12 and 13). In addition, another transition is observed below the melting point, and yet another ^{13}C spectrum is observed from material which has been heated to just above 220 ° C and cooled (Fig. 12).

These studies are all part of the phenomenon known as polymorphism and the drug industry has always been concerned with this subject. There are a number of traditional techniques which can be used to study this subject and solid state NMR should now be added to this list. A further example of polymorphism is exhibited by the compound tripalmitin. This material is known to exist in the three different polymorphs designated α, β and β'. The ^{13}C high resolution solids spectra obtained from each of these forms are shown in Fig. 14 and distinctions can be seen in both the carbonyl region and the region associated with carbons adjacent to oxygen. These differences are the result of variations in the hydrocarbon chain packing in the crystal [7,8] as illustrated in Fig. 15.

Conventional ^{13}C NMR spectra of insoluble materials are difficult to obtain and only partial success has been achieved when swollen gels are examined. The solids spectra of such materials provide a great deal of useful information, however, as can be seen in the following examples. All the resonances of the cross-

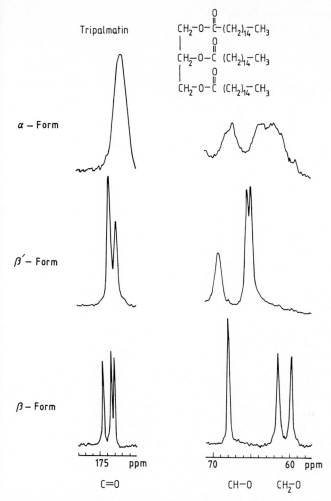

Fig. 14. ^{13}C spectra of various polymorphic forms of tripalmitin

linked epoxides shown in Fig. 16 can be assigned and in this instance it is also possible to observe resonances which are characteristic of the cross-linking molecules. The concentration of cross-links in this example is quite large and it will not always be so easy to find evidence for low concentration.

Another example of an insoluble material is household paint after curing. The solids ^{13}C spectrum of a paint shown in Fig. 17 provides considerable help in identifying the constituents of the material [9]. Cured natural and synthetic rubbers are quite insoluble but can sometimes yield the normal solution ^{13}C spectrum from a swollen gel. This is not very satisfactory however since it is likely that the more immobile regions at the cross-linked sites do not give narrow

Representation of Hydrocarbon Chain Packing
in the α,β' and β form of Tripalmitin.

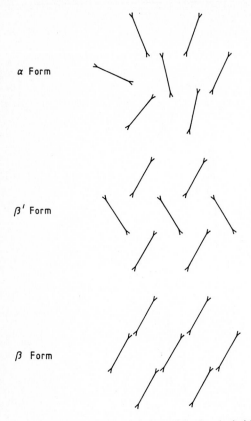

α Form

β' Form

β Form

Fig. 15. Representation of chain packing in tripalmitin polymorphs

signals and remain unobserved. The solid state spectrum can be obtained very easily however and a spectrum of a blend of natural and synthetic rubber is illustrated in Fig. 18. Magic angle rotation and dipolar decoupling are profitably employed, but not cross-polarisation, since the $^1H\,T_{1\rho}$ value is very short indicating considerable molecular mobility, and hence a very low efficiency of cross-polarisation. The resonance lines observed in the spectrum are quite narrow and the assignments are clear. Recent work [10] with high cross-link density natural rubber shows a number of weak resonances which contain information on both the nature of the cross-link sites and the main chain sulphur modifications which are also present (Fig. 19).

The technique of high resolution solids NMR has helped in the structural study of many organic materials, but there are occasions when the required information is not forthcoming. One problem is the inherent linewidths of

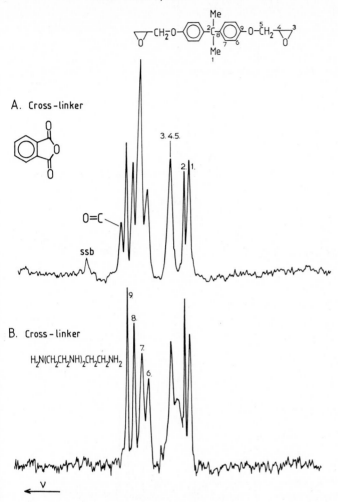

Cross-linked Epoxides

Fig. 16. ^{13}C spectra of crosslinked epoxide resins

resonances. The narrowest lines are of the order of 10 Hz but more commonly the figure is often 20 or 30 Hz. The need to determine the stereoregularity of vinyl polymers has largely been satisfied using ^{13}C NMR of solutions. Polyvinylchloride (PVC) is one such polymer and Fig. 20 shows the considerable information available from the ^{13}C spectrum used to determine the fine structure of this polymer. The solids spectrum shown in Fig. 21 demonstrates the problem of linewidth. All the fine structure seen in solution has been lost within the two broad resonances of the $-CH_2-$ and $-CHCl-$ carbons.

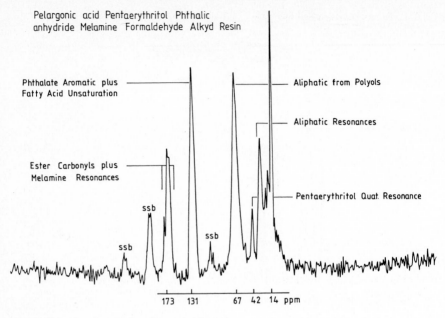

Fig. 17. ^{13}C spectrum of a cured household paint

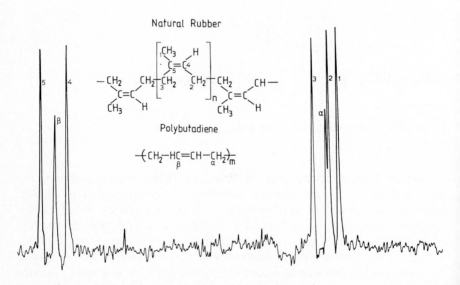

Fig. 18. ^{13}C spectrum of blend of natural rubber with polybutadiene

Sulphur Crosslinked Natural Rubber

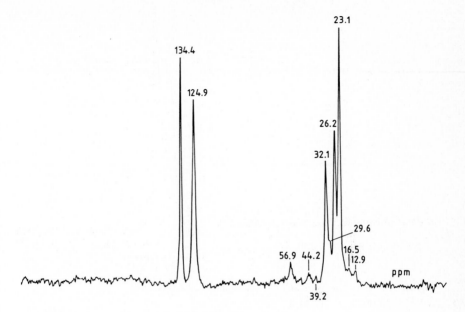

Fig. 19. ^{13}C spectrum of a sulphur-crosslinked natural rubber

POLYVINYL CHLORIDE   Pulsed FT ^{13}C NMR spectra

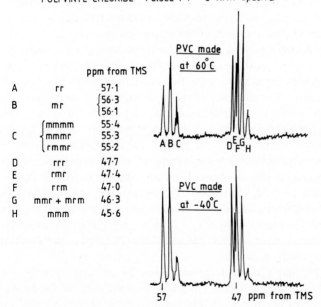

		ppm from TMS
A	rr	57·1
B	mr	56·3 / 56·1
C	mmmm	55·4
	mmmr	55·3
	rmmr	55·2
D	rrr	47·7
E	rmr	47·4
F	rrm	47·0
G	mmr + mrm	46·3
H	mmm	45·6

Fig. 20. Solution high resolution ^{13}C spectrum of polyvinylchloride

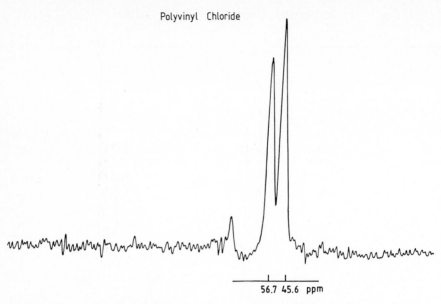

Polyvinyl Chloride

56.7 45.6 ppm

Fig. 21. Solid state high resolution ^{13}C spectrum of polyvinylchloride

A very similar situation arises in the case of low density polyethylene (LDPE). The solution spectrum in Fig. 22 shows resonances which have been assigned to the many types of structures in LDPE. The solids spectrum of a similar polymer is shown in Fig. 23 and although some information on branch type can be seen, the detail is quite hidden by the linewidth of the major resonance.

Many of the polymeric materials referred to previously are structurally heterogeneous and contain both disordered (amorphous) and ordered (crystalline) regions. Such a material will exhibit two or more proton $T_{1\rho}$ relaxation times as discussed earlier. The efficiency of cross-polarisation is clearly going to vary with the type of material under examination. This effect can be turned to advantage since it should be possible to discriminate between these regions by "fine tuning" the pulse sequence used. In general, the amorphous regions (or regions of high molecular mobility) will not respond well to cross-polarisation since the proton relaxation time is short. Thus, an experiment can be devised where cross-polarisation is delayed so that the great majority of the protons in the amorphous regions will have relaxed and are therefore not available for cross-polarisation and enhancement of the ^{13}C signal. Under these conditions the signal from more ordered regions, which generally have longer proton relaxation times, will be preferentially enhanced.

An example of this experiment is shown in Fig. 24 and concerns the ^{13}C spectrum of polyethylene terephthalate. The pulse sequence used for the delayed contact experiment is shown in Fig. 25. The spectra obtained indicate that the delayed contact produces sharper resonances. This is because the environment of

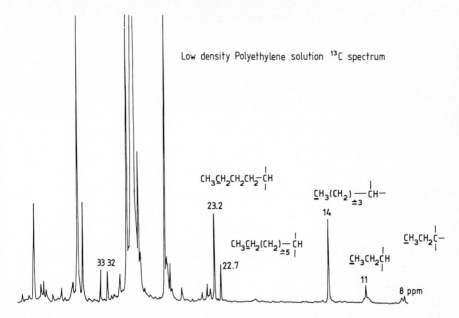

Low density Polyethylene solution ^{13}C spectrum

$CH_3\underline{C}H_2CH_2CH_2\text{--}\overset{|}{\underset{|}{C}}H$

23.2

$\underline{C}H_3(CH_2)_{\geq 5}\text{--}\overset{|}{C}H$

22.7

$\underline{C}H_3(CH_2)_{\geq 3}\text{--}\overset{|}{C}H\text{--}$

14

$\underline{C}H_3CH_2\overset{|}{C}H$

11

$\underline{C}H_3CH_2\overset{|}{\underset{|}{C}}\text{--}$

33 32

8 ppm

Fig. 22. Solution high resolution ^{13}C spectrum of low density polyethylene

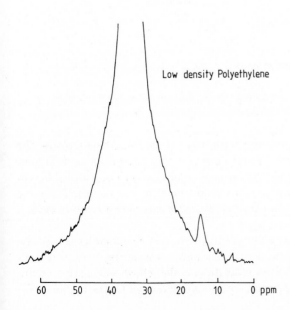

Low density Polyethylene

60	50	40	30	20	10	0 ppm

Fig. 23. Solid state high resolution ^{13}C spectrum of low density polyethylene

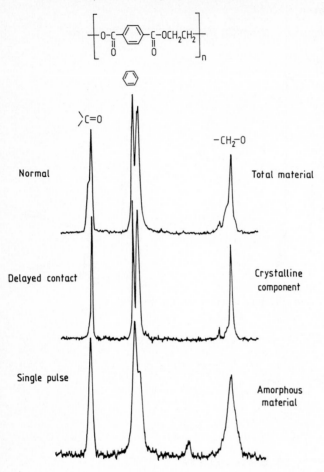

^{13}C NMR spectra of Polyethylene Terephthalate

Fig. 24. ^{13}C spectra of polyethylene terephthalate under different experimental conditions

the individual carbons is more uniform throughout the crystalline regions. The amorphous regions give the broader resonances for the converse reason. That is, the chemical environments of individual carbons are likely to be a little different throughout the disordered regions, thus broadening the resonances. The pulse sequence in operation for the so-called single pulse experiment requires proton decoupling only.

The ^{13}C solids spectrum of an ethylene/vinyl acetate copolymer serves as a further example of the use of different pulse sequences to display various regions of a heterogeneous system. Proton relaxation studies show that three domains with different mobilities are present, which can be designated as $T_{1\rho}$ long, $T_{1\rho}$

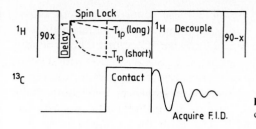

Fig. 25. Pulse sequence for delayed contact condition

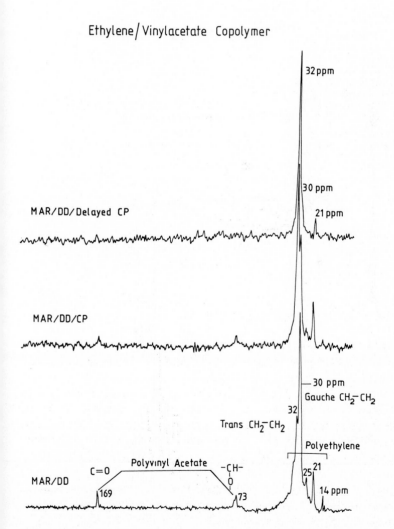

Fig. 26. ^{13}C spectra of an ethylene/vinyl acetate copolymer under different experimental conditions

intermediate and $T_{1\rho}$ short. The use of a pulse sequence without cross-polarisation enhances the resonances obtained from the more mobile regions of the system where cross-polarisation is inefficient, (i.e. $T_{1\rho}$ short). This produces the spectrum shown as MAR/DD in Fig. 26 (magic angle rotation with dipolar decoupling). The presence of polymerised vinyl acetate is apparent. Additional resonances are due to disordered polyethylene sequences and short chain hydrocarbon branches. The pulse sequence (which includes cross-polarisation (CP)) gives the spectrum MAR/DD/CP, and under these conditions the $T_{1\rho}$

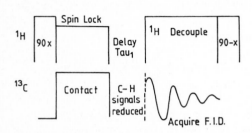

Fig. 27. Pulse sequence for non-quaternary suppression condition

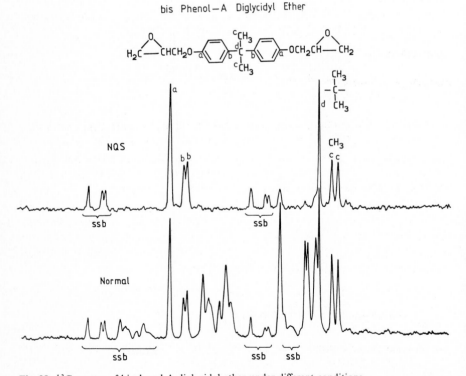

Fig. 28. ^{13}C spectra of bisphenol-A diglycidyl ether under different conditions

intermediate domains predominate. The vinyl acetate sequence is now much less obvious and the majority of short chain hydrocarbon branches have been lost. The polyethylene sequences now show as two distinct conformations with the trans sequence indicating that at least some ordered regions are present. Conditions for the third spectrum (MAR/DD/delayed CP) are deliberately chosen to enhance the response from the $T_{1\rho}$ long domains of the copolymer. The spectrum hardly shows any vinyl acetate, virtually all the ethylene is present as trans sequences and there is little evidence of short chain branching. It is probable that the ethylene trans sequences are present in crystalline regions but other techniques such as X-ray diffraction or differential scanning calorimetry must be used to confirm this proposal.

A further experiment which has proved to be valuable in the assignment of resonances concerns the use of a pulse sequence which discriminates in favour of quaternary carbon atoms. The pulse sequence for non-quaternary suppression is shown in Fig. 27 and differs from the normal cross-polarisation sequence by the delay TAU_1 before the acquisition. During this period when the proton decoupler is inactive the $C - H$ dipolar interactions rapidly dephase the carbon magnetisation leaving mainly the quaternary carbons to contribute to the observed spectrum. The mobility of methyl groups at normal temperatures leads only to an attenuation of the signal and not complete suppression. The usefulness of such an experiment is illustrated in Fig. 28 where the assignment of resonances in complex molecules has been facilitated.

3   Techniques Based on other Nuclei

As described in Sect. 2 the use of high power proton decoupling in combination with cross-polarization and magic angle rotation to obtain high resolution ^{13}C spectra for a wide variety of solid organic materials and substances has become almost routine and a wealth of publications exists. Diversification towards a multinuclear approach high resolution solid-state NMR, akin to the similar development in high resolution solution-state NMR, has taken place in recent years. The problems encountered in order to obtain high resolution solid-state spectra for nuclei other than the magnetically dilute spin $\frac{1}{2}$ ^{13}C are often very different and may prevent a multinuclear approach to a particular sample containing two or more NMR active nuclei.

3.1   Dilute Spin $\frac{1}{2}$ Nuclei other than ^{13}C

The best example of a dilute spin $\frac{1}{2}$ nucleus, ^{13}C, is magnetically dilute due to its low natural abundance of 1.1%, and thus experiences spin dilution by the 98.9% of NMR-inactive ^{12}C (+ ^{14}C) nuclei. Spin dilution can occur by other means, e.g.

chemical dilution of a high naturally abundant spin $\frac{1}{2}$ nucleus in a large molecule, such as a polymer with isolated ^{19}F end groups, or physical dilution by so-called matrix isolation of a nuclide in molecules in which it is otherwise absent. In practice the working definition of a dilute nucleus may be taken as one for which the homonuclear dipolar interactions are weak enough for MAR at readily attainable rates to suffice for averaging. In the case of ^{13}C such dipolar broadening is the order of tens of hertz and so MAR at rates of 3 kHz is more than adequate. MAR also reduces the contribution of chemical shift anisotropy (CSA) to the linewidth. The MAR rate chosen also depends on the CSA (see Sect. 2). Interestingly, calculation of [31]P–[31]P dipolar interactions predicts that these too fall in the region where spinning at such rates should produce averaging and this is indeed found to be the case. [31]P may therefore be considered to be a dilute spin $\frac{1}{2}$ nucleus even though it has a natural abundance of 100%. [31]P linewidths in MAR spectra are an area of much interest as the widths achieved are not always as narrow as predicted and experiments such as CRAMPS (Combined Rotation And Multi-Pulse Spectroscopy), have been undertaken to try and obtain further narrowing [11]. Recent developments in MAR techniques [12,13] have made it possible, in favourable cases, to achieve high enough spinning rates to narrow lines of abundant spin $\frac{1}{2}$ spectra. Rates as high as 23 kHz have been reported but these are for rotors machined out of the material under study. Such rates have not yet become routine for packed rotors. Until such times it will be necessary to discuss and treat dilute and abundant spin $\frac{1}{2}$ nuclei as separate cases.

After ^{13}C the most widely studied nucleus in this general case is [29]Si which is 4.7% abundant. Much work has been performed in the industrially important areas of zeolite catalysts [14] and cements [15,16]. Investigations of proteins, binding of phosphorous containing molecules to proteins and pH-dependent phenomena have been undertaken by observation of [31]P [17].

An interesting case of selective coupling was found during a study of a series of simple inorganic phosphates used as model compounds for more complex biological molecules. Penta-sodium tripolyphosphate hexahydrate, $Na_5P_3O_{10} \cdot 6H_2O$, shows three resonances at 3.20, 0.94 and -6.41^1 ppm (Fig. 29a). The two end phosphates are inequivalent and if the decoupler power is reduced all resonances broaden (Fig. 29b). If no proton decoupling is employed one of the end resonances broadens to such an extent that it is no longer observable (Fig. 29c). The crystal structure [18] confirms that the two terminal phosphates are inequivalent and that one is involved in hydrogen-bonding to five water molecules whereas the other hydrogen-bonds to only one. It seems reasonable to assign the collapsed resonance to the highly hydrogen-bonded end phosphate. Hence ^{31}P solid-state NMR is able to distinguish one end of this molecule from the other. Similar results have been published by another group [19].

---

[1]with respect to external 85% phosphoric acid

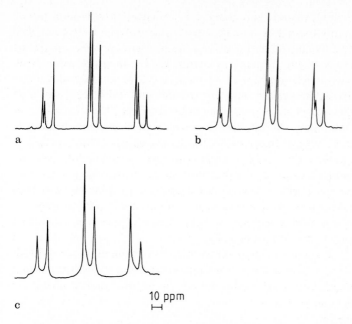

10 ppm

Fig. 29. 80.96 MHz ^{31}P MAR solid-state NMR spectra of penta sodium tripolyphosphate hexahydrate acquired under different levels of proton decoupling field: (**a**) 60 kHz, (**b**) 40 kHz (**c**) no decoupling

When recording spectra of dilute spin $\frac{1}{2}$ nuclei it is not always necessary or indeed desirable to decouple and/or cross polarise (CP). The CP advantage $(\gamma H/\gamma_{obs})$ is less with large γ_{obs}, so for ^{31}P the improvement is approximately half that of ^{13}C, although it may be advantageous if the relaxation times of the observation nucleus is long or in order to provide further information as in the non-quaternary suppression experiment [20]. Decoupling is clearly unnecessary if protons are absent or if the observation nucleus-to-proton distances are large, e.g. where there is tetrahedral or octahedral nearest neighbour coordination by oxygen. Nuclei with large chemical shift anisotropies, e.g. ^{31}P and ^{195}Pt, are better observed at low field due to the large number of sidebands produced, unless very high spinning rates are available. Usually the number of chemically inequivalent sites is low and assignment is easily carried out, recording the spectra at different spinning rates to confirm isotropic resonances if necessary.

3.2   Quadrupolar Non Integer/Odd Halves Spin Nuclei

It was thought for some time that MAR would be unsuccessful for most quadrupolar nuclei, firstly due to the magnitude of the quadrupolar interaction and secondly due to second order effects which cannot be averaged to zero by

MAR. However, it was realised [21] that in the "high field approximation", where the Zeeman term predominates and the quadrupolar term is treated as a perturbation of the Zeeman, the central $m = \frac{1}{2} \leftrightarrow m = -\frac{1}{2}$ transition is unaffected (to first order: both levels are lowered in energy but by an equal amount). In cases where the electric field gradient at the nucleus is low, second order effects can be reduced by MAR together with any dipolar and anisotropic shielding interactions. The second-order line-broadening is inversely proportional to the applied B_0 field strength whereas the chemical shift dispersion is directly proportional to B_0. It follows that in order to observe chemical shift effects it is desirable to work at the highest field possible; conversely quadrupolar effects are best studied at low field. Anisotropic interactions often prevent transitions other than the central $m = \pm \frac{1}{2}$ being observed and at centres of symmetry lower than octahedral or tetrahedral electric field gradient interactions can prevent detection altogether, i.e. in MAR spectra of this class of nuclei not all the sites in a particular sample will necessarily be observable [22].

Examples of such quadrupolar nuclei which have been studied by MAR NMR include ^{23}Na, ^{27}Al, ^{11}B and ^{17}O which are present in such important classes of materials as glasses [23] and zeolites [24]. Another industrial materia

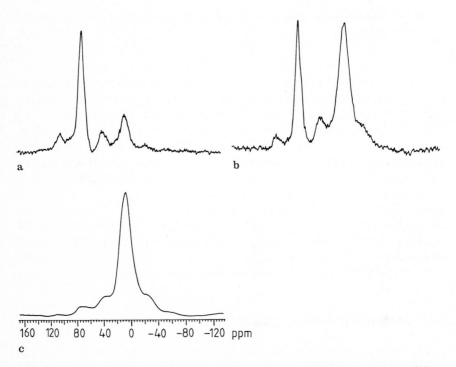

Fig. 30. 104.1 MHz ^{27}Al MAR solid-state NMR spectra of Secar 71 high alumina cement at different stages of hydration: (**a**) partially hydrated; (**b**) semi-hydrated; and (**c**) fully hydrated

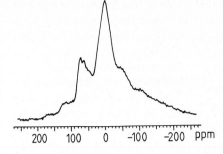

200   100   0   −100   −200   ppm

Fig. 31. 52.1 MHz ^{27}Al MAR solid-state NMR spectrum of the same Secar 71 high alumina cement sample as Fig. 30(b)

which contains aluminium is hydraulic cement. High alumina cement (HAC) is used for monolithic furnace linings [25] and industrial catalyst supports. Although the technical utilisation of cements is in general reasonably well controlled, much of the chemistry of hydration has yet to be understood. The anhydrous cement contains Al tetrahedrally co-ordinated by oxygen, giving Al chemical shifts in the 70–80 ppm range. However, in the hydration products the Al is octahedrally co-ordinated by oxygen with chemical shifts in the region 0–10 ppm. Solid-state ^{27}Al NMR can therefore be used to follow the hydration of HAC by monitoring the change in aluminium co-ordination from tetrahedral to octahedral. Figure 30 (a–c) shows the 104.1 MHz ^{27}Al MAR spectra of partially, semi- and fully-hydrated HAC samples, respectively. In these samples, all the aluminium intensity is recorded and quantitative hydration profiles can be obtained for the aluminium co-ordination change [26]. Figure 31 shows the same sample as Figure 30 (b) recorded at half the B_0 field (i.e. 52 MHz) and exemplifies by comparison with Figure 30b, the benefit of working at the higher field by the greatly improved resolution.

3.3   Abundant Spin $\frac{1}{2}$ Nuclei

The two common examples of this general case are ^{1}H and ^{19}F, both virtually 100% abundant with large magnetrogyric ratio. These nuclei often occur with high chemical abundance, most obviously in the case of protons in most organic substances and polymers. Here the dipolar interaction is usually so large (of the order of tens of kHz) that MAR would have to be performed at impractically high rates. As mentioned in Sect. 3.1 improvements in spinning rates have given useable resolution for some important polymers containing abundant spin $\frac{1}{2}$ nuclei. Practicable spinning rates are also increasingly feasible for loaded rotors (spin rate approx. 17 kHz) which can be sufficient, in certain cases, to give useful resolution for this class of nucleus. Generally, however, spinning samples are only studied in special cases where the dipolar term is small, e.g. due to chemical dilution. Proton or fluorine rich samples are usually studied by static spectra or not at all.

A special case where ^{1}H solid-state NMR studies can be useful even with samples which give very broad lines is where the chemical shift range is greatly increased beyond the normally narrow range (0–10 ppm). This occurs for metals where the interaction with the conduction band electrons gives rise to the well-known Knight shift [27], which can give shifts in percentage as opposed to ppm values. The Knight shift is usually observed for the resonance signal of the metal itself but greatly increased shifts may also be given for nuclei chemisorbed on the metal surface. Hence solid-state ^{1}H NMR has proved to be a useful tool for surface science studies of metal catalysts [28].

A well characterised catalyst [29] is Pt on a support of SiO_2. Figure 32 shows the static 60 MHz ^{1}H NMR spectra of (a) the background hydroxyls of the SiO_2 support, (b) the same sample after exposure to a hydrogen equilibrium pressure of 50 Pa, and (c) the difference spectrum of (b) − (a). Despite a linewidth in the region of 3 kHz for the support hydroxyls the chemisorbed protons are sufficiently shifted, between 30–60 ppm (depending on the exposure pressure), to be observed as a shoulder on the background spectrum. The difference spectrum (c) shows that hydrogen is also adsorbed onto the support on exposure, and that the proton linewidth for the hydrogen chemisorbed on the Pt is somewhat lower, approx. 1.3 kHz, than that of the hydroxyl protons on the SiO_2. The partioning of the hydrogen between the metal and the support may be part of a dynamic exchange

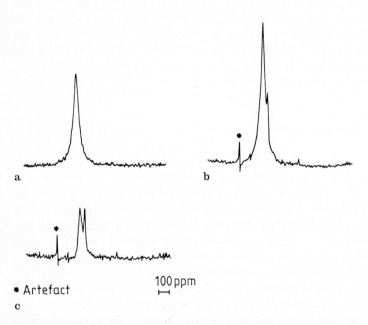

* Artefact   100 ppm

Fig. 32. 60 MHz ^{1}H static solid-state spectra of: **(a)** background hydroxyl hydrogens on SiO_2 supporting a divided Pt catalyst: **(b)** the same sample after exposure to a hydrogen equilibrium pressure of 50 Pa; **(c)** the difference spectrum of b)–a)

which is currently being investigated by variable temperature, different applied static field strengths and multiple pulse methods. Further resolution of these resonances may be possible by utilising multiple pulse sequences such as WAHUHA [30] which is an alternative technique to MAR, well documented in the literature, for achieving line narrowing in the homonuclear dipolar case. The two techniques may also be combined in the CRAMPS experiment as mentioned in Sect. 3.1 above.

Acknowledgements

The authors wish to acknowledge the earlier input of expertise and spectra in this area from Professors R.K. Harris and K.J. Packer. They are also grateful to Dr. G.R. Moore and H. Khodr for permission to use the ^{31}P spectra, D. Lennon for provision of the ^{1}H spectra, Dr. D.J. Greenslade for stimulating an interest in high alumina cement (HAC) chemistry, Dr. C.H. Fentiman of the Lafarge Aluminous Cement Co. Ltd. for providing the HAC samples, A.J. Strike for his enthusiastic support of the instrumentation, Kim Smith for assistance in obtaining many of the spectra, and Angela Saunders for preparing the diagrams.

4   References

1. Andrew ER (1981) Internat. Rev. Phys. Chem. 1: 195
2. Pines A, Gibby MG, Waugh JS (1973) J. Chem. Phys. 59: 569
3. Hartmann SR, Hahn EL (1962) Phys. Rev. 128: 2042
4. Schaefer J, Stejkal EO (1976) J. Amer. Chem. Soc. 98: 1031
5. Bunn A, Cudby MEA, Harris RK, Packer KJ, Say BJ (1982) Polymer, 23: 694
6. Bunn A, Cudby MEA, Harris RK, Packer KJ, Say BJ: J. Chem. Soc., Chem. Comm. 1981: 15
7. Lutton ES (1972) J. Amer. Oil. Chem. Soc. 49: 1
8. Bociek SM, Ablatt S, Norton IT (1985) J. Amer. Oil. Chem. Soc. 62: 1261
9. Marshall GL, Cudby MEA, Smith K, Stevenson TH, Packer KJ, Harris RK (1987) Polymer 28: 1093
10. Komoroski (ed) (1986) High resolution NMR spectroscopy of synthetic polymers in bulk, VCH, New York (Methods in stereochemistry analysis, vol. 7)
11. Harris RK, Jackson P, Wilkes PJ (1987) J. Mag. Reson. 73: 178
12. Dec SF, Wind RA, Maciel GE, Anthonio FE (1986) J. Mag. Reson. 70: 355
13. Langer V, Dauguard P, Jakobson HJ (1986) J. Mag. Reson. 70: 472
14. Fyfe CA (1983) Solid state N.M.R. for chemists (C.F.C. Press, Guelph, Canada, p 337
15. Barnes JR, Clague ADH, Clayden NJ, Dobson CM, Hayes CJ, Groves GW, Rodger SA (1985) J. Mater. Sci. Lett. 4: 1293
16. Clayden NJ, Dobson CM, Hayes CJ, Rodger SA : J. Chem. Soc., Chem. Commun. (1984): 1396
17. Khodr H, Moore GR, Strike AJ, Williamson DJ (1988) Biochem. Soc. Trans. 16: 839
18. Wiench DM, Jansen M, Hoppe R (1982) Z. Anorg. Allg. Chem. 488: 80
19. Griffiths L, Root A, Harris RK, Packer KJ, Chippendale AM, Tromans FR : J. Chem. Soc. Dalton Trans. (1986): 2247
20. See Fig. 27.
21. Meadows MD, Smith KA, Konsey RA, Rothgels TM, Skarjune RP, Oldfield E (1982) Proc. Natl. Acad. Sci. USA 79: 1351
22. Fyfe CA (1983) Solid state N.M.R. for chemists, C.F.C. Press, Guelph, Canada, p 536
23. Oldfield E, Kirkpatrick RJ (1985) Science, 227: 1537

24. Fyfe CA, Klinowski J, Gobbi GC (1983) Angew Chem. Intl. Ed., 22: 259
25. George CM (1983) In: Barnes P (ed) Structure and performance of cements, Applied Science, Barking, Essex
26. Müller D, Rettel A, Gessner W, Scheler G (1984) J. Mag. Reson. 57: 152
27. Slichter CP (1980) Principles of magnetic resonance Springer, Berlin Heidelberg New York, p 106
28. Wang PK, Slichter CP, Sinfelt JA, (1984) Phy. Rev. Lett. 53: 82
29. Pearce HA, Sheppard N (1976) Surface Sci. 59: 205
30. Waugh JS, Huber LM, Haeberlen V (1968) Phys. Rev Lett. 20: 180

CHAPTER 8
Principles and Techniques of Laser Spectroscopy

D.L. Andrews and M.R.S. McCoustra

1 General Principles of Laser Operation

To understand the concepts of laser action, we first need to appreciate the nature of the stimulated emission process on which it is based. Molecules in excited states generally have very short decay lifetimes (often between 10^{-7}s and 10^{-9}s) and by releasing energy they rapidly undergo relaxation processes. In this way, they undergo transitions to more stable states of lower energy; there are many different mechanisms for the release of energy, some of which are radiative, in the sense that light is emitted, and some of which are non-radiative. However, although chemical distinctions can be made between different types of radiative decay such as fluorescence and phosphorescence (see Chap. 5), the essential physics is the same – photons are emitted which match the energy difference between the initially excited state and the final state involved in the transition. Since this kind of photon emission occurs without any external stimulus, it is referred to as spontaneous emission.

Now suppose that a beam of light is directed into a system of excited molecules, with a frequency to exactly match the gap between the excited state and some state of lower energy. Each molecule can relax by emitting another photon of the same frequency as the supplied radiation; however it turns out that the probability of emission is enhanced if other similar photons are already present. Moreover, emission occurs preferentially in the direction of the applied beam, which is thereby amplified in intensity. This behaviour contrasts markedly with the completely random directions over which spontaneous emission occurs when no beam is present. This type of emission is therefore known as stimulated emission: it is emission which is stimulated by other photons of the appropriate frequency. The two types of emission are illustrated in Fig. 1.

Many of the fundamental practical issues involved in the construction of a laser follow from the nature of the stimulated emission process on which the device is based. The primary requirement is for a suitable substance in which stimulated emission can take place, in other words an active medium. An external stimulus is also required to promote atoms or molecules of this medium to an appropriate excited state from which emission can occur. The active medium can

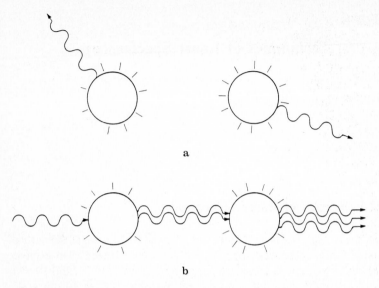

Fig. 1. Spontaneous (**a**) and stimulated (**b**) emission by excited molecules

take many forms, gas, liquid or solid, and the substance used is determined by the type of output required. Each substance has its own unique set of permitted energy levels, so that the frequency of light emitted depends on which levels are excited, and how far these are separated in energy from states of lower energy. The first laser, constructed by Maiman in 1960, had as its active medium a rod of ruby, and produced deep red light. More common today are the gas lasers, in which gases such as argon or carbon dioxide form the active medium; the former system emits various frequencies of visible light, the latter infra-red radiation.

There is a great diversity in the means by which the initial excitation is created in the active medium. In the case of the ruby laser, a broadband source of light such as a flashlamp is used; in gas lasers an electrical discharge provides the stimulus. Even chemical reactions can provide the necessary input of energy in certain types of laser. Two general points should be made, however, concerning the external supply of energy. First, if electromagnetic radiation is used, then the frequency or range of frequencies supplied must be such that the photons which excite the laser medium have an energy equal to or larger than that of the laser output. Secondly, because of heat and other losses, no laser has anything like 100% efficiency, and the energy output is always less than the energy input. One other point is worth mentioning at this juncture. So far, it has been implicitly assumed that the excitation precedes laser emission; indeed this is true for many kinds of laser system. However, for a continuous output, there is a requirement to sustain a population of the generally short-lived excited molecules, the condition known as population inversion.

In general, it is necessary to arrange for the multiple passage of light back and

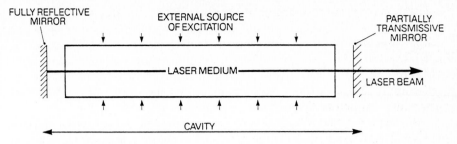

Fig. 2. Essential components of a laser

forth through the active medium in a resonator, so that with each traversal the intensity can be increased by further stimulated emission. In practice this can be arranged by placing parallel mirrors at either end of the laser medium, so that light emitted along the axis perpendicular to these mirrors is essentially trapped, and bounces backwards and forwards indefinitely, growing in intensity all the time. By contrast, photons which are spontaneously emitted in other directions pass out of the active medium and no longer contribute to stimulated emission. A usable output of laser light can then be obtained by having one of the end-mirrors made partially transmissive. Fig. 2 schematically illustrates the basic components of the laser as discussed above.

2 Properties of Laser Light, and their Applications

It has been recognized since the earliest days of laser technology that laser light has characteristic properties which distinguish it from that produced by other sources. In this section we shall look at how these properties arise from the nature of the lasing process, and briefly consider examples of how they are utilized for particular laser applications. Such a discussion obviously has to be highly selective, and those examples which are presented here have been chosen more to illustrate the diversity of laser applications than for any other reason.

2.1 Beamwidth

Since stimulated emission produces photons with almost precisely identical directions of propagation, the end-mirror configuration results in selective amplification of an axial beam which is typically only 1 mm in diameter. The laser thus emits a narrow, and essentially parallel beam of light from its output mirror, usually with a Gaussian distribution of intensities across the beam. A typical figure for the angle of beam divergence is 1 mrad, which would illuminate an area only one metre across at a distance of one kilometre; the exciplex lasers now available have beam divergences of less than 200 microradians. Although the extent of beam divergence is initially determined by the diffraction limit of the output aperture, the little divergence that there is can to a large extent be corrected by suitable optics. A striking illustration of how well collimated a laser beam can be is provided by the fact that it has been possible to observe the reflection of laser light from reflectors placed on the surface of the Moon by astronauts during the Apollo space programme.

One highly significant application of the collimation and narrow beamwidth of lasers is optical alignment in the construction industry, for example in tunnel boring. Pulsed lasers can also be used for tracking and ranging, and for atmospheric pollution monitoring. In the latter case it is the narrow beamwidth that makes it possible to monitor from ground level, by spectroscopic analysis of the scattered light, gases escaping from high factory chinmeys. By measuring the time delay between emission of each laser pluse and detection of the scattered signal, the range of pollutants can thus be accurately monitored; this technique is known as lidar (light detection and ranging) [1].

2.2   Intensity

The property most commonly associated with laser light is a high intensity, and indeed lasers do produce the highest intensities known on Earth. Since a laser emits an essentially parallel beam of light in a well-defined direction, rather than in all directions, the most appropriate measure of intensity is the irradiance, defined as the output power divided by the cross-sectional area of the beam. To put things into perspective when we look at typical laser irradiances, we can note that the mean intensity of sunlight on the Earth's surface is of the order one kilowatt per square metre, i.e. $10^3 \mathrm{W\,m}^{-2}$.

Let us consider first a moderately powerful argon laser which can emit something like 10 W power at a wavelength of 488 nm. Assuming a cross-sectional area for the beam of $1 \mathrm{mm}^2$, this produces an irradiance of $(10\,\mathrm{W})/(10^{-3}\mathrm{m})^2 = 10^7 \mathrm{Wm}^{-2}$. In fact we can increase this irradiance by focussing the beam until we approach a diffraction limit imposed by the optics. In this respect, too, laser light displays characteristically unusual properties, in that by focussing it is possible to produce intensities that exceed that of the source itself; this is not generally possible with conventional light sources. As a rough guide, the minimum radius of the focussed beam is comparable with the wavelength, so that in our example a cross-sectional area of $10^{-12}\mathrm{m}^2$ would be realistic, and give rise to a focussed intensity of $10^{13}\mathrm{Wm}^{-2}$.

It is in the pulsed lasers which first accumulate energy as a population inversion is built up, and then rapidly release it through emission of a pulse of light, that we find the highest output intensities, though we have to remember that the peak intensity is obtained only for a very short time. A good Q-switched ruby laser, for example, which emits 25 ns pulses (1 ns $= 10^{-9}$s) at a wavelength of 694 nm, can give a peak output of 1 GW $= 10^9$W in each pulse, though typically in a somewhat broad beam of about 500 mm^2 cross sectional area. The mean irradiance of each pulse is thus approximately 2×10^{12}Wm^{-2}, which can easily be increased by at least a factor of 10^6 by appropriate focussing. Because the peak intensity from a pulsed laser is inversely proportional to its pulse duration, there are various methods of reducing pulse length so as to increase the intensity. The most important are Q-switching, which produces pulses typically of nanosecond (10^{-9}s) duration, and mode-locking, which can produce pulses of picosecond (10^{-12}s), and in some cases sub-picosecond duration.

Many applications of lasers hinge on the high intensities available. A fairly obvious example from industry is laser cutting and welding. For such purposes, the high-power carbon dioxide and Nd-YAG lasers, which produce infra-red radiation, are particularly appropriate. Such lasers can cut through almost any material, though it is sometimes necessary to supply a jet of inert gas to prevent charring, for example with wood or paper; on the other hand, an oxygen jet facilitates cutting through steel. A focussed laser in the 10^{10}Wm^{-2} range can cut through 3 mm steel at approximately 1 cm s^{-1}, or 3 mm leather at 10 cm s^{-1}, for example. Applications of this kind can be found across a wide range of industries, from aerospace to textiles, and there are already several thousand laser systems being used for this purpose in the USA alone.

2.3   Coherence

Coherence is the property which most clearly distinguishes laser light from other kinds of light, and it is again a property which results from the nature of the stimulated emission process. In the laser, by contrast with any conventional light source, photons emitted by the excited laser medium are emitted in phase with those already present in the cavity. There are surprisingly few applications of laser coherence; the main one is holography, which is a technique for the production of three-dimensional images. The process involves creating a special type of image, known as a hologram, on a plate with a very fine photographic emulsion. Unlike the more usual kind of photographic image, the hologram contains information not only on the intensity, but also on the phase of light reflected from the subject; clearly, such an image cannot be created using an incoherent light source. Subsequent illumination of the image reconstructs a genuinely three-dimensional image. One of the main problems at present is that only monochrome holograms can be made, because phase information is lost if a range of wavelengths is used to create the initial image; although holograms which can be viewed in white light are now quite common, the colours they display are only the result of interference, and not the colours of the original subject.

2.4 Monochromaticity

The last of the major characteristics of laser light, and the one which has the most relevance for chemical applications, is its essential monochromaticity, resulting from the fact that all of the photons are emitted as the result of a transition between the same two atomic or molecular energy levels, and hence have almost exactly the same frequency. Nonetheless, there is always a small spread to the frequency distribution which may cover several discrete frequencies or wavelengths satisfying a standing-wave condition; this situation is illustrated in Fig. 3. The result is that a small number of closely separated frequencies may be involved in laser action, so that an additional means of frequency selection has to be built into the laser in order to achieve the optimum monochromaticity. What is generally used is an etalon, which is an optical element placed within the laser cavity and arranged so that only one well-defined wavelength can travel back and forth indefinitely between the end-mirrors.

With lasers which have a continuous output, it is quite easy to obtain an emission linewidth as low as $1 \, \text{cm}^{-1}$, and in frequency-stabilized lasers the linewidth may be four or five orders of magnitude smaller. One of the important factors which characterize lasers is the quality factor, Q, which equals the ratio of the emission frequency v to the linewidth Δv, i.e. $Q = v/\Delta v$. The value of the Q-factor can thus easily be as high as 10^8, a point which is clearly of great significance for high resolution spectroscopy. Spectroscopists generally prefer to have linewidth expressed in terms of wavelength or wavenumber units, where the latter represents the number of wavelengths of radiation per unit length, usually per centimetre $(v = 1/\lambda)$. We then find that $\Delta v = \Delta\lambda/\lambda^2$, where $\Delta\lambda = \lambda/Q$.

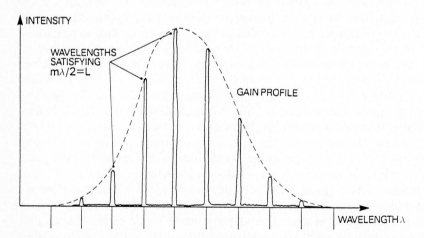

Fig. 3. Emission spectrum of a typical laser. Emission occurs only on wavelengths within the fluorescence linewidth of the laser medium which experience gain and also satisfy the standing-wave condition

Another important area of application associated with the high degree of monochromaticity of a laser source lies in laser isotope separation. Since molecules which differ in isotopic constitution generally have slightly different absorption frequencies, use of a very narrow linewidth laser enables a mixture of the compounds to be selectively excited, and then separated by other means. Not surprisingly, there is a great deal of interest in this type of application within the nuclear industry.

3   Laser Sources of Spectroscopic Importance

Laser light sources can be broadly classified into two groups, namely fixed-frequency and tunable. In the former, laser output occurs predominantly at a single frequency, or over a limited range of frequencies. Although such lasers give good coverage of the electromagnetic spectrum as demonstrated in Fig. 4, their spectroscopic applications are somewhat limited, and they mostly find application in routine Raman spectroscopy. However, for the most part, modern laser spectroscopy is carried out using lasers that can be continuously tuned over a wide range of wavelengths.

The classification into fixed-frequency and tunable lasers represents one general classification. A second division is possible when one considers the mode

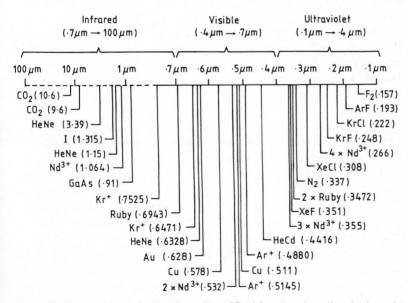

Fig. 4. The laser spectrum showing the variety of fixed frequency laser lines in the region from the near-UV to the mid-IR

of laser operation, i.e. whether it is pulsed or continuous wave (CW). This difference is largely responsible for determining the ultimate resolution with which it is possible to record spectra using lasers, and for determining the particular spectroscopic technique to be employed. Typically, for a CW dye laser operating in the visible, bandwidths of the order of 500 kHz ($0.00015\,\mathrm{cm}^{-1}$) are routinely achievable and represent the limit set by the mechanical and optical inaccuracies and instabilities of the current generation of devices. Clearly, such devices are applicable only in what could be termed ultra-high resolution spectroscopy. In contrast, for pulsed lasers, the limiting bandwidth possible is set simply by the Fourier transform of the pulse-width. Thus, for example, a laser with a 10 ns pulse-width could, given the correct laser cavity design, achieve a bandwidth of approx. 100 MHz ($0.03\,\mathrm{cm}^{-1}$). However, multiple-mode operation in most commercial pulsed dye lasers increases the operating bandwidths to approx. $0.2\,\mathrm{cm}^{-1}$ unless further line-narrowing (mode-selecting) optical elements are introduced into the laser cavity. Given the vast range of laser systems available, space does not permit a description of them all. In the following, what could be considered the more important laser systems are briefly described.

3.1   Fixed-Frequency Lasers

3.1.1   The Ruby Laser

In 1960 the ruby laser was the first system in which visible laser action was observed. The first visible laser simply consisted of a synthetic ruby rod, with polished and mirrored end surfaces, excited by a helical flashlamp positioned around the crystal. Today, ruby laser design has changed little, although more modern pumping chamber design allows more efficient coupling of the pump radiation into the ruby rod. Although now largely surpassed for most applications by Nd-YAG lasers, ruby lasers still find use in studies of plasma formation and in laser flash photolysis, and off-the-shelf systems are routinely capable of producing multi-Joule energies per pulse.

Typically, ruby lasers output between 0.03 and 20 J per pulse on their fundamental wavelength of 694.3 nm; Q-switching ensures pulse durations of the order of 20-30 ns. As a consequence, peak powers of approx. 1 MW to 1 GW are available, making second harmonic generation (frequency doubling) a routine option. Pulse repetition rates are generally low, however, both as a result of heating directly from the pump flashlamps and also from the relaxation of the excited chromium(III) ions responsible for the laser action. Nonetheless, in general it can be said that the lower the pulse energy of the laser, the higher the repetition rate obtainable. Thus repetition rates of approx. 50 Hz can be obtained from small ruby lasers.

3.1.2 The Nd-YAG laser

The Nd-YAG laser is in many aspects physically similar to the ruby laser. Laser action is obtained by flashlamp pumping a YAG (yttrium aluminium garnet) rod doped with Nd(III) to approx. 1–2% by weight. As a consequence of the energy level system involved, fundamental laser emission occurs at 1064 nm, with pulse energies up to approx. 1.2 J per pulse available from standard commercial laser systems. By modification of the support matrix, the fundamental laser wavelength can be shifted by several nm, and lasers based on other YAG-like crystals are now available. However, the difficulty in growing large YAG and YAG-like crystals will continue to prevent the production of large YAG-based Nd(III) lasers. As a result most high-energy Nd(III) lasers are based on glass rods, and pulse energies in excess of 100 kJ per pulse have been obtained with research lasers of this type. Typically, such high power lasers are based in central facilities, especially those concerned with nuclear fusion research.

While the pulse-width from a standard Nd-YAG laser is normally approx. 1–2 μs, Q-switching allows pulse-width reduction to 10–20 ns. The high peak powers consequent upon this make for high efficiency in doubling to 532 nm, tripling to 355 nm and even quadrupling to 266 nm. It is here that the usefulness of the Nd-YAG laser is apparent. From a single infrared laser, it is possible to obtain discrete wavelengths in the UV/visible range which are invaluable for dye laser pumping (532 and 355 nm) and for UV photochemistry (355 and 266 nm). Although ns pulsed Nd-YAG lasers probably represent the major application of this laser system, the low laser threshold means that CW operation in possible, and powers of up to 25 W at 1064 nm can be obtained. By modelocking such CW lasers, it is possible to produce pulse-trains with a typical pulse-width of about 100 ps at both the fundamental and the second harmonic wavelength. The latter has proved valuable in the production of tunable ps and fs (10^{-15} s) laser pulses from modelocked dye lasers. More recently, a number of manufacturers have begun to employ a technique known as injection seeding with Nd-YAG lasers. In this, pulsed or CW radiation at 1064 nm from a single mode diode laser is optically injected into the Nd-YAG rod as it is being pumped. The Nd-YAG laser then behaves simply as an optical amplifier for the seed light. Such injection-seeded lasers promise true single mode behaviour (ultra-narrow bandwidth), very high beam quality (very low divergence) and, very possibly, an improvement on the already excellent reliability record of Nd-YAG lasers.

3.1.3 The Nitrogen Laser

The nitrogen laser was the very first UV gas laser to come into commercial production. Pumped by electric discharge, laser emission occurs on the $C \rightarrow B$ electronic transition of molecular nitrogen at around 337 nm. This wavelength is ideal for dye laser pumping and for photochemical applications. Although output pulse energies are somewhat small, typically 1 mJ, the nitrogen laser does have the advantage over larger exciplex laser systems that it naturally produces sub-ns pulse-widths. Larger nitrogen lasers do exist, producing 4 or 5 times more output. However to gain an increase in output energy the short pulse-width is sacrificed; high pulse energy nitrogen lasers have typical pulse widths of 10–20 ns.

The short pulse-width of the nitrogen laser, particularly when used to pump a short pulse dye laser, make it ideal for photophysical measurements where the expense of a modelocked ion or Nd-YAG pumped dye laser could not be justified. However, in contrast to such laser systems, the repetition rates of nitrogen-pumped dye lasers are usually very low, typically 20–30 Hz.

3.1.4 Exciplex (Excimer) Lasers

In the last 10 years, the growth rate of the gas exciplex laser field has been astonishing, and probably represents the fastest growing sector of laser sales at present. A number of exciplex gas lasers exist and provide good coverage of the UV; the more routinely used of these are listed in Table 1. This variety of wavelengths gives the exciplex laser its wide applicability, from dye laser pumping with XeCl and XeF through photolithography to fusion research with huge multi-Joule KrF lasers.

As with the nitrogen laser, rare gas exciplex lasers are pumped by electric discharge. Discharge through a gas mixture containing a rare gas Rg and a halogen X, leads to the formation of bound, electronically excited species (RgX) termed exciplexes (often incorrectly called excimers). The emission from these bound electronically excited states to the unbound ground state forms the basis for the laser action. Output pulse energies vary markedly and depend strongly on the exciplex being used. Thus for example it is not unusual to obtain up to 400 mJ per pulse from a XeCl exciplex, while KrF is capable of producing 2 or 3 times this output in pulse-widths of around 10–20 ns. Recently, there has been a trend to produce longer pulse-widths while maintaining the output pulse energy. This trend has been initiated in an effort to reduce the problem of damage associated with such high energy UV pulses in some medical applications.

Although gas mix lifetime still represents a major problem with exciplex lasers, recent advances have pushed the fill lifetime to 10^6–10^7 shots. Using cryogenic cleaning and halogen topping-up techniques this limit can be extended

Table 1 Exciplex Laser Systems

Exciplex	Output Wavelength nm	Typical Output Pulse Energy mJ[a]
XeF	351	80
XeCl	308	150
KrF	248	250
KrCl	222	na
ArF	193	200
F_2	157	na

na Generally not specified

[a] For a medium sized exciplex laser, Lambda Physik EMG101MSC

further. The current aim of exciplex laser manufacturers is to produce a product that is modular, reliable and easy to use for both industrial and scientific users.

3.1.5   The Carbon Dioxide Laser

Like the nitrogen and exciplex lasers, the carbon dioxide laser is a discharge pumped gas laser. In contrast to the former, however, laser action in carbon dioxide occurs through collisional excitation of vibrationally excited states of the active medium from excited species, particularly nitrogen, in the buffer gas of the laser. Laser output is then dominated by infrared output at approx. $10.6\,\mu m$, corresponding to a transition from a state having one quantum of the antisymmetric stretch to a state having two quanta of the bending mode, and also at around $9.6\,\mu m$, corresponding to a transition from the same initial state to a state with one quantum of the symmetric stretch. The output can be either broadband or, as is more usual, can be made tunable over the various rotational lines within the vibrational band. Further tunability is achieved by the use of isotopically substituted carbon dioxide.

Operation of the laser can either be in pulsed or CW mode. In pulsed applications, output pulse energies of up to 2 kJ are readily achievable and pulse-widths are typically a few μs. Pulse clipping allows the production of pulses of a few 10's of ns wide, though with a consequent loss of output. In CW mode, powers of $> 50\,W$ are readily achievable from commercial laser systems, and recent trends suggest that higher power systems will soon become available.

Given the high infrared output power of the carbon dioxide laser, it is only natural that the applications of such lasers are dominated by their use as cutting tools both in engineering and medicine. The high power means that, like exciplex and nitrogen lasers, carbon dioxide lasers have also found wide application in optical pumping, although at far-infrared as opposed to visible wavelengths. This represents one method of obtaining tunable laser radiation in the far-infrared, a region traditionally regarded as difficult to work in due to the weak nature of most conventional far-IR sources.

Direct spectroscopic applications of the carbon dioxide laser are restricted by the limited tunability of the output. Those applications that do exist mostly rely on Stark or Zeeman tuning of the molecular energy levels into resonance with a suitable laser line, and hence are usually restricted to CW sources. Pulsed carbon dioxide lasers, however, find widespread use in photochemistry. The high peak powers of these lasers means that multiple infrared photon absorption is possible. As a consequence, this has led to the development of infrared multiphoton dissociation (IRMPD) and the birth of infrared laser photoche-mistry. [2, 3].

3.1.6   The Atomic Iodine Laser

Laser action in the atomic iodine laser occurs on a spin-orbit transition in the ground electronic state of atomic iodine and results in the emission of $1.315\,\mu m$

radiation. Population inversion is produced by photodissociation following flashlamp excitation of a gaseous perfluoroalkyl iodide. This represents the only commercial example of a photodissociation laser, although numerous other media, such as nitric oxide (NO) from the photodissociation of nitrosyl chloride (NOCl), are known to be viable.

Output powers of up to 3 J are obtainable from commercial systems, in pulses 2–6 μs wide; Q-switching allows reduction of the pulse to approx. 30 ns. Given the high powers of such lasers, it is not surprising to find applications in non-linear optics, often as a pump laser for optical parametric oscillators, the infrared and near-infrared equivalent of the pulsed dye laser (see below). Primarily, however, iodine lasers are used in the study of fast relaxation kinetics in aqueous media. e.g. biochemical kinetics, using the temperature-jump technique based on the strong absorption of water in this region. Large research atomic iodine systems, like Nd-YAG lasers, also find application in laser-assisted fusion research.

3.1.7 Metal Vapour Lasers

Metal vapour lasers now represent one of the most rapidly developing fields of laser science. These lasers are unique in providing very high repetition rate (up to 20 kHz) pulsed operation, with pulse-widths of < 60 ns, in the visible and near-infrared. A number of metals are used, including manganese, lead and barium, but most favoured are copper, with outputs at 510.6 nm and 578.2 nm, and gold, with output at 627.8 nm. Average output powers of up to 40 W are readily obtainable from such lasers. The laser medium in such lasers is a low pressure gas of monatomic metal vapour produced by discharge heating in a ceramic laser tube. Typical temperatures required are 1200 K, and as a result larger metal vapour lasers have a considerable warm-up period of up to an hour, during which there is little or no laser output.

In spectroscopy, there are two major applications of metal vapour lasers. The first is in dye laser pumping, particularly for red and near-infrared dye output. Similarly, metal vapour lasers are now finding widespread use as high repetition-rate sources for ps and fs pulse amplifiers in photophysics and ps/fs spectroscopy. The high repetition rate enhances the rate of data acquisition by 2 or 3 orders of magnitude over traditional Nd-YAG or exciplex-pumped pulse amplifiers. Finally, the use of metal vapour lasers as light sources in Raman spectroscopy is also possible, again the principal advantage being the high repetition rates and consequent enhancement of signal acquisition.

3.1.8 Rare Gas Ion Lasers

The argon and krypton ion lasers represent what could be termed the classic examples of a CW laser. The laser medium is a plasma of the appropriate rare gas generated by low voltage, high current discharge through the gas. In both cases, laser action occurs on a number of electronic transitions in the ion. Thus the argon ion laser principally outputs radiation on 10 lines between 454 and 529 nm, with the 488.0 and 514.5 nm lines dominating, while the kryton laser outputs on 9 lines between 476 and 799 nm, with the 647.1 nm line dominating. Both systems additionally provide a number of UV lines at high discharge currents.

Output powers of up to 30 W can be obtained from commercial argon ion lasers, with up to 4 W in the UV lines. The output from krypton lasers is generally much smaller, up to 1 or 2 W. Pulsed operation is possible by use of mode-locking techniques, and this forms the basis for the generation of very high repetition-rate (approx. 80 to 100 MHz) ps laser pulses from the dye lasers. Firstly, as CW dye laser pumps, ion lasers provide the basic radiation for the generation of continuously tunable CW laser radiation from around 380 nm to beyond 800 nm. The second application, which is perhaps more widespread, is the use of these lasers in Raman spectroscopy. Ion lasers have for the most part replaced the lower power, longer wavelength helium-neon laser as the excitation light in modern Raman spectrometers for reasons both of power and wavelength.

3.2   Tunable Lasers

The lasers discussed above can be simply regarded as fixed-frequency light sources. As such they are largely inapplicable to most spectroscopic techniques, with the notable exception of Raman spectroscopy. The real power of lasers as spectroscopic tools lies in the use of tunable lasers, both in the UV/visible and in the infrared.

3.2.1   Dye Lasers

The dye laser is perhaps the single most important tool for the laser spectroscopist. Pulsed operation is based on pumping either by flashlamps or by a fixed-frequency source such as a nitrogen or exciplex laser or the second or third harmonic of a Nd-YAG laser. Such lasers provide narrow bandwidth, continuously tunable laser radiation from about 380 nm to 900 nm. Typical output pulse energies are of the order of 10–15% of the pump laser pulse energy, with pulse-widths comparable to those of the pump light and repetition rates limited only by the pump source. Energies of the order of 2 or 3 J can be obtained from flashlamp-pumped systems in 1–2 μs pulses, although Q-switching is often used to reduce the pulse-width. For CW applications, pumping normally involves an ion laser, and average powers can exceed 1 W.

In all cases, the medium in which laser emission occurs is a solution of a large dye molecule such as a rhodamine or a coumarin, and population inversion is achieved by a combination of optical pumping and collisionally assisted vibrational relaxation. As a consequence, lasing typically occurs on the S_1–S_0 electronic transition. Without the use of wavelength-selective optics such as a diffraction grating in the laser cavity, output is generally broadband, approx.

$10-100 \, cm^{-1}$, and centred around the maximum of the fluorescence spectrum of the dye. The introduction of dispersive elements into the laser cavity results in a narrowing of the bandwidth, typically to $0.1-1 \, cm^{-1}$. Although a narrower bandwidth can be obtained by the introduction of further elements, in pulsed applications it is ultimately limited by the Fourier transform of the laser pulse-width. This latter restriction is removed in CW dye lasers, which represent the highest resolution spectroscopic devices available. Ring dye lasers in which light is constrained to travel around a cyclic path between a series of mirrors can achieve ultimate resolutions of the order of $10^{-6} \, cm^{-1}$.

While a considerable quantity of spectroscopic information can be obtained from spectra in the visible and near-UV, the majority of samples absorb only in the region below 350 nm. Using non-linear optical techniques such as frequency-doubling and sum-frequency generation in materials such as potassium dihydro-gen phosphate (KDP) or β-phase barium borate (BBO), extension of the wavelength range down to 185 nm is routinely possible for pulsed dye lasers with pulse energies of approx. 1 mJ. Further extension into the VUV and XUV regions is possible using frequency-tripling and related techniques, although output pulse energies are low. Using a related technique, difference-frequency generation, it is possible to shift the laser output from the visible to the infrared. Pulsed infrared radiation down to $20 \, \mu m$ has been obtained in μJ quantities using multiple stages of this method. As an alternative to second harmonic generation etc., frequency shifting both to the infrared and the UV can be achieved using Stimulated Raman Scattering (see below), particularly in molecular hydrogen.

3.2.2   Diode Lasers

Semiconductor diode lasers represent as powerful a tool for infrared spectroscopy as the dye laser does for UV/visible spectroscopy. Diode lasers, which are essentially specially constructed light-emitting diodes, provide continuously tunable, very narrow bandwidth pulsed or CW infrared radiation. By judicious choice of diode, excitation current and temperature, extensive coverage of the near-and mid-infrared is possible. However until recently, diode lifetime and tuning range of about $100 \, cm^{-1}$ for each diode have required the user to maintain a large library of diodes for coverage of any extended regions of the infrared. With the rapid development of the optical communications industry, these factors are sure to be markedly improved in the next few years.

3.2.3   Optical Parametric Oscillator (OPO)

In many respect the OPO is an infrared analogue of the visible dye laser. Typically the device consists of a non-linear crystal such as lithium niobate $(LiNbO_3)$ located in an optical cavity similar to that of a dye laser and pumped by the fundamental of a high power Nd-YAG laser. By a process which is essentially the reverse of sum-frequency generation, two frequencies are created from a single pump frequency, v_0; the signal, v_{SIG} and the idler, v_{IDL}, such that $v_0 = v_{SIG} + v_{IDL}$. Output on either of these infrared frequencies is controlled by wavelength-selective elements in the cavity, and the actual frequency determined by the orientation of the crystal with respect to the pump beam and the crystal temperature. Output pulse energies of several mJ can be obtained from such devices. However, their application is severely limited at present by the low damage thresholds of the majority of the current generation of OPO crystals.

More recently, the operation of a visible OPO pumped at 355 nm by the third harmonic of a pulsed Nd-YAG laser has been demonstrated. If the promise of such solid state, tunable, visible light sources is fulfilled, we may see at least a partial replacement of the pulsed tunable dye laser as a spectroscopic tool.

4   Absorption Spectroscopy

Laser-based systems for absorption spectroscopy can be based either on fixed-frequency or tunable laser sources. Fixed-frequency lasers providing emission at only one, or a few discrete wavelengths, are not at all amenable to the usual absorption methods which require scanning over a continuous range of wavelengths, and thus call for specialised techniques. Two such methods, both developed in the late 1960s, namely laser magnetic resonance [4] and laser Stark spectroscopy [5] are still used for high-resolution research applications. However, the introduction of tunable dye lasers at about the same time made it possible to obtain an absorption spectrum by scanning the source itself across the appropriate wavelength range. This kind of approach has many advantages, and obviates the need for any monochromator. Although the following account concentrates on electronic spectroscopy in the visible range using a dye laser, the general principles discussed here are of much wider application, and include infra-red spectroscopy with diode lasers.

A simple setup for absorption spectroscopy using a dye laser is illustrated in Fig. 5. This arrangement produces a spectrum by monitoring the transmission through the sample as a function of wavelenth; the attenuation relative to a reference beam provides a direct measure of the absorption, as in conventional spectrometry. However, relative absorbances as low as 10^{-5} can be detected in

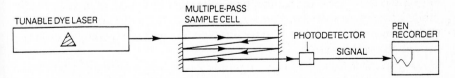

Fig. 5. Schematic diagram showing the setup for measurement of an optical absorption spectrum using a tunable laser source. As usual, the signal received by the photodetector is ratioed with that obtained from a reference beam (not shown) derived from the same source, but which does not pass through the sample

the laser configuration. Since the absorption signal is proportional to the distance travelled through the sample by the radiation, long path-lengths through the sample are often used for laser spectroscopy; the figure illustrates one simple means of obtaining a long path length by multiple passes of radiation through the sample medium. With a cell 20 cm in length, for example, it is fairly straightforward to generate 50 traversals, resulting in an effective path length of 10 m. In this way, an absorption band producing a decrease in intensity per traversal of as little as one part in 10^4 ultimately produces a drop in intensity of 0.5%, which is easily measurable. The technique is, of course, particularly well suited to studies of gases, which have very low absorbance; however, it is also of use in obtaining the spectra of components in dilute solutions, in which absorption features due to the solute have to be distinguished from the often very much more intense features due to the solvent. In such a case, the reference beam would be passed through a multiple-pass cell containing pure solvent. An alternative method of obtaining long path lengths is to use long hollow glass or quartz fibres filled with sample fluid.

A different technique which is also used to increase the absorption signal from sample is intracavity enhancement. As the name indicates, this phenomenon relates to the apparent increase in the intensity of absorption displayed by samples placed within the laser cavity. However, it should be noted that the mechanisms for this enhancement are such that there is no simple relationship between laser output and sample absorption. Finally, sensitivity can be improved by amplitude- or frequency-modulation of the incident light; here, the absorption signal is obtained by phase-sensitive detection of the oscillation intensity of the transmitted light; usually the latter type of modulation is more successful. It is worth noting, however, that strikingly good results can be obtained without any of the refinements described above.

Whilst "traditional" absorption spectroscopy with lasers has proved particularly valuable in the infrared using low power CW diode lasers, studies in the visible prove more difficult. As a consequence of the high average powers of visible dye lasers, severe non-linearities can be introduced both in the absorption process itself and in the detection step. This is particularly true when using pulsed dye lasers for measurements of absorption, where pulse energies as low as 10–100 μJ can lead to saturation of an optical transition. Furthermore, with such high power lasers, the measurement of weak absorption can be difficult if not impossible, since one is then attempting to measure very small changes in a large initial intensity. Few detectors are sensitive enough to do this job well, especially since there is normally a 1–5% laser intensity fluctuation superimposed on the desired absorption signal.

4.1   Specialised Absorption Techniques [3]

As seen above, a common problem with transmittance methods is the difficulty of detecting weak absorption features, since the signal is generally very small compared to the background intensity. There are, however, several alternative, but highly sensitive measurement techniques particularly suited to laser spectroscopy. These methods are all based on the monitoring of physical processes which take place subsequent to the absorption of radiation. Before examining these in detail, however, it is worth emphasising that all the methods to be described in this section involve precisely the same absorption process in the initial excitation of the sample, and also produce spectra through analysis of the dependence on excitation frequency; hence they may all properly be described as types of absorption spectroscopy.

4.1.1   Excitation Spectroscopy

The deactivation of atoms and molecules excited by the absorption of visible or ultraviolet light can often involve the emission of light at some stage. In the case of atomic species, fluorescent emission generally takes place directly from the energy level populated by the excitation. In the case of molecular species there are usually a number of different decay pathways which can be followed, of which spin-allowed fluorescence from the electronic state initially populated provides the most direct, and usually the most rapid, means of deactivation. However, radiationless decay processes which occur in the vibrational levels prior to fluorescence result in emission over a range of wavelengths, so that even if the initial excitation is at a single fixed wavelength, the spectrum of the emitted light may itself contain a considerable amount of structure and so provide very useful information. This is the basis for fluorescence spectroscopy, discussed later in Sec. 5. Excitation spectroscopy, or laser-excited fluorescence, by contrast, is concerned not with the spectral composition of the fluorescence, but with how the overall intensity of emission varies with the wavelength of excitation. Figure 6(a) illustrates the various transitions giving rise to the net fluorescence from an atomic or simple molecular species with discrete energy levels.

The sensitivity of this absorption method stems from the fact that the signal is detected relative to a zero background; every photon collected by the detection system (usually a photomultiplier tube) has to arise from fluorescence in the sample, and must thus result from an initial absorptive transition. If every photon absorbed results in a fluorescence photon being emitted, in other words if the quantum yield is unity, then in principle the excitation spectrum should accurately reflect both the positions and intensities of lines in the conventional absorption spectrum. In practice, collisional processes may lead to non-radiative decay pathways and thus a somewhat lower quantum yield, although this problem may to some extent be overcome in a gaseous sample by reducing the pressure. Also, the imperfect quantum efficiency of the detector has to be taken into consideration, along with the fact that a certain proportion of the fluorescence will not be received by the detector because it is emitted in the wrong direction. Nevertheless, it has been calculated that fractional extinctions as low as 1 part in 10^{14} are measurable by commercially available instrumentation using the method of excitation spectroscopy. Amongst other applications, the high

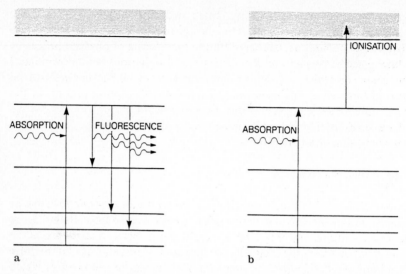

Fig. 6. Energy level diagrams for (**a**) excitation spectroscopy, and (**b**) ionisation spectroscopy of a simple species with discrete energy levels

sensitivity of this method makes it especially well suited to the detection of short-lived chemical species.

4.1.2 Ionisation Spectroscopy [6]

Another specialised technique which is used to monitor absorption in the UV/visible range is ionisation from the electronically excited states populated by the absorption, as illustrated in Fig. 6(b). Here, the spectrum is obtained by monitoring the rate of ion production as a function of the irradiation frequency. Various methods can be used to produce the ionisation, but clearly the process has to be sufficiently selective that ground state species are not ionised. For excited states close in energy to the ionisation limit, ionisation can be induced either by application of an electric field, or by collisions with other atoms or molecules. Alternatively, a photoionisation technique can be employed, using either the laser, or any other suitable frequency light source to produce photons with enough energy to bridge the gap to the ionisation continuum. The ions or free electrons so produced can be detected by electrical methods, which often have close to 100% efficiency; ionisation spectroscopy is thus one of the most sensitive methods of detecting absorption, and under ideal conditions virtually every atom or molecule excited by the laser radiation is subsequently ionised and detected [7]. The simplest arrangement is where laser radiation provides the energy for both the initial excitation and the subsequent ionisation.

4.1.3   Thermal Lensing Spectroscopy [8]

In this form of spectroscopy, and also in photoacoustic spectroscopy discussed below, the optical absorption from a tunable laser beam is monitored through an effect based on the heating produced by absorption. The initial heating itself results from the decay of electronically excited states, and may be termed a photothermal effect. In the case of thermal lensing spectroscopy, the specific phenomenon utilised is the temperature-dependence of the refractive index, which results in non-uniform refraction around the laser beam as it passes through any absorbing gas or liquid.

The mechanisms at work in this kind of spectroscopy, also known as thermal blooming spectroscopy, are in many respects more complicated than in any other absorption method. In the first place, the production of heat as the result of absorption depends on the detailed kinetics of relaxation in the sample molecules. The localised change in temperature which then ensues depends on the bulk heat capacity, and the extend of both conduction and convection in the sample; consequently, liquids are much more amenable to study by this method than gases. The characteristics of the laser beam also play a crucial role. Since the spatial distribution of intensity across the laser beam should normally conform to a well-defined Gaussian profile, the small volume of sample traversed by the centre of the beam experiences a greater intensity and thus exhibits a greater temperature rise than sample at the outside edge of the beam. The localised refractive index gradient which is consequently established is a function of the sample's thermal expansion properties, and leads to a defocussing of the laser beam which is ultimately measured in the setup illustrated in Fig. 7. Sensitivity can be increased by chopping the beam, and detecting the modulation in the detected signal.

The instrumentation required for thermal lensing spectrometry places constraints both on the kind of laser which can be employed, and on the nature of the sample which can be studied. It is a method which is more appropriate for absorption in the visible range than in the infra-red where direct heating effects occur, and it calls for a tunable laser with a consistent spatial intensity cross-section. Since the beam has to be focussed close to the sample in order for the defocussing to be observed, only a very small volume of sample is required to

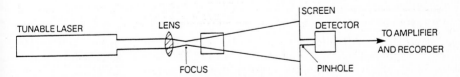

Fig. 7. Diagram of the apparatus used in thermal lensing spectroscopy. The defocussing which arises as the laser is tuned through an absorption band of the sample results in a broadening of the transmitted beam, and is observed as a decrease of intensity through the pinhole

generate the spectrum, and the intensity of the signal does not increase in proportion to any increase in the path-length. Samples must be carefully filtered, to overcome false signals due to light scattering impurities: also, the method is basically inappropriate for flowing samples, unless the flow is very slow indeed (less than 1 ml per minute). The dependence on the thermal properties of the sample make it a far more sensitive method for organic solvents than for water; in the former case, relative absorbances as low as 10^{-7} can be detected. This indicates that a detection limit as low as 10^{-11} M may be achievable for solutions of strongly coloured solutes. Indeed, perhaps the most outstanding feature of thermal lensing spectroscopy is that it represents one of the cheapest means of obtaining high sensitivity in analytical absorption spectroscopy.

4.1.4 Photoacoustic Spectroscopy [9]

As mentioned above this form of spectroscopy, also known as optoacoustic spectroscopy, is similar to thermal lensing spectroscopy, in that both hinge on the heating effect produced by optical absorption. In this case, however, the specific effect which is measured results from the thermally produced pressure increase in the sample, and makes use of the fact that if the laser radiation is modulated at an acoustic frequency, pressure waves of the same frequency are generated in the sample, and can thus be detected by a piezoelectric detector or a microphone. Since the intensity of sound must depend on the amount of heating, it reflects the extent of absorption; hence a spectrum can be obtained by plotting the sound level against the laser frequency or wavelength. Once again, the method produces spectra of very high resolution (see also Chapter 3).

The instrumentation for laser photoacoustic spectroscopy is shown in Fig. 8. Visible or infra-red lasers can be used for the source, so that either electronic or vibrational (or in some cases vibration-rotation) spectra can be obtained. The laser light first passes through a chopper which provides the acoustic-frequency modulation of the beam incident upon the sample; a modulation signal is also derived at this point from a beam-splitter and optical detector (not shown on the diagram). The acoustic signal generated as the beam passes through the sample cell, or spectraphone, is then sent to a phase-sensitive amplifier locked into the

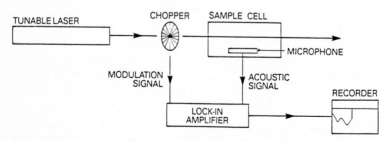

Fig. 8. Schematic diagram of the apparatus used in the photoacoustic spectroscopy of a gas

modulation signal, and produces an output which is plotted by a pen recorder. Alternatively, a pulsed laser can be used in conjunction with an averaging system.

In contrast to thermal lensing spectrometry, photoacoustic absorption methods are easily applicable to both liquids and gases, and can also be adapted to solid samples; in the latter case, a microphone can be used to pick up an acoustic signal from the gas above the sample. Flowing fluid samples are also relatively easy to study, provided the acoustic noise generated by the flow is electronically filtered out of the microphone signal. In the case of liquids, it is again the case that solutions in organic solvents provide a much greater sensitivity than aqueous solutions, due to the high heat capacity of water. Nonetheless, it is important to choose a solvent which is comparatively transparent over the spectral range being examined, or else there may be little improvement over the spectra obtainable by conventional absorption methods.

Gases are the simplest to work with, and optimum sensitivity is obtained if the sample cell dimensions support resonant oscillations of the acoustic waves, which in practice means tailoring the modulation frequency to the particular sample being studied. Relative absorbances of 10^{-7} are measurable in both liquids and gases, although in gases the detection limit may be as much as two orders of magnitude lower. Sensitivity is in fact comparable to that of a good gas chromatography-mass spectrometry system, and concentrations of gases in the ppb (parts per billion) range can be detected over a wide range of pressures, from several atmospheres down to 10^{-3} atmospheres. The technique is thus particularly well suited to investigations of gaseous pollutants in the atmosphere, and there are few other absorption techniques which match photoacoustic spectroscopy for sensitivity.

4.1.5   Optogalvanic Spectroscopy [10]

Another method which is commonly used to measure absorption spectra with a tunable laser source is based on the optogalvanic effect, which is essentially a change in the electrical properties of a gas discharge (or in some cases a flame) when irradiated by certain frequencies of light. In contrast to the other specialised methods discussed so far, this type of spectroscopy provides information on atomic or ionic species, rather than molecules. Also, the optical transitions involved are not only those originating from the ground state, since excited states are also appreciably populated.

The electrical current passing through a gas discharge results in ionisation of the atoms, and the efficiency of this process depends on the energy they already possess, being most effective for atoms in highly excited states near to the ionisation continuum. The current is thus a sensitive function of the various atomic energy level populations. In the presence of a beam of light with suitable frequency, absorption processes result in atomic transitions to states of higher energy, so that the relative populations of the various levels are changed. This effect can be registered as a change in the voltage across the discharge, which may

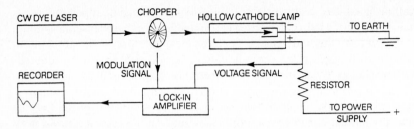

Fig. 9. Apparatus used for optogalvanic spectroscopy in a gas discharge

be either positive or negative. Plotting the variation in discharge voltage against the irradiation frequency thus provides a novel kind of absorption spectrum, and completely obviates the need for an optical detection system.

The apparatus used in discharge photogalvanic spectroscopy is shown in Fig 9. In practice, the voltage change across an external resistor is monitored as a dye laser focussed into the discharge in a hollow cathode lamp is tuned across its spectral range. To increase the sensitivity, the laser radiation may be chopped, and the modulation of the signal at the chopping frequency detected with a lock-in amplifier and recorder. By using a lamp containing uranium or thorium, whose spectra are already known with high precision, optogalvanic spectroscopy can incidentally be used as a method of wavelength calibration for tunable laser sources.

The primary analytical applications of optogalvanic spectroscopy arise elsewhere however, in flame methods similar to those already widely used in atomic absorption and fluorescence spectrometry. Samples are usually in solution form, and are introduced as a narrow jet into flames of an ethyne (acetylene)/air mixture. However, the optogalvanic instrumentation additionally requires the coupling of the burner head to the positive side of a power supply, and the insertion of high-voltage cathode plates on either side of the flame. The remainder of the setup is then similar to that shown in Fig. 9, with the burner assembly and cathode plates replacing the hollow cathode lamp. One alternative arrangement is to replace the continuous dye laser with one which is flashlamp-pumped, using a photodiode to detect each pulse and trigger a signal-averaging amplifier.

In flames, it is generally found that the optogalvanic effect results in an increase in the extent of ionisation between the electrodes each time an absorption line is encountered. For this reason, the alternative designation laser-enhanced ionisation spectroscopy has also gained usage. Compared to the optical detection methods used in conventional flame spectrometry, this method suffers none of the problems usually caused by the flame background and scattering of the laser light. It also has improved sensitivity, both in spectral resolution and in its detection limits for certain elements. The best example of this sensitivity is its use in the analysis for lithium, which has been detected in concentrations as low as $1\,pg/ml$ $(10^{-12}\,g/ml)$ by this method, representing an improvement over the detection limit of most flame analytical methods by several powers of ten [10].

4.1.6 Photofragment Excitation Spectroscopy [11, 12]

As discussed above, there are often difficulties associated with recording direct absorption spectra using pulsed lasers. Moreover, when the species of interest does not emit light during relaxation from its excited state, excitation spectroscopy is impossible. However, often in such circumstances the species of interest is known to dissociate upon electronic excitation either by a direct or an indirect (predissociation) mechanism. Under such circumstances, the technique of Photofragment Excitation (PHOFEX) spectroscopy can be applied.

The basic technique behind this form of spectroscopy is the idea of what is normally referred to as a pump-probe experiment. Using a tunable, high-power pulsed laser (the pump) the species of interest is photodissociated. A second tunable, pulsed laser (the probe) detects a particular photofragment, generally a diatomic, by either excitation of ionisation spectroscopy after a short (nanosecond) delay. By fixing the wavelength of the probe laser to a particular feature of the excitation or ionisation spectrum of the fragment and tuning the pump laser, the PHOFEX spectrum is generated. Whether or not this spectrum exhibits structure is largely determined by the dissociation mechanism of the species of interest.

One important use to which this technique has been put is in the trace analysis of N-nitrosamines [13], powerful carcinogens found in meats treated with nitrites. Using the third harmonic of a Nd-YAG laser, it is possible to photodissociate every molecule within the focus of the laser beam. A delayed probe laser then detects the nitric oxide fragments using laser-induced fluorescence. Concentrations as low as $10^{-12}\,mol\,cm^{-3}$ have been detected and improvements in the probe technique may improve this to $10^{-5}\,mol\,cm^{-3}$ or lower.

4.1.7 Polarisation Spectroscopy [6]

Like PHOFEX, polarisation spectroscopy is a pump-probe technique. In this case, however, it requires only a single laser. A simplified experimental layout is shown in Fig. 10. The output from a tunable laser is split into a weak probe beam and a strong pump beam, the latter being circularly polarised by passing it through a quarter-wave plate. This light is then directed into the sample where its absorption introduces a degree of anisotropy. As a consequence of this anisotropy, the sample becomes birefringent and can rotate the plane of polarisation of the plane-polarised probe beam. This rotation can be easily detected by the placement of crossed polarisers on either side of the sample.

The great advantage of polarisation spectroscopy over direct absorption is

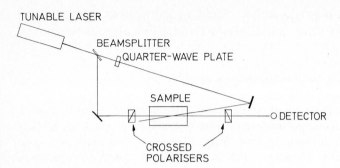

Fig. 10. Schematic diagram of the apparatus used in measuring polarisation spectra of gases and liquids

then clear. In contrast to direct absorption where small changes in a large background are detected, here small changes on an essentially zero background are measured (cf. excitation spectroscopy). As a consequence, this technique is considerably more sensitive than most other forms of absorption spectroscopy.

5   Fluorescence Spectroscopy [14]

In each type of laser spectroscopy examined so far, the essential mechanism for introducing spectral discrimination has been the absorption of radiation by sample atoms or molecules. We now consider an alternative but equally well established branch of spectroscopy based on the emission of radiation by the sample (see also Chapter 5). The laser provides a very selective means of populating excited states, and the study of the spectra of radiation emitted as these states decay is generally known as laser-induced fluorescence. There are two main areas of application for this technique; one is atomic fluorescence, and the other molecular fluorescence spectrometry.

5.1   Laser-Induced Atomic Fluorescence [15]

This technique is now increasingly being used for trace elemental analysis. It involves atomising a sample, which is usually in the form of a solution containing the substances for analysis, in a plasma, a furnace, or else in a flame, and subsequently exciting the free atoms and ions to states of higher energy using a laser source. Since the atomisation process populates a wide range of atomic energy levels, the emission spectrum is complex, despite the monochromaticity of the laser radiation. The fluorescence frequencies are nonetheless highly characteristic of the elements present, and whilst detection capabilities are typically in the 10^{-11} g/ml range, concentrations two orders of magnitude smaller have been measured by this technique, corresponding to a molarity of 10^{-19} M. In fact, it has been demonstrated that laser-induced atomic fluorescence has the ultimate capability of detecting single atoms, although this is in an analytically rather unuseful setup based on an atomic beam.

A wide variety of lasers can be utilised for atomic fluorescence measurements. The most important factor to take into consideration is the provision of a high intensity of radiation in the range of absorption of the particular species of interest. For this reason, although individual measurements are made with a fixed irradiation wavelength, it can be helpful to employ a tunable source in order to provide a detection facility for more than one element. For example use of a frequency-doubled dye laser in conjunction with the traditional air/acetylene flame has been shown to offer dramatically improved detection limits for several precious metals [16]. In common with several other types of spectroscopy considered previously, sensitivity is often increased by either modulating a cw laser with a chopper, detecting the signal with a phase-sensitive lock-in amplifier, or else by integrating the signal using a pulsed laser, as shown in Fig. 11. The principles of laser-induced atomic fluorescence may also be employed in a microprobe configuration. Here, material is vapourised from the surface of a heterogeneous solid sample by a pulsed laser, and its elemental composition is characterised by its atomic fluorescence.

Another variation on this theme is laser-induced breakdown spectroscopy. Here, high intensity pulses of laser light, as for example from a Nd:YAG laser, are focussed into the sample and result in the formation of a spark through a process of dielectric breakdown. Once again, elemental analysis can then be carried out by measurement of the resultant atomic fluorescence. This method, which has evolved from the field of electric spark spectroscopy, has the advantage of

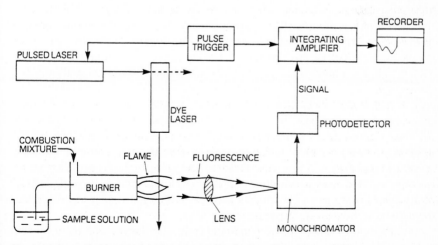

Fig. 11. Schematic diagram of the instrumentation for laser-induced atomic fluorescence measurements using a dye laser pumped with a pulsed laser, and with flame atomisation of the sample solution

obviating the need for electrodes, and hence removes any possibility of spectral interference associated with atomisation of electrode surfaces. It also has the advantage of speed over many other analytical techniques since it requires little or no sample preparation. This kind of procedure enables ppb concentrations to be detected in optimal cases. Applications to tissue analysis and the identification of trace constituents of blood and sweat have been demonstrated, making the technique an attractive alternative to many of the methods more traditionally used in medical and forensic laboratories.

5.2   Laser-Induced Molecular Fluorescence [17]

Compared to atomic fluorescence, this method suffers from much poorer sensitivity due to the much broader lines in the emission spectra of all but the smallest polyatomic molecules. In general, then, the high monochromaticity and narrow linewidth of a laser source does not result in a comparable resolution in laser-excited molecular emission spectra. However, it does present some advantages over the more traditionally used black-body radiation sources. Fluorescence spectra are often complicated by features which owe nothing to fluorescence but rather to Raman scattering in the sample, a process which results in the appearance of frequencies shifted by discrete amounts from the irradiation frequency (see the following section). The practical advantage of laser-induced fluorescence measurements is that when monochromatic light is used, the Raman features occur at discrete frequencies, and can easily be distinguished from the broad fluorescence bands. It is also worth noting that for certain analytical applications fluorescence measurement may be more selective than absorption spectroscopy, since even when two different compounds absorb at the same frequency, their fluorescence emission frequencies may be quite different.

The energetics of molecular excitation and decay are such that fluorescence spectra can only be collected at wavelengths longer than the irradiation wavelength; subject to this restriction, however, there is comparative freedom over the choice of laser wavelength. Only in the spectroscopy of diatomic and other small molecules does the facility for selectively exciting a particular state play a significant role. However, fluorescence spectroscopy is not solely concerned with chemically stable systems; many important applications concern study of the photolytic processes which can occur in a sample through the interaction with laser light. When the energy absorbed by sample molecules is sufficient, fragmentation processes can occur, resulting in the formation of short-lived transient species which are often strongly fluorescent. For example laser photolysis of cyanogen C_2N_2 produces free CN radicals which can be detected through collection of the fluorescence spectrum.

Another useful feature of laser-induced molecular fluorescence is that with a pulsed source, it is possible to monitor the time-development of the emission process. This provides a further means of discriminating against the Raman scattering which takes place only during the period of irradiation, since the fluorescence is associated with decay processes subsequent to absorption. Different radiative transitions can in fact be discriminated by their different decay constants in some cases. Perhaps more useful in connection with the study of photochemical reactions, however, is the possibility of measuring excited state and transient lifetimes by this method, using either a Q-switched or a mode-locked source. The methods of time-resolved laser measurement are thus particularly pertinent to the study of chemical reaction kinetics [3].

Both continuous-wave and pulsed lasers can be used for laser-induced molecular fluorescence measurements. Helium-cadmium and argon lasers are the most popular continuous sources; nitrogen lasers and nitrogen-pumped dye lasers are usually adopted for pulsed sources. As usual, the latter are generally operated with a signal-integrating detection system. With such a setup, detection limits for solution samples can be as low as 10^{-13} M, or 10^{-5} ppb. With comparatively complex samples where one is analysing for a particular component in admixture with a large concentration of other species, a detection limit of 1 ppb is more realistic.

Laser-induced fluorescence in increasingly being used for the analysis of eluant from liquid chromatography [18]. It is also possible to 'tag' suitable biological species with fluorescent compounds, so as to facilitate laser fluorimetric detection [19]. Mention should also be made of the novel analytical method which, although not strictly spectroscopic, is nonetheless based on fluorescence and is known as remote fibre fluorimetry. Quite another field of application lies in the remote sensing of oil spills at sea, using an airborne ultraviolet laser source and telescoping detection system to scan the surface of the water for the characteristic fluorescence. Finally, laser fluorescence spectroscopy is widely employed for combustion diagnostics, since it enables the concentrations and temperatures of transient species within a flame to be ascertained with a high degree of spatial resolution, without interfering with the gas flow or chemistry. Intermediates which have been detected by this method range from atomic oxygen and nitrogen to comparatively large fragments such as CH_3O, although OH is most commonly monitored since it provides a relatively straightforward measure of the degree of reaction.

6 Raman Spectroscopy [3]

Raman spectroscopy is based on a phenomenon involving the inelastic scattering of light by molecules (or atoms). The term 'inelastic' denotes the fact that the scattering process results in either a gain or loss of energy by the molecules responsible, so that the frequency of the scattered light differs from that incident upon the sample. The energetics of the process are illustrated in Fig. 12. The two types of Raman transition, known as Stokes and anti-Stokes, are shown in (a) and (b); the former results in an overall transition to a state of higher energy, and the

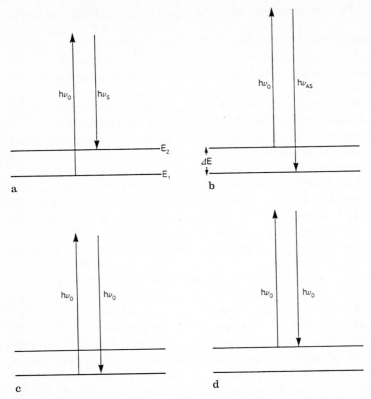

Fig. 12. Diagrams illustrating Raman and Rayleigh scattering. Only the energy levels directly involved are depicted: (**a**) shows a Stokes Raman transition, and (**b**) an anti-Stokes Raman transition; (**c**) and (**d**) show Rayleigh scattering from the two different levels

latter a transition to a state of lower energy. The elastic scattering processes represented by (c) and (d), in which the frequencies absorbed and emitted are equal, are known as Rayleigh scattering.

From the diagrams of Fig. 12, it is clear that the Stokes Raman transition from level E_1 to E_2 results in scattering of a frequency given by $v_s = v_0 - \Delta E/h$, and the corresponding anti-Stokes transition from E_2 to E_1 produces a frequency $v_{AS} = v_0 + \Delta E/h$, where $\Delta E = E_2 - E_1$, and v_0 is the irradiation frequency. For any allowed Raman transition, then, two new frequencies usually appear in the spectrum of scattered light, shifted to the negative and positive sides of the dominant Rayleigh line by the same amount, $\Delta v = \Delta E/h$. For this reason, Raman spectroscopy is always concerned with measurements of frequency shifts, rather than the absolute frequencies. In general, of course, a number of Raman transitions can take place involving various molecular energy levels, and the spectrum of scattered light thus contains a range of shifted frequencies. In the particular case of vibrational Raman transitions, these shifts can be identified

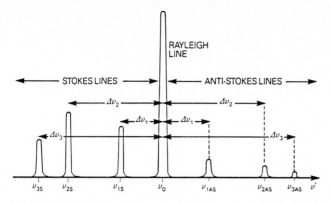

Fig. 13. Model Raman spectrum showing the intensity of light scattered with frequency, ν'

with vibrational frequencies in the same way as absolute absorption frequencies in the infra-red spectrum. The theory is described in detail in Chapter 9.

One other point is worth raising before proceeding further. Although the Stokes and anti-Stokes line in a Raman spectrum are equally separated from the Rayleigh line, they are not of equal intensity, as illustrated in Fig. 13. This is because the intensity of each transition is proportional to the population of the energy level from which the transition originates; the ratio of populations is given by the Boltzmann distribution, and the anti-Stokes line is invariably weaker in intensity. There is also a fourth-power dependence on the scattering frequency, as the detailed theory shows. The ratio of intensities of the Stokes line and its anti-Stokes partner in a Raman spectrum is thus a temperature-dependent quantity, a fact which can be made good use of in certain applications, for example in the non-invasive determination of flame temperatures.

The Raman effect is a very weak phenomenon; typically only one incident photon in 10^7 produces a Raman transition, and hence observation of the effect calls for a very intense source of light. Since the effect is made manifest in shifts of frequency away from that of the incident light, it also clearly requires use of a monochromatic source. Not surprisingly, then, the field of Raman spectroscopy was given an enormous boost by the arrival of the laser, so that Raman spectroscopy is today virtually synonymous with *laser* Raman spectroscopy. A typical setup is shown in Fig. 14. Light scattered by the sample from a laser source (usually an ion laser) is collected, usually at an angle of 90°, passed through a monochromator, and thereby produces a signal from a photodetector. The Raman spectrum is then obtained by plotting the variation in this signal with the pass-frequency of the monochromator. Even with a fairly powerful laser, Raman spectroscopy requires a very sensitive photodetection system capable of registering single photons to detect the weakest lines in the spectrum.

All phases of matter are amenable to study by the Raman technique, and there are several advantages over infra-red vibrational spectroscopy in the particular

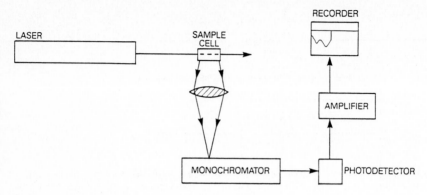

Fig. 14. Instrumentation for laser Raman spectroscopy

case of aqueous solutions. Firstly, because only wavelengths in the visible region are involved, conventional glass optics and cells can be used. Secondly, since water itself produces a rather weak Raman signal, spectra of aqueous solutions are not swamped by the solvent. For these reasons, Raman spectroscopy is especially well suited to biological samples. As with most types of laser spectroscopy the spectrum is derived from a relatively small number of molecules because of the narrow laser beamwidth, and even with a comparatively insensitive spectrometer only a few ml of sample is sufficient. Raman spectroscopy is thus also a useful technique for the analysis of the products of reactions with low yield. One further point worth noting is that compared to infra-red spectroscopy sample heating is much reduced by using visible radiation. The only case where heating does cause a real problem is with strongly coloured solids, where it has become a common practice to spin the sample so that no single spot is continuously irradiated; this is particularly important in the case of resonance Raman studies, discussed below. The whole field of Raman spectroscopy has recently been given a fresh impetus with the development of Fourier Transform techniques (see Chapter 9).

An application that is gaining popularity is known as the laser Raman microprobe [20], and is principally used for heterogeneous solid samples, for example geological specimens. Two different methods have been developed based on microprobe principles. In one method illustrated in Fig. 15, various points on the surface of the sample (or indeed if it is transparent various regions within the sample) are irradiated with a laser beam and the Raman scattering is monitored. Since any particular chemical constituent should produce Raman scattering at one or more characteristic wavelengths, it is possible to filter out the emission from the surface at one particular wavelength, and hence produce a map of the surface concentration of the substance of interest. Used in conjunction with a visual display unit and image-processing techniques, this method undoubtedly has enormous potential as a fast and non-destructive means of chemical analysis.

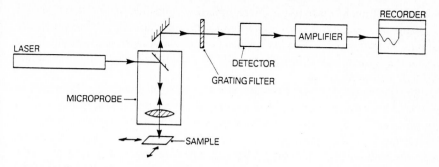

Fig. 15. Raman microprobe (point illumination) instrumentation. In the alternative global illumination configuration, a larger surface is illuminated, and the detection equipment is replaced with an image intensifier phototube and camera

A less sensitive, but simpler technique involves global irradiation of the sample. In this way a small area of the surface can be imaged directly, and the entire image can be passed through a filter to a camera system.

6.1   Specialised Raman Techniques

As with absorption spectroscopy, there is a wide range of modifications to the standard methods of Raman spectroscopy. To some extent, these are modifications to the Raman process itself, rather than simply different means of detection; nonetheless, all the techniques discussed below involve essentially the same types of Raman transition.

6.1.1   Resonance Raman Spectroscopy [21]

The frequency of radiation used to induce Raman scattering need not in general equal an absorption frequency of the sample. Indeed, it is generally better that it does not, in order to avoid possible problems caused by the interference of absorption and subsequent fluorescence. However, there are certain special features which become apparent when an irradiation frequency is chosen close to a broad intense optical absorption band (such as one associated with a charge-transfer transition) which make the technique a useful one despite its drawbacks. Quite simply, the closer one approaches the resonance condition, the greater is the intensity of the Raman spectrum. The selection rules also change, so that certain transitions which are normally forbidden become allowed, thus providing extra information in the spectrum. Lastly, in the case of large polyatomic molecules where any electronic absorption band may be due to localised absorption in a particular group called a chromophere, the vibrational Raman lines which experience the greatest amplification in intensity are those involving vibrations of nuclei close to the chromophore responsible for the resonance. In principle, this facilitates deriving structural information concerning particular sites in large molecules. This feature has been much utilised, for example, in the resonance Raman vibrational spectroscopy of biological compounds containing strongly coloured groups.

Resonance Raman spectroscopy can be performed with any laser source provided it emits a wavelength lying within a suitable broad absorption band of the sample. Clearly for application to a range of samples, a tunable dye laser is to be preferred. Unfortunately, resonance fluorescence occurs with increasing intensity, and over an increasingly broad range of emission wavelengths, as the resonance condition is approached. However, this problem may be overcome by use of a mode-locked source providing ultrashort (picosecond) pulses. Since there is no detectable time-delay for the appearance of the Raman signal, while resonance fluorescence is associated with a lifetime typically in the nanosecond range, the signal from the photodetector can be electronically sampled at suitable intervals and processed so that only the true Raman emission is recorded.

6.1.2 Stimulated Raman Spectroscopy [22]

As noted above, the Raman effect is essentially a very weak one, and it is in general only under conditions of resonance enhancement that the scattering intensity could be regarded as sizeable. However, there is one other way in which the effect can be enhanced, which is as follows. If a sufficiently intense laser is used to induce Raman scattering, then despite the initially low efficiency of conversion to Stokes frequencies, Stokes photons which are emitted can stimulate the emission of further Stokes photons through Raman scattering of laser light by other sample molecules in the beam. This process evidently has the self-amplifying character always associated with stimulated emission, and is therefore most effective for Stokes scattering in approximately the direction of the laser beam through the sample. In fact with a giant pulse laser this stimulated Raman scattering can lead to generation of Stokes frequency radiation in the 'forward' direction with a conversion efficiency of about 50%.

Once a strong Stokes beam is established in the sample, of course, it can lead to further Raman scattering at frequency $v'_s = v_s - \Delta v$, and this too may be amplified to the point where it produces Raman scattering at frequency $v''_s = v'_s - \Delta v$, and so on. Hence a series of regularly spaced frequencies $(v_0 - m\Delta v)$ appears in the spectrum of the forward-scattered light. Because of the self-amplifying nature of the process, it is usually the case that only the vibration producing the strongest Stokes line in the normal Raman spectrum is involved in stimulated Raman scattering. The effect does not, therefore, have the analytical utility of conventional Raman spectroscopy.

There is more than one spectroscopic method based on the principle of stimulated Raman scattering. One of the possibilities is to simultaneously irradiate the sample with both a pump laser beam of frequency v_0 and with another tunable probe laser beam. When the frequency of the latter beam is tuned through a Stokes frequency it stimulates the corresponding Raman transition, and thus experiences a gain in intensity. The Raman spectrum is thus obtained by plotting the intensity of the probe beam after passage through the sample against its frequency; this method is known as Raman gain spectroscopy.

6.1.3   CARS Spectroscopy [23]

Next we consider coherent anti-Stokes Raman scattering spectroscopy, usually abbreviated to CARS. The mechanism for this process is a so-called four-wave interaction between two laser beams of frequency v_0 and v_1 directed into the sample. The frequency v_1 is tuned across the frequency range below v_0, and the four-wave interaction produces coherent emission at a frequency $v' = 2v_0 - v_1$, so that when v coincides with any Stokes Raman frequency v_s, we have $v' = 2v_0 - v_s$ $= 2v_0 - (v_0 - \Delta v) = v_0 + \Delta v = v_{AS}$, corresponding with the anti-Stokes frequency.

Although the strongest lines in the CARS spectrum occur when v_1 strikes resonance with a Stokes frequency, this is usually seen against a background emission due to a relatively frequency-insensitive non-resonance mechanism. For this reason, CARS spectroscopy is not well suited to applications in trace analysis. However since it does produce a spectrum typically $10^4 - 10^5$ times more intense than the normal Raman effect, but subject to the same selection rules, there are other important areas of application. While CARS is a two-beam method, part of the output from a single pump laser may be frequency-converted in a dye laser cavity to produce the second beam, as shown in Fig. 16. In principle, the CARS emission frequency is completely determined by the two incident frequencies, and the spectrum may be obtained by plotting the intensity of light received by the detector against the frequency difference $v_0 - v_1$. Occasionally a monochromator is placed before the detector to cut out stray frequencies resulting from light scattering and fluorescence processes in the sample. However, it is often sufficient to use a simple optical filter with a cut-off just above v_0 for this purpose, making use of the fact that only higher frequencies can result from the CARS process.

An unusual feature of CARS is that it depends quadratically on the density of

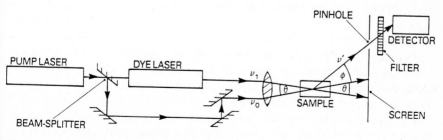

Fig. 16. Instrumentation for CARS spectroscopy [3]

the sample. Hence whilst the vibrational spectra of liquids can be obtained with CW lasers, giant pulse lasers producing megawatt power levels are required in order to obtain the vibration-rotation CARS spectra of gaseous samples; a frequency-doubled Nd:YAG source is a common choice. It is, nonetheless, a most useful technique for samples obtainable only in very small quantities; since the CARS signal is produced only by molecules at the intersection focus of the two applied beams, only μl volumes of liquid are required, or gas pressures of 10^{-6} atmospheres.

One special area in which CARS spectroscopy has found widespread application, by virtue of its particular suitability to highly luminescent samples, is in combustion and other high-temperature reaction diagnostics. The main advantage over traditional Raman spectroscopy is the fact that a coherent high-frequency output beam is produced, which can be much more easily detected against a strong fluorescence background. Here the CARS technique often provides results which usefully complement the data on chemical transients available from laser-induced fluorescence measurements. By making the two applied laser beams intersect within the flame or reaction volume to be studied, the CARS spectrum not only helps to reveal the chemical composition in any small region, but also the temperature can be accurately determined from the relative intensities of rotational lines. The method of temperature determination not only surpasses that of any standard thermocouple device in its accuracy, but also in that measurements are possible well above the usual thermocouple limit of about 2500 K. The technique is moreover non-instrusive, and hence in no way affects the process being studied. A good industrial example of this type of application lies in the analysis of gas streams in coal gasification plants.

One other specialised area of Raman spectroscopy, Raman optical activity is described in Chapter 10.

7   Free Jet Expansions and Laser Spectroscopy [24–26]

Molecular beams or jets have had a long history of application in molecular spectroscopy. However, traditionally such sources have been effusive in nature, providing essentially a Boltzmann distribution of states in the species under study. While adequate for atomic sources, the generally high temperatures employed in effusive sources severely limit application to polyatomic molecules.

In contrast to an effusive source, the supersonic beam or jet has only recently found application in spectroscopy, although having been applied in gas-phase reaction dynamics for many years. The major advantages of a supersonic source over effusive sources are the cooling associated with adiabatic expansion of a high pressure gas into a high vacuum, and also the directed nature of the expansion. Essentially two methods exist for the generation of supersonic expansions, providing either continuous or pulsed jets. The generation of a continuous expansion can be regarded as a brute force approach. Typically, large high-speed diffusion or turbomolecular pumps are required to maintain the pressure in the expansion chamber below 10^{-4} mbar. This clearly represents a major investment both in equipment and in the space required to house it. In contrast, the pumping requirements for a pulsed expansion are more modest, typically employing a 10–25 cm diffusion pump backed by a suitable rotary pump. In general pulsed expansions are more intense by 10 to 100 times than continuous expansions obtained under similar conditions. This combination of factors has led to a rapid

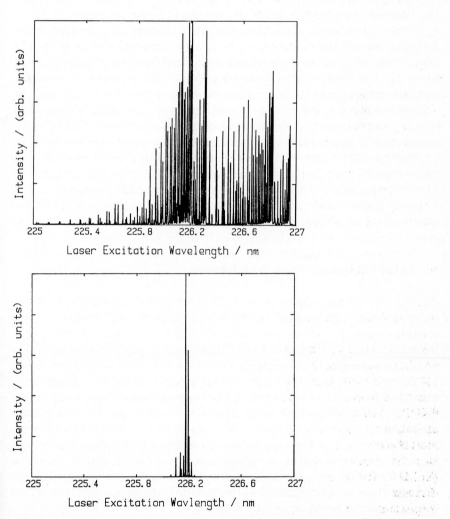

Fig. 17. A comparison of the room temperature laser-induced fluorescence excitation spectrum of NO with the spectrum of 1% NO co-expanded with argon through a pulsed molecular beam valve

growth of interest in the spectroscopy of jet-cooled molecules using pulsed free jet sources.

As mentioned above, the major advantage of using a supersonic beam is the cooling effect associated with the adiabatic nature of the expansion. Using a few percent of the species of interest (the seed) in mixture with a rare gas, typically helium or argon, cooling of the seed gas to temperatures as low as 2 or 3 K is routinely possible. This leads to a dramatic simplication of the spectrum of the seed molecule, as illustrated by the comparison in Fig. 17. Such cooling can also have the side-effect of encouraging the formation of weakly bound van der Waal's molecules, particularly in the case of argon-based expansions, and of the condensation of the seed molecule into clusters of 2 or more molecules.

For analytical applications, the low effective temperature associated with a molecular beam has two consequences. Firstly, the distribution of quantum state populations, which is normally spread over a large number of states at room temperature, is compressed into just a few levels. This can lead to a sensitivity enhancement of approx. 10^3 over spectroscopy at room temperature. Thus for diatomic molecules as few as 10^6 molecules/cm^3 can be detected in free jets using laser-induced fluorescence excitation spectroscopy as a probe. Secondly, the appearance of broad relatively unstructured spectra, for example those of polycyclic aromatic hydrocarbons, may be reduced to a few narrow vibronic features which are distinctive of the species of interest. As a consequence, the free jet enhances the selectivity of laser spectroscopy. More details of the generation of molecular beams are to be found in the references. Analytical applications are discussed in detail in Ref. [26].

8   Laser Ionisation Mass Spectrometry [27, 28]

The last topic we shall discuss is one which is bringing about a revolution in the well-established field of mass spectrometry. This technique involves sample ionisation using a tunable laser rather than the traditional electron impact methods. Since ionisation of most molecules requires energies in excess of that which can be supplied by a single photon, unless very short wavelengths below 150 nm are used, laser ionisation generally has to involve a multiphoton excitation process; hence the term multiphoton ionisation mass spectrometry (MUPI – MS or MPI – MS) is also applied to this method. The multiphoton excitation can proceed via a number of mechanisms, as shown in Fig. 18, but is most effective if the excitation proceeds via a resonant intermediate state. In this case the process is known as resonance-enhanced multiphoton ionisation (REMPI). The advantage of using a mass spectrometer is that it enables the different types of ion produced by multiphoton absorption and subsequent fragmentation to be distinguished.

An early laser mass spectrometer was typically based on a pulsed nitrogen pumped dye laser. More recently, Nd-YAG and exciplex pumped lasers have

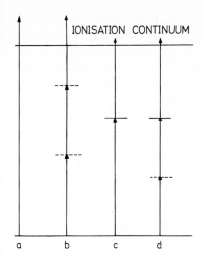

IONISATION CONTINUUM

a   b   c   d

Fig. 18. Schematic representation of a number of non-resonant and resonant multiphoton ionisation schemes; **(a)** 1-photon ionisation, **(b)** 3-photon multiphoton ionisation (MPI), **(c)** (1 + 1) Resonance enhanced multiphoton ionisation (REMPI), **(d)** (2 + 1) REMPI

become commonplace alternatives. Many details of the ionisation cell are determined by the nature of the sample. Samples are often introduced into the system in the gas phase, at vapour pressures usually not less than 10^{-6} atmospheres, but in some cases as high as 50 atmospheres. Even liquid samples, such as hplc eluants, and solid samples, including supported samples, can be studied directly. Compared to a conventional instrument, the ionisation efficiency may be several orders of magnitude better, approaching 10% for the ground state molecules. However, the ionisation volume is much smaller, again by several orders of magnitude, in a laser setup. The ionisation volume can of course be increased by using an unfocussed laser, but this has to be played off against the reduction in multiphoton ionisation efficiency.

Detection methods vary, as in other mass spectrometers, but one of the most effective is time-of-flight (TOF) analysis. Here, the various ion fragments are electrically accelerated, and pass along an evacuated flight tube. Traditionally, thus tube is linear; however, recent work [29–31] has demonstrated the higher mass resolution (up to 10,000) of a Reflectron time-of-flight mass spectrometer. Here, an electrostatic reflector is used to reflect the ion fragment bunches along the second half of the flight tube, and more importantly to "focus" the bunches to improve mass resolution. The time taken by each ion to reach the detector is determined by its mass, and is typically measured in microseconds. Since laser pulses of only nanosecond duration are used to produce the ionisation, the mass spectrum can be obtained by monitoring the ion current as a function of the time elapsed since the laser pulse. Commercial instruments applying either linear time-of-flight or Reflectron time-of-flight principles are now available.

One of the most important aspects of laser ionisation mass spectrometry is the fact that the mass spectrum changes both with laser wavelength and laser power. Indeed if it were possible, the mass spectrum could be represented as a plot in four dimensions of ion current against laser wavelength, laser power and ion mass. More usually the laser power is fixed, and the mass spectrum is represented as a three dimensional plot of ion current versus laser wavelength and mass. At low laser power, soft ionisation is possible, producing only the parent ion peak and little fragmentation. Increasing the laser power increases the degree of fragmentation until ultimately simple atomic ions are produced in very intense laser fields. Since multiphoton ionisation generally produces fragmentation patterns quite different from those of normal mass spectra, this method additionally provides a useful insight into the dynamics of multiphoton ionisation. Finally, there is the possibility of obtaining further information by varying the laser polarisation. Whilst these methods can be used directly on chemically complex samples, it is possible to treat samples by conventional "wet chemistry" techniques to produce solutions for laser ionisation mass spectrometry. This kind of approach is particularly expedient in analysis for inorganic elements; for example, Fe atoms can be detected by their resonant two-photon ionisation at a wavelength of 291 nm. By varying the irradiation wavelength, almost any element can be selectively ionised and quantitatively measured in the mass spectrometer [32]. The use of ultra-high resolution laser spectroscopic techniques could eventually extend this to any isotope of any element.

Perhaps the most interesting recent development in laser ionisation mass spectrometry has been the increasing use of free jets in such systems to pre-cool the species of interest before ionisation. Under such conditions, selective ionisation of one component of a complex molecular mixture is feasible. Thus for example, it is possible to selectively ionise a particular xylene isomer in a jet-cooled mixture of isomers. In a further development of the jet-cooling technique in conjunction with soft pulsed infrared laser desorption and Reflectron time-of-flight mass analysis, (Fig. 19) it has been possible to obtain parent ion mass spectra for large biomolecules such as chlorophyll, small peptides, nucleotides

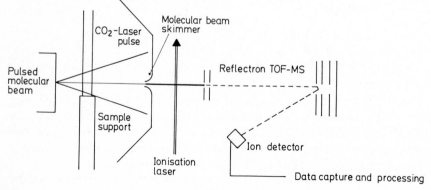

Fig. 19. Schematic representation of the combined laser desorption/molecular beam coding reflection time-of-flight laser photoionisation mass spectrometer due to Boesl and Schlag [29–31]

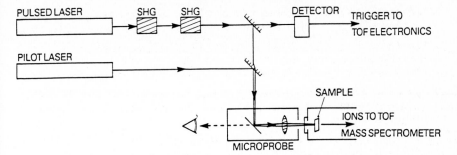

Fig. 20. Laser microprobe mass analysis apparatus for transmission sampling

and sugars [29–31]. This would be nigh impossible in a traditional electron impact mass spectrometer.

Another type of laser mass spectrometry more directly aimed at analytical applications should be mentioned in conclusion. This is known both as laser microprobe mass analysis (LAMMA), or laser-induced mass analysis (LIMA), and is based on the microprobe concept discussed earlier. As shown in Fig. 20, high-power pulses of light from a frequency doubled Nd-YAG laser are focussed onto small areas of sample, typically vapourising volumes of about 1 μl per shot. A pilot HeNe laser follows a collinear optical path onto the sample surface so as to facilitate visual location of the focal region. Ions released from the sample are subsequently analysed in a time-of-flight mass spectrometer; the method illustrated is appropriate for transmission samples, but laser desorption in a reflection mode is now practicable.

The reflection method is very well suited to the characterisation of different areas of inhomogeneous surfaces; it can also be applied to the subsurface analysis of inhomogeneous solids, since successive laser pulses vapourise and release ions from progressively deeper layers within the sample. Thus it has already established extensive applications ranging from geology to microelectronics. This method is particularly well suited to the study of fibres, environmental particles and metal corrosion processes, where it frequently out-performs conventional microanalytical techniques. Apart from microsampling capability, the main advantage of laser mass analysis are speed of analysis (a complete mass spectrum can be obtained in a matter of microseconds), a molecular elemental selectivity on the ppm scale or better, and a potential mass range of 1–10,000 mass units. It can thus also offer a powerful method for the determination of organic and biological residues or contaminants. Commercial laser mass analysis systems based on these principles are now in production.

9   References

1. Sharp BL (1982) Chem. Brit 18: 342
2. Letokhov VS (1983) Nonlinear laser chemistry, Springer, Berlin Heidelberg New York

3. Andrews DL (1986) Lasers in Chemistry, Springer, Berlin Heidelberg New York
4. Thrush BA (1981) Acc. Chem. Res. 14: 116
5. Hollas JM (1982) High resolution spectroscopy, Butterworths, London
6. Demtroder W (1982) Laser spectroscopy, Springer, Berlin Heidelberg New York
7. Hurst GS (1987) Phil. Trans. Roy. Soc. Lond. A323: 155
8. Fang HL, Swofford RL (1983) In: Kliger DS (ed) Ultrasensitive laser spectroscopy, Academic, New York, p 176
9. Tam AC (1983) In: Kliger DS (ed) Ultrasensitive Laser Spectroscopy, Academic, New York, p 2
10. Travis JC, DeVoe JR (1981) In: Heiftje GM, Travis JC, Lytle FE (eds) Lasers in chemical analysis Humana, Clifton, p 93
11. Bitto H, Guyer DR, Polik WF, Moore CB (1986) Faraday Discuss. Chem. Soc. 82, Paper 8
12. McCoustra MRS, Pfab J (1987) Chem. Phys. Lett. 136: 231
13. McCoustra MRS, Pfab J (1987) The relevance of N-nitroso compounds to human cancer: Exposure and mechanisms, IARC–WHO Scientific Publications, Lyons, France, p 228
14. Harris TD, Lytle FE: p 369 in Ref. [8]
15. Weeks SJ, Winefordner JC: p 159 in Ref. [10]
16. Kachin SV, Smith BW, Winefordner JD (1985) Appl. Spec. 39: 587
17. Wright JC: p 185 in Ref. [10]
18. Yeung ES: p 273 in Ref. [10]
19. Strojny N, de Silva JAF: p 225 in Ref. [10]
20. Rosasco GJ (1980) Adv. Infrared Raman Spec. 7: 223
21. Rousseau DL, Friedman JM, Wilson PF (1979) In: Weber A (ed) Raman spectroscopy of gases and liquids. Springer, Berlin, Heidelberg New York, p 203
22. Esherick P, Owyoung A (1982) Adv. Infrared Raman Spec. 9: 130
23. Nibler JW, Knighten GV: p 253 in Ref. [21]
24. Smalley RE, Wharton L, Levy DH (1977) Acc. Chem. Res. 10: 139
25. Levy DH, Wharton L, Smalley RE (1977) In: Moore CB (ed) Chemical and biological applications of lasers vol 2, Academic, New York, p 1
26. Hayes JM (1987) Chem. Rev. 87: 745
27. Parker DH: p 234 in Ref. [8]
28. Lichtin DA, Zandee L, Bernstein RB: p 125 in Ref. [10]
29. Walter K, Boesl U, Schlag EW (1986) Int. J. Mass Spec. Ion Proc. 71: 309
30. Boesl U, Grotemeyer J, Schlag EW (1987) Anal. Instrum. 16: 151
31. Grotemeyer J, Schlag EW (1988) Angew. Chem. Int. Ed. Engl. 27: 447
32. Singhal R, Ledingham K (1987) New Scientist, 116 (1588): 52

CHAPTER 9

Raman Spectroscopy

S.F.A. Kettle

There are four – somewhat inelegant – questions to be asked about any spectroscopy: What is it? How does it occur? How do you measure it? What can you learn from it? Inevitably, the answers to these questions are inter-linked but I will try to separate them as I deal with each in turn.

1 What Is Raman Spectroscopy ? [1–9]

In its usual form, Raman spectroscopy is a study of the vibrations of molecules or crystals. In order to achieve this objective, Raman Spectroscopy either uses visible (laser) light or, in Fourier Transform Raman Spectroscopy (FTRS), infrared (laser) light. The frequency of each vibration is given as the difference between the frequency of light emitted by the sample and the frequency of the incident light (that of the laser). Raman spectroscopy, then, consists of measuring with high resolution a very weak light emitted by a sample when the latter is in a laser beam. This emission should not be confused with fluorescence, although the latter can occur and, when it does, may well swamp the Raman-emitted light, making the latter unobservable.

2 How Does the Raman Effect Occur?

When a light wave (electromagnetic wave) is incident on a molecule its effect is to distort the molecule. This is most simply seen if one, hypothetically, thinks of a single molecule being held in the electric field between the two plates of a parallel plate capacitor which has an oscillating potential across the plates (this arrangement mimics the effect of the electric vector of the electromagnetic radiation on a single molecule). The electrons and nuclei of the molecule will be displaced in opposite directions by the electric field, the displacements oscillating with the frequency of the oscillating potential applied to the capacitor plates. In the case of the electromagnetic field of a light wave one sometimes speaks of such a molecule being in a "virtual state" i.e. one which is not achievable by the molecule in the absence of the field. We may note that when the oscillating frequency coincides with that of a natural frequency of the molecule resonance occurs and the molecule absorbs energy from the incident light. This is the classical description of the origin of the absorption of infrared, visible or ultraviolet light, absorption which gives rise to the corresponding spectroscopies. The normal Raman effect is a bit more complicated because in it the molecule does not have a natural frequency which coincides with that of the incident radiation. Figure 1 illustrates and enlarges the above discussion.

To understand the Raman effect we have to look at the above description in more detail. If, again hypothetically, we rotate the molecule with a constant potential on the parallel plate capacitor, then, for an anisotropic molecule, the distortion of the molecule will vary with the orientation of the molecule with respect to the electric field. If we were to explore all orientations of the molecule and to measure the distortions then we could plot the distortions as a function of the molecular orientation. The surface so constructed, which would look something like a rugby or American football, is a measure of polarisability of the molecule. Note, however, that because an anisotropic molecule will have an 'easy' direction, we cannot assume that the distortion induced in the molecule will be exactly collinear with the field that induces it. To actually compute and plot the polarisability ellipsoid of a molecule this non-coincidence has to be taken into account. Suppose now that we were to physically distort the molecule, to move the atoms slightly – as happens in a vibration. For convenience, let us regard the atoms as fixed at the extremities of their vibrational displacement, physically unreasonable as this may be. If we now, even more hypothetically, measure the polarisability ellipsoid of this distorted molecule we cannot guarantee that we will obtain the same ellipsoid as for the undistorted molecule. We conclude that associated with some (normal) vibrations there will be a change in polarisability of a molecule (the change in polarisability ellipsoid from that of the molecule in its equilibrium configuration will be essentially proportional to the displacement of

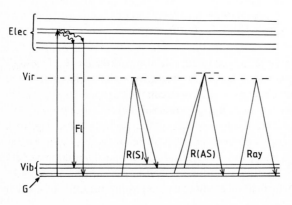

Fig. 1. Fluorescence (Fl) Stokes Raman R(S), Anti Stokes Raman R(AS) and Rayleight Scattering (Ray), involving the Ground state (G), Vibrationally excited states (Vib), Electronically excited states (Elec) and a laser frequency-dependent – Virtual state (Vir)

the atoms along the normal coordinate). This change in polarisability can be an increase or a decrease – it can be positive, negative or zero. Accordingly, it is more difficult to draw a picture of it and so it is usual to handle it mathematically and talk in terms of a (derived) polarisability tensor.

In the Raman effect, a laser polarises molecules which emit light of a different frequency. The incident laser light polarises the molecule and, if we treat the process of emitting light as reversible, and consider the molecule in the field of the emitted light then the molecule will again be polarised. However, the polarisation of the molecule associated with the emitted light may well differ from that associated with the incident light. Let us suppose there is, indeed, a difference in polarisation of the molecule in the two cases. This change is accommodated by the excitation of a vibration which is one of those which changes the polarisation of the molecule. This is a simplified, semi-classical, pictorial description of the Raman effect. The reader is warned that the simplified mathematical account commonly given, and which we repeat in an Appendix, whilst helpful in presenting a different viewpoint, also has limitations: in the Appendix we indicate its main defects.

3   How Is a Raman Spectrum Measured?

3.1   'Conventional' Raman Spectroscopy

For the majority of spectroscopies there is a similarity in the experimental set up:

Source of————Sample————Energy-resolver————Detector
Radiation   (Monochromator)

For the Raman experiment, shown schematically in Fig. 2, a laser is now invariably used as the source. There are many types of laser available for conventional Raman spectroscopy. For some higher-order Raman-related spectroscopies (which we shall discuss later) a high-energy pulsed laser is used. More commonly, a continuous power of between 10 milliwatts and 1 watt is used, the range 50–200 mW being the most common. At the upper limit an argon-ion (gas) laser is usually chosen although a kryption-ion (gas) laser is an alternative (the krypton laser has a somewhat lower power but can be made to lase at a wider range of frequencies). A mixture of argon and krypton is also used. Such mixed gas lasers are now less available commercially but have an advantage for the Raman spectroscopist. Although the power available in any one lasing line is lower than with either pure gas, a total of about ten different lasing lines, extending from near infrared to near ultra-violet, is available (although the extremes require special optics). In theory, choice of wavelength is important; quantum mechanics shows that the intensity of the Raman scattered light varies as the fourth power of the laser frequency for constant input power. Blue laser light is to be preferred. However, many compounds are photolysed by blue light

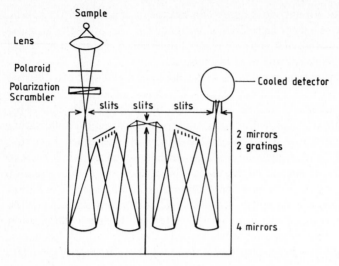

Fig. 2. A typical Raman Spectrometer using a double (grating) monochromator

and even more fluoresce, so some compromise is often necessary (red light tends to cause less photolysis and, certainly, less fluorescence). In practice, the v_0^4 dependence does not seem as important as one might expect and it is commonly ignored.

For experiments, such as the Resonance Raman measurements which we shall discuss later, in which a really continuous range of laser wavelengths is important a dye laser is used. In these, organic dyestuffs in solution absorb light at one wavelength (frequently from an argon-ion laser) and emit at another. The process is akin to fluorescence; by suitable adjustment of a grating or prism, however, the dominant emission is made to occur in a single, narrow, band. A wide selection of dyes can be used which, between them, cover the visible and near infrared regions (see Chap. 8).

Any gas laser, in particular, emits light at a variety of frequencies – those corresponding to atomic transitions within the gas species. These non-lasing lines must be removed or else they will appear in the final spectrum (scattered by the Rayleigh mechanism, shown in Fig. 1). An interference filter, prism or custom-built grating pre-monochromator is therefore invariably used to clean-up the laser beam before it strikes the sample. Even so, weak 'non-lasing line' peaks are sometimes seen in the spectrum. They are always sharp – and so useful as frequency markers. Unlike genuine Raman lines, they are not reproduced when a different laser exciting line is used. Raman spectroscopists usually have at hand a list of those non-lasing lines which occur with their experimental set-up.

The sample may be gaseous, liquid or solid. In order to increase the scattering from gases or liquids mirrors are positioned around the sample both to reflect the laser beam back along its own path and to reflect into the collecting optics scattered light which would otherwise be lost. The angle between the incident laser beam and the spectrometer axis (that along which light is collected) can be 90° (most common), 0° or 180°: for these latter cases the laser beam usually comes in at 90° but is reflected into the 0° or 180° orientation by a tiny prism or mirror. Here 180° means that the laser beam is co-axial with the collecting optics but is directed by the prism into a direction away from the spectrometer; 0° means that the laser beam is directed towards the spectrometer but encounters a stop in front of the collecting optics.

Liquid samples may be held in any transparent container – even a test tube may be used – but best results are obtained when optical flats are used on all principal light paths. The same is true for gases, but these are often studied in cells which are compatible with a carefully aligned pair of mirrors which pass the laser beam back and forth across the axis of the entrance optics. Powder samples are usually held in a tiny capillary, a diluent – such as KBr – seldom being used. A few milligrams can give a good spectrum; if necessary, the sample can be either heated or cooled or even held under high pressure. Matrix-isolated species have been the object of much recent study. Thermal decomposition of the sample can often be avoided by pressing it into a circular groove and spinning about the axis of this groove so that a fresh part of the sample is constantly presented to the laser; similar techniques are often employed in Resonance Raman spectroscopy. Some typical sample-holder arrangements are shown in Fig. 3.

A lens with a large f number (usually ca f1) is used to collect the scattered radiation, which is focused onto the entrance slit of a spectrometer. In order to distinguish the scattered light from general stray and background light a single

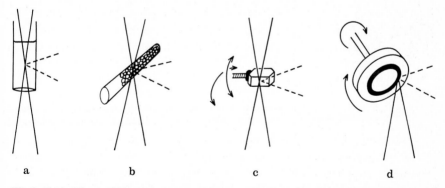

a   b   c   d

Fig. 3. Typical Raman 90° sample arrangements. The *solid lines* indicate the focused laser beam and the *dotted lines* represent the light scattered from the sample into the spectrometer (see Fig. 2)
(a) Liquid samples. The bottom of the liquid cell is an optical flat; mirrors may be placed around the samples to increase both the laser and scattered light intensity
(b) Powdered solids held in a capillary (small amounts of liquids may be similarly studied)
(c) Single crystals. The crystal is mounted on goniometer arcs; ideally the crystal faces involved in the light paths are optically flat and perpendicular
(d) Spinning cell arrangement for dark or heat sensitive samples. A spinning cell can be used for liquids, but this is an adaptation of (a), the cell being spun about a vertical axis and the laser beam being off-centre

monochromator is not adequate. A double monochromator is most common, although for low-frequency studies (vibrations down to a few wavenumbers) a triple monochromator will probably be necessary. The advent of holographic gratings, however, has brought about a resurgence of the double mono- chromator – the problems of optical alignment can be less with the double than with the triple monochromator. Prisms are never used. Because gratings diffract light polarised along and perpendicular to the direction of the grooves with different and wavelength-dependent efficiencies, it is necessary to depolarise all light entering the spectrometer. This is done using a polarisation scrambler. This usually consists of a wedge of natural quartz which rotates the plane of polarised light by an amount that depends on its thickness (so that, for a wedge, the full range of rotation is covered, irrespective of the polarisation of the incident light). Bonded to the wedge of natural quartz, is one of fused quartz, tapering in the opposite direction, which has no effect on polarisation but which compen- sates for the varying pathlength in the natural quartz.

For the highest resolution, a photomultiplier is used as a detector. It is essential that it is sensitive to near-infrared light (or else a red laser line could not be used). This means that a suitable photocathode must be used. The most popular is currently a gallium arsenide photocathode because it has a near- constant response over the visible and near infrared regions right up to its cut-off point. An alternative is the S20 'trialkali – Cs, Na, K/Sb photocathode. In either case the photomultiplier is cooled by about $30°$ C to reduce spontaneous electron emission and thus enhance the signal-to-noise ratio.

Almost invariably photon-counting electronics are used in the output circuit. Suitable gating circuits enable the rejection of the low-amplitude pulses arising from electrons which are emitted from one of the secondary anodes of the photomultiplier (the avalanche effect guarantees a minimum size pulse from electrons originally generated at the photocathode). Other gating circuits remove pulses that have too great an amplitude to result from photoemission at the photocathode (they could result from background cosmic radiation, for instance).

Control of the spectrometer (but not of the laser) is invariably by dedicated computer. This allows digital manipulation of the spectrum co-adding, subtrac- tion, smoothing and so on. A point to watch for when high resolution spectra are to be obtained is that until recently the RAM available on the standard computers supplied with the instrument may have been such as to permit manipulation of only part of the spectrum at any one time. A computer also enables the use of alternative, lower resolution, detectors (but which are adequate for all but those spectra requiring resolution of approx. $0.1\,cm^{-1}$ or better). Typical of these is a diode array detector[10] – a detector which may be thought of as akin to the sensitive element in a television camera, which enables a ca. $100\,cm^{-1}$ spectral range to be scanned simultaneously (each detector contains ca. 1000 separate channels). Such detectors also enable such a spectral range to be scanned rapidly and repeatedly, enabling a study of time-dependent phenomena

[11,12]. For optimum use of such detectors the image of the entrance slit has to be in focus on a flat field (different frequencies coming to a focus at different points in the field). If this is to be obtained it is normally done by building it into the design of the monochromator and most workers therefore choose to be guided by the spectrometer manufacturer's recommendations on diode array detectors. To enhance the sensitivity of such flat-plate detectors image intensifiers may be added in front of them. These image intensifiers are related to those used in infrared night vision viewers. Another development is in the use of charge-coupled semiconductor devices [13]. These are even more similar to the television camera detector because it is possible to record different spectra in each 'line' of the detector – and each detector contains about four hundred 'lines'. Charge-coupled devices have a high sensitivity and low noise, giving an excellent signal-to-noise ratio. When used with visible lasers their performance in this respect is better than that of most photomultipliers. However, this is achieved only at the additional complication of a need to cool the detector to $-100\,^\circ$C or below.

3.2   Fourier Transform Raman Spectroscopy [14]

Conventional Raman spectroscopy suffers from disadvantages. Firstly, accurate optical alignment is vital. Considerable time may be needed to make sure that a narrow, focused, laser beam is incident on a solid sample in a capillary tube at a point which is not only on the optic axis of the entrance optics but is such that the collecting lens gives a focussed image on the entrance slit. Secondly, as mentioned above, many materials fluoresce in the laser beam, giving a broad band in the spectral region in which a Raman spectrum is expected. This fluorescence, which may well not be visible to the naked eye, often swamps the Raman signal. The fluorescence problem may be alleviated either by careful purification of the sample or by photobleaching – leaving the sample in the beam for several hours, when the source of the fluorescence may be photolytically removed. Often, however, the fluorescence cannot be removed and thus to Raman spectrum obtained. The problem is particularly severe for commercial products: thus over 90% of polymers fail to give Raman spectra, not because the spectra do not exist but because their fluorescence is too strong to enable Raman spectra to be measured.

Fourier Transform Raman Spectroscopy, a technique which is still in its infancy, offers the prospect of removing both of the above problems. A typical experimental set-up is shown in Fig. 4. Basically, it consists of using an infrared laser to excite the Raman spectrum, which then falls in the spectral region covered by a normal FTIR spectrometer. The interferometer and detector system of the FTIR are then used to obtain the Raman spectrum. The possibility of a joint FTIR/FTRS spectrometer is obvious and, indeed, models are already commercially available.

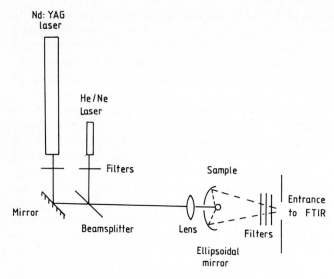

Fig. 4. Schematic arrangement for a typical FT Raman spectrometer

The fluorescence problem largely disappears because fluorescence involves excitation into electronic excited states and these seldom fall in the infrared. The large – 'barn door' – optical arrangement of an interferometer, in contrast to the slits of a conventional spectrometer, largely alleviate the sensitivity to optical alignment. However, there are problems. Firstly, there is the v_0^4 dependence of the magnitude of the Raman effect on frequency: a weak effect is expected to be yet weaker (by a factor of about 20 compared to the visible). Secondly, there is the need to work with an (invisible) infrared laser. Use of a co-axial HeNe alignment laser can, of course, overcome this. A third problem is laser noise. The neodymium YAG laser commonly used for excitation – and most of the alternative sources – are rather noisy and do not have the long-term intensity stability of the gas lasers used in the visible. This instability reappears as noise in the spectrum. Fourthly, the need to avoid saturating the detector with laser light, (which is six orders of magnitude more intense than the scattered light), means that some very effective way has to be found to prevent the infrared laser light from reaching the detector. Currently, multiple wavelength-optimised interference rejection filters are used, but there also block out the low-frequency region – spectra can only be obtained from ca $300\,\mathrm{cm}^{-1}$ upwards (or, at best, $200\,\mathrm{cm}^{-1}$). Alternative filter arrangements are currently under active development and it seems certain that before long it will be possible to make measurements down to around $40\,\mathrm{cm}^{-1}$. Fifthly, the detectors are relatively poor for a variety of reasons but here, again, improvement is to be expected. Neither photomultipliers nor charge-coupled devices have any sensitivity in the infrared. Ultimately, the signal-to-noise (S/N) ratio achieved in FTRS is in some considerable measure determined by the detector. The change from conventional Raman to FTRS is accompanied by a change from a shot noise limitation of the S/N ratio to a detector noise limitation.

Three other advantages of FTRS should be noted. Firstly, some of the wavelength accuracy of FTIR is now available in Raman, enabling improved frequency comparisons – with conventional infrared and Raman, peaks up to approx $4\,cm^{-1}$ apart have to be regarded as 'coincident' – although this advantage of FT methods can be over-emphasised (because of off-axis light paths, for instance, errors of about $2\,cm^{-1}$ are not uncommon). Thus, it is reported that moving the focus of the Raman light scattered from the sample across the entrance aperture moves the spectrum by about 3–$4\,cm^{-1}$. However, provided that the sample is carefully constrained to the same position, highly reproducible spectra can be obtained so that careful calibration allows removal of systematic errors. Secondly, unlike conventional Raman, because of the absence of slits there is no need to accurately focus the laser on the sample. One result of this is that powers up to approx. $500\,mW$ are routinely used – in conventional Raman, the use of such power levels would burn many samples. Finally, the throughput (Jacquinot) advantage is very important – it compensates for many of the disadvantages of FTRS.

At the time of writing, FTRS has clear advantages for 'difficult' samples and for 'ordinary' samples if run by the non-expert. More advanced techniques involving depolarisation, single crystal and microscope measurements, for example, are best performed on conventional instrumentation. The fact that the lines currently obtained from a gas laser are significantly narrower than from an infrared one means that conventional instrumentation will be the choice when high resolution is important.

4   What Can Be Learned from Raman Spectroscopy?

Although both Raman spectroscopy and infrared spectroscopy study molecular vibrations so that in general appearance the spectra obtained are similar (see Fig. 5), the two techniques are often complementary. Firstly, the selection rules are different. The infrared selection rule requires that $(\partial\mu/\partial Q_i)$ be non-zero where μ is the molecular dipole amount and Q_i is the ith normal coordinate whereas, as we have seen, the Raman selection rule is that the polarisability change with respect to normal coordinate, $(\partial\alpha/\partial Q_i)$, be non-zero. In centrosymmetric molecules this infrared-Raman selection rule difference reaches the limit in that those vibrations which are infrared active vibrations are symmetry required not to be Raman active and vice versa. Note that it does not automatically follow that infrared and Raman spectra will have no frequencies in common in such cases (differences in frequency may be less than the instrumental errors), although such frequency exclusion is common for small molecules. Secondly, not only intensities but also band contours are often rather different in the two

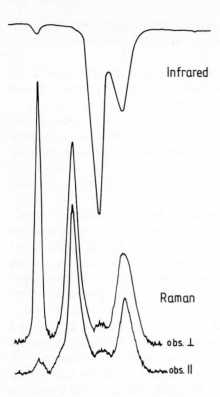

Infrared

Raman

obs. ⊥

obs. ‖

Fig. 5. Infrared and Raman spectra of $Mn(CO)_5Br$ in the $\nu(CO)$ region, showing that for this compound peaks strong in the infrared tend to be weak in the Raman and vice versa. A totally symmetric mode (extreme left) is identified in the Raman by its polarization characteristics (see Fig. 6)

spectroscopies. Quite often sharp bands in the Raman may correspond to broad infrared features and vice versa. Thirdly, there are solvent differences. In particular water, which, even with FTIR, is a difficult solvent to use in infrared studies, is excellent for use as a solvent in conventional Raman, although less so in FTRS, for just the same reasons that is not a good infrared solvent – it has broad and intense absorption bands in the infrared. Finally, studies of single crystals (and, thus, the study of vibrational phenomena associated with the crystalline state) are very much easier in the Raman. It is usually difficult to get crystals thin enough for infrared study (before they are ground thin enough to transmit a significant amount of infrared light they tend to craze, crack and disintegrate). In summary, then, that which we learn from Raman is complementary to that learnt from infrared studies. The one technique is, in a sense, incomplete without the other.

However there are aspects of Raman spectroscopy which set the technique apart from infrared and which are behind the current increase in commercial interest in Raman, particularly FTRS. Firstly, Raman studies can usually be performed with no sample preparation. Apart from the fluorescence problem two main sample difficulties arise. The first lies in sample absorption of the laser light – this can lead to sample destruction and/or weak signals. However, it also leads to the Resonance Raman effect, which we shall describe in the next section.

Secondly, very small particle size leads to scattering of the laser light with consequent high background and low signal; milky solutions and fine powders tend to be 'difficult'.

The next aspect of Raman arises from the fact that the incident laser light is polarised. Suppose that this light is incident on an isotropic molecule – an octahedral or tetrahedral species (both of which are cubic – for each $x \equiv y \equiv z$) for example. Because of this isotropy, the polarisation of the molecule induced by the electric vector of the light wave is parallel to the electric vector (in contrast to the more general situation described in our earlier discussion). Suppose the vibration excited by the Raman effect has a symmetry which preserves the molecular isotropy – it is a 'breathing' mode for example. In this case the scattered light is also fully polarised. We have, then, a simple test for vibrations which preserve the molecular symmetry (so-called totally symmetric vibrations). A sheet of polaroid inserted into the collecting optics will, in one orientation, have no effect on the intensity of the peak associated with a totally symmetric vibration but, rotated through 90°, will remove it from the spectrum. An example of this is shown in Fig. 6. Of course, the majority of molecules are not cubic but the general principle remains – totally symmetric vibrations retain more of the inherent laser polaris-ation than do vibrations which change, and usually reduce, the molecular symmetry. This means that a simple augmentation of the Raman experiment for liquids and gases – the inclusion of a sheet of polaroid – can provide information on the symmetry of molecular vibrations. The polaroid film also finds use in the study of the spectra of single crystals, but here it is also important to identify and control the orientation of the crystal axes with respect to those of the spectrometer (and, thus, of the polaroid film). In favourable cases, by control of crystal-spectrometer axes, direction of laser polarisation and the orientation of the polaroid film, all of the various Raman-active symmetry species can be distinguished (but for the symmetry of the crystal rather than that of the molecule).

Another attraction of Raman spectroscopy is that it is possible to measure Raman spectra through a microscope (or, at least, a specially adapted micro-

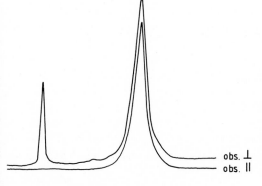

obs. $\perp$
obs. $\parallel$

Fig. 6. $\nu(CO)$ spectrum of $Mo(CO)_6$ showing that the peak associated with the a_{1g} mode (*left*) is polarized – and so can be stopped by a suitably oriented film of polaroid – but that the e_g peak (*right*) is not

scope) although it has to be admitted that in this respect infrared is doing its best to catch up! As mentioned in Sect. 3.2, although it is possible to make microscopic measurements using a Fourier Transform system, the throughput of the microscope is only about a hundredth of that of the FT system, reducing the S/N ratio accordingly. The conventional Raman set-up, which works with a sample which is almost an effective point source, is much better adapted to measurements with a microscope. The experimental arrangement is shown in Fig. 7. The microscope is adjusted manually so that a tiny speck of impurity in a polymer, for example, is at the centre of the field of view. The laser is then switched on and the beam brought through the microscope optics and focused on the impurity particle. The Raman scattered light from this particle is collected through the microscope objective (i.e. the 180° arrangement) and then reflected into the spectrometer. The Raman spectrum of the impurity can thus be obtained. As an alternative, where the sample under study contains a number of components, each with a characteristic Raman frequency, the spectrometer can be set at each frequency in turn. If the sample is then tracked back and forth under the microscope so that a grid is traced out – akin to the grid traced by the electron beam in a television tube – then the distribution of each component across the entire sample can be determined.

Although successful studies have been carried out, the inherent weakness of the Raman effect makes studies of surfaces and adsorbed species 'difficult' with one exception. This arises from the observation that molecules adsorbed onto the

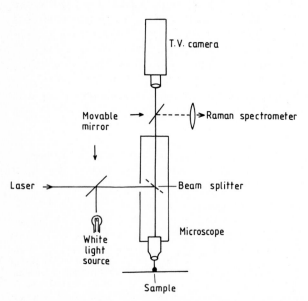

Fig. 7. Schematic Raman microprobe arrangement. Only a limited number of commercial microscopes can be adapted for this use

surfaces of some noble metals such as silver give spectra which are several orders of magnitude stronger than might be expected (up to $\times\ 10^6$) [16,17]. The origin of this Surface-Enhanced Raman Scattering (SERS) is not well understood. It probably arises through a coupling of fields within the metal to molecules via surface roughness. On the basis of this model not all metals are expected to be equally effective – it depends on their electronic structure – and those that are expected to show the strongest effect are those that indeed do. However, it seems agreed that there is at least one other mechanism operative, more chemical in nature. Some argue that surface metal-ligand bond coordination, or a similar effect, is dominant. Whatever its origin, the effect is currently exploited in the study of electrode surfaces and of the interaction of biomolecules with surfaces.

5 Other Raman Spectroscopies

In the past twenty years there have been several spectroscopies introduced which are related to Raman spectroscopy, although, as yet, only one has made any major impact. This important advance has been in the field of Resonance Raman Spectroscopy [18]. In the discussion above on the mechanism of the Raman effect it was assumed that the laser light used is not of a frequency which corresponds to an (electronic) transition within the molecule. It is when there is just such a coincidence (and dye lasers may be used to ensure it) that the Resonance Raman effect occurs. A Resonance Raman spectrum is characterised in two ways. Firstly, the intensities observed are often much greater than those of the normal Raman effect. Secondly, only a limited number of vibrational modes – perhaps one or two – both show this enhancement and also appear not just as fundamental vibrations (frequency shift v_i) but also as overtones $2v_i$, $3v_i$, $4v_i$ etc. Evidently, the most spectacular results will be shown when a blue laser is used (otherwise the detector may run out of sensitivity before the Resonance Raman spectrum is completed). In favourable cases up to ca $15\,v_i$ have been detected. Such measurements enable vibrational anharmonicities to be determined and approximate dissociation energies obtained. An example of a Resonance Raman spectrum is shown in Fig. 8. An area where the Resonance Raman effect has found special application is in the field of biochemistry [19–21]. Biological molecules are usually large and so have very rich and complicated vibrational spectra. From this maze, the Resonance Raman effect selects out just a few vibrations. By choosing a laser frequency that coincides with an electronic transition of a transition metal ion in the molecule, for example, the modes enhanced by the Resonance Raman effect will be selected from those that have high amplitudes close to the transition metal ion. It is thus possible to selectively tease out of the spectrum vibrations sensitive to the environment of an individual transition metal ion.

A technique of – as yet relatively unfulfilled – promise is CARS (Coherent Anti-Stokes Raman Scattering) [22]. Anti-Stokes scattering is illustrated in

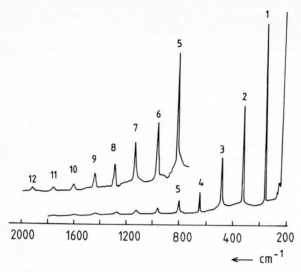

Fig. 8. Resonance Raman spectrum of TiI_4. The order of harmonic is indicated for each peak

Figure 1; in contrast to the Stokes scattering normally studied in Raman spectroscopy, it arises when a vibrationally excited molecule undergoes deactivation by the Raman process – the light emitted is at a higher frequency than that of the exciting laser (the process is also referred to in the Appendix). In CARS two lasers are used, one of which is a tunable dye laser. The scattered light is at higher frequency than either. Scattering occurs whenever the difference in frequency of the two lasers coincides with a molecular vibrational frequency, v_i. With one laser held at fixed frequency, v, and the other over tuned over a frequency range, peaks are observed at the various frequencies $v + v_i$ corresponding to molecular vibrations. Further, the scattering occurs along the line of the two (near-superimposed) incident laser beams so that an intense beam may result. The main problem of the method is the achievement of an adequate mechanical, frequency and intensity stability. Further details are given in Chap. 8.

Finally, we may mention the inverse Raman effect [22, 23]. It is called 'inverse' because, unlike the normal Raman effect, which is manifest as an emission of light, the inverse effect is an absorption. Specifically, one irradiates the molecule with light from a giant-pulse laser and, simultaneously, with white light covering the spectral region between the wavenumber of the laser and up to ca $4000\,cm^{-1}$ higher in energy. The inverse Raman effect is manifest by absorption peaks in the continuum, each corresponding to a molecular frequency. This is a difficult technique. Not only has synchronisation between the laser pulse and continuum to be achieved but also the whole spectrum has to be scanned in approx. $10^{-7}\,s$ (the duration of the pulse).

Table 1. Number of papers at a recent conference[f(iii)]

How does the Raman-effect occur?	
Raman Theory	6
Band Shapes and Dynamics	36
Band Intensities and Molecular Conformation	11
How do you measure a Raman Spectrum?	
New Techniques	21
What do we learn from Raman Spectroscopy?	
Vibrational Analysis and Molecular Structure	24
Inorganic Materials and Matrices	17
The Solid State	30
Semiconductors and Superconductors	29
Phase transitions and the effects of	
temperature and pressure	31
Low dimensional and amorphous solids	12
Macromolecules and Proteins	28
Lipids and Biomembranes	9
Time-resolved Raman and Transient Species	14
Raman Microscopy	15
SERS, Surface and Interfacial Phenomena	51
Other Raman Spectroscopies	
Resonance Raman	38
Biological Pigments	22
Biological Systems and Medical Applications	22

6 The Future

The present article is an attempt to give a readable overview of the Raman field as it has developed over the recent past. What of the future? Whilst an attempt has been made to assess future trends at various points in the text, an independent assessment can be gained from the distribution of contributed papers at a recent conference on Raman Spectroscopy (f(iii)). These are given in Table 1, listed roughly according to the section of the present article to which they have the greatest affinity.

Acknowledgement

The author is indebted to Professor H.A. Willis for helpful comments.

Appendix

An electric field E incident on a molecule induces a polarisation P. If we assume a simple proportionality between them, then:-

$$P = \alpha E \tag{1}$$

where α is the polarisability. The field E is generated by the oscillating incident light wave and so may be written in the form

$$E = E_0 \sin 2\pi v_0 t \tag{2}$$

Combining (1) and (2)

$$P = aE_0 \sin 2\pi v_0 t \tag{3}$$

Now the polarisability of a molecule will depend on the molecular vibrations since these involve excursions of nuclei from their equilibrium positions. If we represent a vibration by Q then we may expect that as Q varies (i.e. as the molecule vibrates) there will be a corresponding, approximately linear, change in α. In its most general form we may write the proportionality constant as $\partial\alpha/\partial Q$ – since this form pre-supposes no pre-knowledge about either α or Q. For the total of i vibrations of a molecule, therefore, we have

$$\alpha = \alpha_0 + \sum_i \left(\frac{\partial\alpha}{\partial Q_i} \right) Q_i \tag{4}$$

Substituting this into Eq. (3) and using the fact that Q_i is an oscillating quantity

$$Q_i = Q_0 \sin 2\pi v_i t \tag{5}$$

where v_i is the corresponding natural frequency, we have

$$P = \left[\alpha_0 + \sum_i \left(\frac{\partial\alpha}{\partial Q_i} \right) Q_0 \sin 2\pi v_i t \right] E_0 \sin 2\pi v_0 t$$

$$= \alpha_0 E_0 \sin 2\pi v_0 t + \sum_i \left(\frac{\partial\alpha}{\partial Q_i} \right) Q_0 E_0 (\cos 2\pi (v_0 - v_i)t - \cos 2\pi (v_0 + v_i)t)$$

The first term in this expression, $\alpha_0 E_0 \sin 2\pi v_0 t$, represents scattering of the incident laser beam. This is called Rayleigh scattering and is the dominant effect. For the other – Raman – term to be non-zero we have the requirement that the term $(\partial\alpha/\partial Q_i)$ is non-zero for some of the i vibrations. Only for these vibrations will one see a Raman effect. This, then, is the Raman selection rule. In this from the selection rule is of no great immediate help but it can be re-expressed in terms of a much more useful symmetry-based selection rule – only vibrations which have the same symmetry properties as a product of coordinate axes give rise to Raman bands.

The term in $\cos 2\pi (v_0 - v_i)t$ gives rise to the Raman effect. The frequency $(v_0 - v_i)$ is the frequency of the weak emitted light which is studied in Raman spectroscopy. As is evident from the discussion in this chapter, and also from this derivation, it is the frequency difference between the weak emitted light and the laser frequency, v_0, which gives the molecular vibrational frequency. This scattering is often called 'Stokes scattering'. The final term, $\cos 2\pi (v_0 + v_i)t$, is called "Anti-Stokes scattering". It corresponds to the destruction of a vibration in a molecule and the addition of its energy to that of the incident laser beam with consequent increase in frequency – and, therefore, of energy.

Weaknesses in this derivation are that in Eq. (1) it is implied that P and E are collinear, so that α is a simple proportionality constant. Secondly, Eq. (5) implies that the molecule is already vibrating. This aspect is made explicit in the discussion of Anti-Stokes scattering at the end of the previous paragraph.

7 References

1. Herzberg G (1945) Infrared and Raman spectra, Van Nostrand, New York
2. Gibson TR, Hendra PJ (1970) Laser Raman spectroscopy, Wiley, London
3. Loader J (1970) Basic Raman spectroscopy, Heyden, London
4. Tobin MC (1971) Laser Raman spectroscopy, Wiley-Interscience, New York
5. Koningstein JA (1972) Introduction to the theory of the Raman effect, D. Reidel, Dordrecht
6. Anderson A (ed) (1973) The Raman effect, M. Dekker, New York
7. Colthup NB, Daly LH, Wiberley SE (1974) Introduction to infrared and Raman spectroscopy, Academic, New York
8. Woodward LA (1972) Introduction to the theory of molecular vibrations and vibrational spectroscopy, Oxford University Press, Oxford
9. Long DA (1977) Raman spectroscopy, McGraw Hill, New York
10. The diode-array advantage in UV/visible spectroscopy, (Hewlett-Packard, Waldbronn, FRG, 1989)
11. Laubereau A, Stockburger M (eds) (1985) Time resolved vibrational spectroscopy, Springer, Berlin Heidelberg New York
12. Atkinson GH (ed) (1987) Time resolved vibrational spectroscopy, Gordon and Breach, New York
13. Batchelder DN (1988) European Spectroscopy News 80: 28
14. Bergin FJ, Shurvell HF (1989) Applied Spectroscopy 43: 516
15. Manufacturers literature is helpful in this area; examples of 'do it yourself' attachments are to be found in the literature, for example several of the papers in Sec. 18 of the conference proceedings detailed in the Table
16. Chang RK, Furtak TE (eds) (1982) Surface Enhanced Raman scattering, Plenum, New York
17. Cardona M, Güntherods G (eds) (1984) Light scattering in solids, vol 4 Springer, Berlin Heidelberg New York
18. See, for example Clark RJH, vol II of Bibliography Ref. d, p 95
19. Spiro TG (ed) (1987) Biological applications of Raman spectroscopy, Wiley, New York
20. Clark RJH, Hester RE (eds) (1986) Spectroscopy of biological systems vol 13 of Bibliography Ref. d, Wiley, Chichester
21. Stepanek J, Augenbacher P, Sedlacek B (eds) (1987) Laser scattering spectroscopy of biological objects, Elsevier, Amsterdam
22. Harvey AB (ed) (1981) Chemical applications of non-linear Raman spectroscopy, Academic, New York
23. Takayanagi M, Hamaguchi H, Tasumi M (1986) Applied Spectroscopy 40: 1137

8 Bibliography

In addition to the references given in the text, mention should be made of a number of review periodicals each running to several volumes. Even non-current ones contain useful material and so are included.

a) Mathieu JP (ed) Advances in Raman Spectroscopy, Heyden, London
b) Seymanski HA (ed) Raman spectroscopy, theory and practice, Plenum, New York
c) Anderson A (ed) The Raman effect; Principles: applications, Marcel Dekker, New York

d) Clark RJH, Hester RE (eds) Advances in infrared and Raman spectroscopy. The early volumes were published by Heydon, London, later by Wiley, Chichester. Current volumes, whilst retaining the series volume numbers, are being given subject-area names; simultaneously the scope has been widened to cover other spectroscopies
e) Durig JR (ed) Vibrational spectra and structure. Elsevier, Amsterdam
f) Biannual conferences in Raman spectroscopy are held and the proceedings published. The most recent have been:
Proceedings of International Conference of Raman Spectroscopy
  i) IX Tokyo, Japan 1984
  ii) X Eugene, Oregon 1986
  iii) XI London 1988
  iv) XII Columbia, S. Carolina 1990

Publishers of the last two named and the next three volumes are Wiley (Chichester).

A fairly recent review of several advanced aspects (cf. e.g. the Resonance Raman Effect and CARS) is to be found in the series Topics in Current Physics, vol II, Weber A (ed) (1979) Raman spectroscopy of gases and liquids, Springer, Berlin Heidelberg New York.

A very useful series of reviews of Raman spectroscopy and applications including non-linear phenomena, CARS, biological molecules, inorganic species, Raman microscopy and surface chemistry is to be found in Chemistry in Britain 25: 589 (1989).

CHAPTER 10
Natural and Magnetic Circular Dichroism Spectroscopies

A.J. Thomson

1 Introduction

Circular dichroism ($\Delta\varepsilon$) is defined as the difference between the molar absorption coefficients for left (ε_L) and right (ε_R) circularly polarised light. A circular dichroism (CD) spectrum, consisting of a plot of $\Delta\varepsilon$ against wavelength or frequency, can therefore be observed only within absorption bands. CD generally arises from electronic transitions, although recently CD within vibrational bands has been measured. The two forms of CD spectroscopy to be discussed here are natural and magnetically induced.

Natural CD appears only for optically active or chiral molecules, that is, for molecules not congruent with their mirror images. This form of spectroscopy is widely used for the determination of absolute configuration and for the study of bipolymer conformation.

In magnetic CD, molecules become chiral in the presence of a magnetic field applied parallel to the direction of the measuring light beam. The phenomenon arises from the Zeeman effect, and is quite different to that of natural optical activity. The technique has been most widely used for assignment of the electronic spectra of transition-metal compounds. Although organic molecules do give MCD spectra they are often weak and the information content is low. The technique is now finding special application as an optical probe of the paramagnetism of metal ions in proteins and, in this way, is complementary to other techniques such as electron paramagnetic resonance (EPR) spectroscopy (Chap. 12) and magnetic susceptibility measurement.

The two techniques have a close relationship because the instruments used to measure the spectra are the same, except that the measurement of MCD spectra requires the addition of a high magnetic field, usually provided by a super-conducting solenoid. In this review, the origins of each effect will be presented, and some examples will be given to illustrate the applications of the technique.

2   Natural Circular Dichroism

Pasteur first provided the criterion for optical activity to be present, namely, that a molecule should exist in two isomeric forms related as non-superposable mirror images. This rule can be re-stated in terms of three sterochemical symmetry conditions for optical activity. These are that the molecules should be devoid of all the following symmetry elements

a) a centre of symmetry
b) a plane of symmetry
c) an alternating rotation-reflection axis.

The last requirement is the fundamental one since a one-fold alternating axis is equivalent to a plane of symmetry and a two-field alternating axis is identical with a centre of inversion. These conditions for optical activity are equivalent to the symmetry requirements for a transition to have electric and magnetic transition dipole moments with components in a common direction. Hence, electronic CD can occur within an absorption band only if the excitation is associated both with a linear displacement and a rotation of charge. The linear motion of charge generates an electric dipole moment (μ) and the rotation of charge leads to the creation of a magnetic dipole moment ($\mathbf{m}$) perpendicular to the plane of rotation. If the angle between the two transition dipole moments is θ a quantity termed the rotational strength, R, is given by

$$R = \mu m \ \cos \theta$$

A helical displacement of charge involves a rotation about the helix axis and hence the development of a magnetic moment along this axis. There is also a linear displacement of charge along the helix axis so that an electric dipole is created in the same direction. Note that a helix is non-superposable upon its mirror image, that is, it has a handedness.

The rotational strength, R, is related to an experimental spectrum in the following way:

$$R = \frac{3hc^2 \varepsilon_0 \, 10^3 \, \text{In} \, 10}{8\pi^2 N} \int \frac{\Delta \varepsilon}{\lambda} \, d\lambda$$

where all the symbols have their usual meaning. The constant before the integration is equal to 7.637×10^{-4} C.A.m.^{3}. The integration runs over the area of the circular dichroism band. Note that R can have either a positive sign, when θ lies between $\pm 90°$, or a negative sign when θ has a value between 90 and 180°. This implies that R is positive when μ and m are parallel but negative when μ and $\mathbf{m}$ are anti-parallel. Thus the mutual relative orientation of $\mathbf{m}$ and μ for a transition determines the sign of the CD. It is possible then to relate this to absolute configuration in favourable cases, that is, to the handedness of a helix.

2.1 Electronic Circular Dichroism

It is convenient to divide optically active (chiral) molecules into two classes, those which contain an inherently chiral chromophore (Class I) and those whose chromophore is locally achiral but is perturbed by its chiral surroundings (Class II)[1].

Kuhn[2] introduced a wavelength-dependent quantity, g, the dissymmetry factor, which is defined as:

$$g = \frac{2.\Delta\varepsilon}{\varepsilon} = \frac{2(\varepsilon_L - \varepsilon_R)}{(\varepsilon_L + \varepsilon_R)}$$

or

$$g' = \frac{4R}{D}$$

where g' is a number characteristic of a complete absorption band, R and D being the rotational and dipole strengths, respectively. The former is determined from the CD spectrum and the latter from the absorption spectrum. Hence.

$$g' = \frac{4\mu m \cos\theta}{\mu^2} \approx \frac{m}{\mu}$$

Thus the anisotropy factor is of the order of magnitude of the ratio of the magnetic dipole to the electric dipole transition moments. Bands due to transitions which are both magnetic- and electric-dipole allowed, that is Class I molecules, have $g' \approx 10^{-2}$ whereas for molecules of Class II, $g' < 10^{-2}$; g' is usually about 10^{-4}, for a transition which is electric-dipole allowed but magnetic-dipole forbidden. However the value of g' can be between 10^{-2} and 10^{-1} if the transition is magnetic-dipole allowed and electric-dipole forbidden.

2.1.1 Helical Conjugated Molecules – Class I

An example of this class is a twisted diene, Fig. 1[3]. The circular dichroism arises from an electric dipole transition within a π-electron system which is sterically distorted into a helical conformation. A common case would be a steroid diene. The allowed electric dipole transitions are polarised either along the length of the diene (y-polarised) or along the short axis (x-polarised). A twist in the diene chain causes the generation of a magnetic dipole moment along the y-axis and the x-axis. Hence both the long and the short axis polarised transitions gain rotational strength because they have collinear electric and magnetic dipole moments.

The 'cisoid diene' rule states that a cisoid diene skewed in the sense of a right hand helix, (Fig. 1) produces a positive circular dichroism band under the long axis polarized transition, that is, the transition of longest wavelength. The short axis polarized transition gives rise to a negative circular dichroism band for the

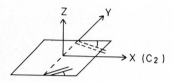

Fig. 1. Butadiene skewed as a *cisoid* diene in the sense of a right-handed helix. The lower half shows the coordinate set referred to in the text

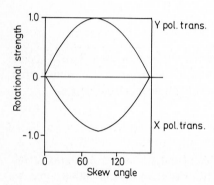

Fig. 2. The rotational strengths of the two electronic transitions of butadiene as a function of skew angle

same sense of helix. This rule has been verified for a large number of diene compounds. Moreover, it has proved possible to relate the magnitude of the experimental rotational strength to the skew angle of the diene, Fig. 2 [4]. The validity of this relation for skew angles greater than 25° is not established.

2.1.2   n-π* Band of Carbonyl Compounds-Class II [5].

The n-π* transition of the carbonyl chromophore involves a rotation of charge due to the transition from the oxygen lone-pair p_y orbital, to a π-anti-bonding orbital, comprising a linear combination of carbon and oxygen p_x orbitals, (Fig. 3.) By the right-hand rule this leads to a magnetic dipole moment along the z-axis. But the electric dipole moment in this direction is zero to a first approximation. Some electric dipole intensity in the z-direction can be 'stolen' from higher energy transitions such as the π-π* transition which is z-polarized or a σ-σ* transition of the C = O band. The mechanism of borrowing may be vibrational. However, this does not introduce optical activity because the carbonyl chromophore possesses a plane of symmetry. On the other hand, if it is surrounded by groups which are chirally arranged or if at least one of the groups

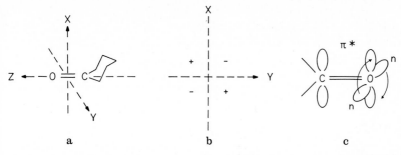

Fig. 3(a). The nodal planes xy and yz of the π^* orbitals. **(b)** The sign of the rotatory power induced in the 300 nm carbonyl absorption by substituents, other than fluorine, placed in the four rear octants defined by the nodal planes. **(c)** the n and π^* orbitals

has an asymmetric centre then optical activity is induced. The disposition of this centre determines the sign of the CD spectrum. Sets of empirical rules, called octant rules, have been devised to relate the sign of the CD band to the disposition of the optically active centre [6].

This illustrates a general phenomenon. A chromophore which has no intrinsic optical activity will invariably become active when wrapped in a chiral environment. This environment could be a protein polypeptide chain, which is of course optically active, or it could be a set of asymmetric centres. The derivation of configurational information from such second-order effects is problematic. In all cases it has been necessary to proceed empirically using molecules of known structure for calibration.

2.1.3 Coupled Oscillators [7].

Two or more chromophores can interact with one another, even though they are not conjugated, when one is excited. The dipole created by the excitation of a transition on one chromophore can interact with another dipole. Expressed in another way the excitation energy becomes delocalised over two or more centres. The simplest form of the interaction is of the dipole-dipole type. If the interacting chromophores are arranged to form a chiral array then a CD spectrum results which can take a simple and interesting form.

Consider the example of a chiral dibenzoate (Fig. 4) 1,2-glycol benzoate [8]. The benzoate chromophores have a very intense absorption band at 230 nm, corresponding to a "charge-transfer transition" between the benzene ring and the COOR moiety. The electric dipole moment μ, of this transition lies approximately parallel to the bond between the benzene ring and the carbon atom of the ester moiety. In the preferred conformation shown in Fig. 4, the electric dipole transition moments are inclined at an angle. Thus there are two possible combinations of these moments called A and B. The resultant electric dipole moments μ_A and μ_B are shown together with the resultant magnetic dipole

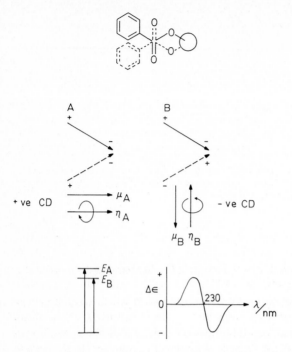

Fig. 4. Coupled oscillators. The chiral dibenzoate chromophore. Two possible combinations A and B of the mutually inclined transition moments are shown giving rise to parallel and anti-parallel electric (μ) and magnetic (**m**) transition moments, respectively. The relative energies, E_A and E_B, of the two resulting excited states are also given. The resulting CD spectrum consists of a couplet of negative sign [Reproduced with permission from Angew. Chem. Int. Ed. Engl. (1979) 18: 363]

moments, $\mathbf{m}_A$ and $\mathbf{m}_B$. The charge rotation created by the combined action of the two partial moments corresponds to a right-hand helix in arrangement A and a left-hand helix in the case of B. The charge rotation can be recognised by envisaging the axis of the resultant electric dipole as a cylinder and the individual components of μ as tangents to this cylinder. In case A the magnetic and electric moments are parallel and the resulting CD is positive, whereas arrangement B gives rise to anti-parallel moments and therefore to a negative CD. Moreover, for arrangement A the induced electric dipoles have their heads close together and thus the dipolar interaction is repulsive, whereas in case B the dipoles are arranged head-to-tail, an attractive interaction. Therefore the excited state due to the arrangement A is at a higher energy, E_A, than the excited state due to case B. The resulting CD spectrum is called a positive couplet. This type of sigmoid curve is characteristic of coupled oscillators and is usually rather intense. A more detailed analysis can give the distance and the angles between the coupled oscillators since the interaction is of the dipolar form, depending upon $(3\cos^2\phi - 1)/R^3$ where R is the distance between the centres of the two dipoles and ϕ the angle between the dipole moments of the interacting oscillators. Such an analysis has been used to determine the relative dispositions of the chromophores in a membrane protein bacteriorhodopsin which contains three retinyl polyene chains interacting in a chiral array [9].

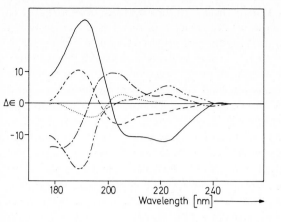

Fig. 5. Spectra for five major secondary structures from 178 to 260 nm: α-helix (——), antiparallel β-sheet (---), parallel β-sheet (-•-•), β-turn (-••-), another (random) structure (•••) [Reproduced with permission from Anal. Biochem (1986) 155: 155]

2.1.4 Proteins

CD spectroscopy between 180-250 nm has been shown to be sensitive to the secondary structure of globular proteins. This region of the optical absorption spectrum contains bands arising from the n-π* transitions of the amide group. The relative disposition of the amide groups is different in an α-helix and a β-sheet. CD arises from the exciton coupling of the transitions of the individual amide molecules [10] and is therefore sensitive to the secondary structure. Figure 5 shows the CD spectra of five major secondary structures, namely the α-helix, the antiparallel β-sheet, the parallel β-sheet, the β-turn and another random structure. The spectra are quite distinctive. However, problems arise in the analysis of the CD spectrum of a protein which may contain contributions from some or all of these structural motifs. One technique has been to use curve-fitting analysis of protein CD spectra assuming a linear relationship between the spectrum of interest and a set of CD spectra corresponding to synthetic polypeptides in known conformation. The validity of such methods has been in doubt. Synthetic polypeptides do not resemble real proteins whose structural elements may differ substantially from ideal models. For example, the intensity of the helical CD spectrum is chain-length dependent. In addition there may be contributions from disulphide bonds and aromatic amino acids. Recent methods analyse protein CD spectra using matrix techniques that automatically consider all parameters, whether recognised or not, avoiding the instability normally associated with large amounts of noisy data. The paper by Compton and Johnson [11] describes in detail the methodology and evaluates its success.

2.2 Vibrational Optical Activity

One of the most rapidly developing branches of the subject is vibrational optical activity (VOA), comprising both vibrational circular dichroism (VCD) and

Raman optical activity (ROA). VCD measured the difference in absorbance of left- and right- circularly polarised infrared light in the region of vibrational absorption bands of optically active molecules. ROA measures the difference in scattered intensity of left- and right-circularly polarised incident laser radiation. Vibrational optical activity is becoming a powerful tool in determining stereochemistry, both conformation and absolute configuration, in chiral molecules. Unlike electronic CD spectroscopy which provides information only about chromophores and their immediate environments, every part of a molecule can contribute to a vibrational CD spectrum. It should be possible to determine both absolute configuration and conformation from the VCD or ROA alone. The primary experimental difficulty is that VOA is very weak, signals being four or five orders of magnitude smaller than the parent effects, that is, infrared absorption and Raman scattering. However within the last ten years great progress has been made in both areas [12–16].

The highest and lowest frequencies at which VCD has been reported using dispersive instruments are $\sim 6000\,cm^{-1}$ and $900\,cm^{-1}$, respectively. Over this range the sensitivity limit in terms of absorbance $A = \varepsilon cl$ (where c is concentration and l the path-length) can be $\Delta A = 10^{-5}$–10^{-6} at a bandwidth of 5–$10\,cm^{-1}$, sufficient to resolve most room temperature liquid-phase VCD spectra. The low-frequency limit of VCD measurements using FT instrumentation is now $\sim 600\,cm^{-1}$. For ROA no such frequency limits exist although the largest effects commonly occur at frequencies below $1000\,cm^{-1}$. Furthermore, the difficulty of obtaining measurable signals in ROA has limited experiments to very concentrated samples, either pure liquids or saturated solutions, where intermolecular effects may be dominating. VCD on the other hand has been measured on sample concentrations of 0.1–0.01 M.

2.2.1   Vibrational Circular Dichroism (VCD)

The observation of optical activity in molecular vibrational transitions presents considerable experimental difficulty since the effects are much smaller than those for electronic optical activity. In the case of VCD the differential absorption between left- and right-circularly polarised infrared radiation is very small, but the absorption coefficients of infrared bands are generally comparable to those of electronic bands [13]. Typical values of $\Delta A/A$ (or $\Delta\varepsilon/\varepsilon$), for vibrational transitions are 10^{-5} to 10^{-4}. Again it is possible to define a dissymmetry factor, g, where for a vibrational transition from vibrational state i to f using Kuhn's definition given earlier

$$g_\phi = 4(\boldsymbol{\mu}_\phi \cdot \mathbf{m}_\phi)/\mu_\phi^2$$

where $\boldsymbol{\mu}_\phi$ and $\mathbf{m}_\phi$ are, respectively, the electric and magnetic dipole transition moments generated by the vibrational mode under consideration. Hence the formal similarity with electronic optical activity is clear.

Typical VCD and absorption spectra of the relatively simple molecule (*trans-*

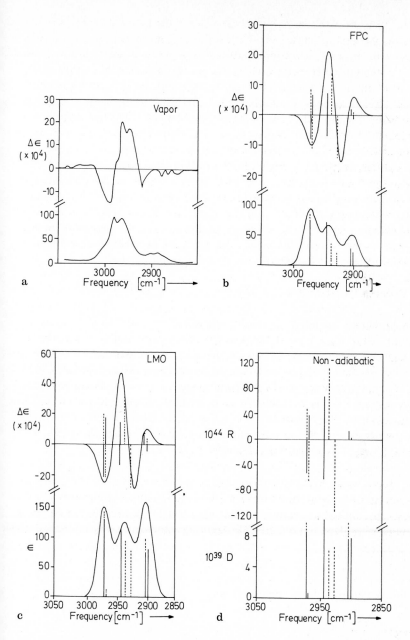

Fig. 6. (a) Absorption and VCD spectra of trans -$1R$, $2R$-$C_4H_6D_2$ in the C–H stretching region; **(b)**–**(d)** Absorption and VCD spectra of the same molecule calculated using different theoretical models; **(b)** fixed partial charge; **(c)** localised molecular orbital approach; **(d)** non-adiabatic theory. The relative dipole and rotational strengths are represented by vertical lines (——) for equatorial and (----) for axial conformers [Reproduced with permission from Ann. Rev. Phys. Chem. (1985) 36: 213]

$1,2\text{-}C_4H_6D_2$) from $3100\,cm^{-1}$ to $900\,cm^{-1}$ have been measured in both the solution and the vapour phase [14]. The results for both phases are qualitatively the same. This molecule has conformational flexibility and exists in two equi-energetic puckered conformations. This fact greatly increases the complexity of the spectrum and its interpretation. Figure 6 shows the experimental spectra in the C–H stretching region together the results of using several different theoretical methods to calculate the spectrum. Stephens [15] has described a general theory of VCD, the so-called non-adiabatic theory, permitting prediction of vibrational rotational strengths and spectra. There are also a variety of heuristic models including coupled oscillators, and fixed partial charges (FPC).

Following the demonstration the VCD spectra can be measured with high reliability, theoretical analysis has advanced considerably. The field is entering a phase of collecting spectra and making comparison with theoretical models. The experimental data require a special home-built instrument for measurement and the analysis is non-trivial. However, the prospect of making unambiguous stereochemical assignments of conformation is promising.

2.2.2 Raman Optical Activity (ROA)

The experimental quantity normally measured is the circular intensity differential, CID, $\Delta_\alpha = (I_\alpha^R - I_\alpha^L)/(I_\alpha^R + I_\alpha^L)$ where I_α^R and I_α^L are the α-components of the scattered intensities in right- and left-circularly polarised incident light. Referring to Fig. 7 one can define polarised and depolarised CID's, Δ_X and Δ_Z, corresponding to intensity components I_X and I_Z being measured through an analyser with its transmission axis perpendicular and parallel to the scattering plane yz. The measurement of the parallel CID is much freer of artifact than is the perpendicular and most effort has concentrated on the former. The first reliable out-of-plane CID spectra were published in 1979 [17].

An example is given in Fig. 8 of the in-plane and out-of-plane polarised Raman spectra of $(-)\text{-}(R)$-3-methyl-1-indanone. Also shown are the polarised Raman spectra. Note that these workers define a differential scattering as $(I^L - I^R)$, a convention adopted to conform to that used in CD spectroscopy. It appears that, almost without exception, the CID bands have the same signs in both the in-

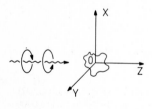

Fig. 7. Geometry for polarized light scattering in ROA

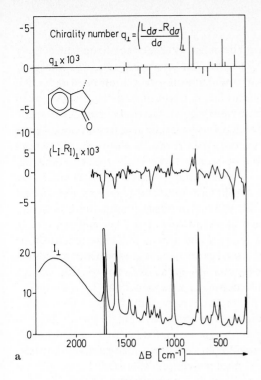

Fig. 8(a). The polarised Raman and CID Raman spectrum of $(-)$-(R)-3-methyl-1-indanone; data not corrected for instrument response

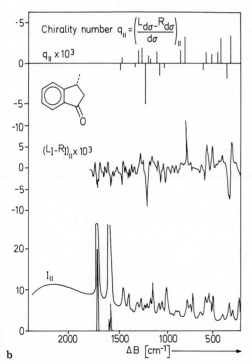

Fig. 8(b). The depolarised Raman and CID Raman spectrum of $(-)$-(R)-3-methyl-1-indanone; data not corrected for instrument response [Reproduced with permission from Tetrahedron (1978) 34: 607]

plane and out-of-plane directions. The most intense Raman optical activity appears to come from bands which are depolarised or only slightly polarised.

The acquisition of data has been the main pre-occupation of workers and much less effort has gone into interpretation [18]. However, spectra are examined for exciton couplets showing a positive and a negative band under a single vibrational peak [19]. These are predicted to be strong signals and possibly the most easy to interpret. Indeed this proves to be the case. Such couplets have been found in the depolarised Raman CID spectrum of tartaric acid and its derivatives. A large couplet is seen at $\sim 500\,\mathrm{cm}^{-1}$ in the depolarized Raman CID of $(2R,3R)$ $(+)$-, $(2S,3S)(-)$-tartaric acid and in $(2R,3R)(+)$-dimethyl tartrate [18]. These bands have been assigned to in-phase and out-of-phase combinations of the in-plane C—C—O deformations. Each of the isomers of tartaric acid can exist in three possible staggered conformations, 1, 2 and 3, (Fig. 9). The simplicity of the Raman spectrum of each of the isomers suggests that one conformer dominates at room temperature. NMR data indicate that species 2 is not present in high concentration, but cannot distinguish between conformers 1 and 3. Conformers 2 and 3 constitute a twisted, highly chiral arrangement of the O—C—C—O unit. Coupling between the in-pase and out-of-phase combinations of the two deformations is expected to generate a Raman optical activity couplet from which the absolute configuration can be deduced if the in-phase and out-of-phase bands can be identified. Conformer 1 does not contain a twisted O—C—C—O unit and is therefore not expected to yield an intense Raman CID.

There are many difficulties to be resolved before this interpretation can be accepted unreservedly. First, it has to be demonstrated that the signals observed are not due to inter-molecular effects. VCD work here suggests great caution is required. Secondly, there is a need for careful assignment of the vibrational modes of such complex molecules, possibly with a full normal co-ordinate analysis being required before the detailed interpretation is accepted. For example, it is possible that the C—C—C—C deformations also lie in the same spectral region.

However, this example brings out the potential of the complementary techniques of VCD and ROA, if good spectra on dilute samples can be obtained over a wide wavelength range. The sensitivity of VOA to configuration and conformation has now been demonstrated by the growing body of experimental data, although the translation of these results into molecular information is still a formidable task.

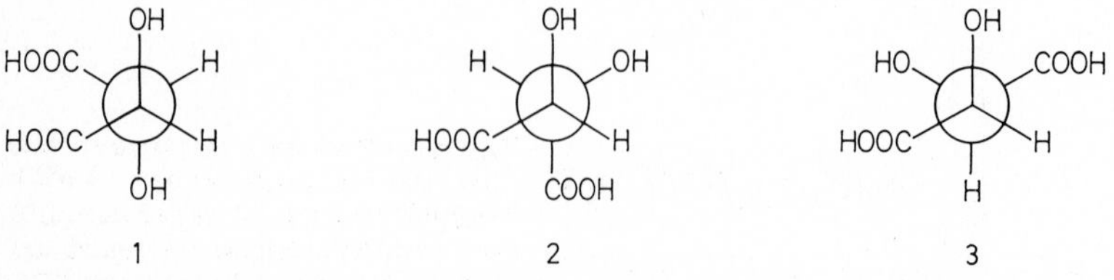

Fig. 9. The three staggered conformers of the $(2R,3R)(+)$ isomer of tartaric acid

3   Magnetic Circular Dichroism (MCD)

Circular dichroism is induced in all matter by a uniform magnetic field applied parallel to the direction of propagation of the measuring light beam. Although phenomenologically similar to natural CD, the molecular origin of the effect is different. Both MCD and CD can be present in an optically active molecule in a magnetic field. The two effects are additive. Within the last 20 years much experimental MCD data have been collected and the theoretical analysis is well founded. MCD spectroscopy presents a host of interesting applications in chemistry, physics and biochemistry [20]. The origin of the effect depends upon two facts.

(a) Degenerate electronic states are split in the presence of a magnetic field (the first-order Zeeman effect) to yield a set of sub-level called Zeeman components. All states are mixed together by an applied magnetic field (the second-order Zeeman effect).

(b) Electronic transitions from the Zeeman sub-levels of the ground state to

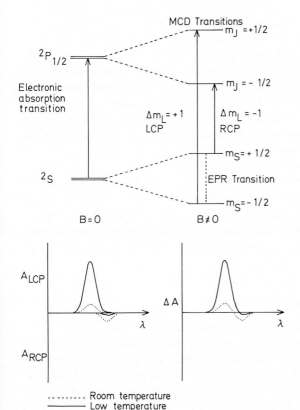

Fig. 10. The origin of the MCD in the atomic transition $^2S \rightarrow {}^2P$. The upper figure shows the allowed optical transitions between the Zeeman sublevels of the ground and excited state. The lower part illustrates the MCD intensity resulting from these transitions

those of an excited state are circularly polarised if the magnetic field is parallel (or anti-parallel) to the direction of the light beam.

The origin of MCD can be illustrated by the following example. Consider an electronic transition from a ^2S ground state, with spin S = 1/2 and zero orbital moment, to a $^2P_{1/2}$ excited state as shown in Fig. 10. When a magnetic field is applied the degeneracies are lifted by the Zeeman effect. The selection rules for electric-dipole allowed transitions between the Zeeman sub-levels of the ground and excited states are $\Delta M_L = +1$ for the absorption of left circularly polarised (LCP) light and $\Delta M_L = -1$ for right circularly polarised light (RCP): $\Delta M_S = 0$. Thus the left- and right-circularly polarised photons impart angular moments of opposite signs to the system. At temperatures of 300 K the two components of the ^2S ground state are almost equally populated since the splitting $g\beta B \ll kT$. The two allowed transitions will occur with equal intensity as shown by the dotted traces on Fig. 10. Also shown is the corresponding measured quantity ΔA ($= A_L - A_R$). When the temperature is lowered population is frozen into the lower component of the ground state and the LCP transition gains intensity at the expense of the RCP intensity. Therefore the MCD intensity, ΔA, of a paramagnet is temperature- (and field-) dependent, increasing in intensity as the temperature is lowered. This contribution to the MCD intensity is referred to as the C-term. Temperature-independent contributions, the so-called A- and B-terms, are described elsewhere [21,22]. At temperatures between 1.5–100 K the C-terms are usually dominant and much stronger than the A- and B-terms. Since these signals are due only to paramagnetic centres the MCD provides a method of directly probing the metal centres of proteins, for example, with little interference from the rest of the molecule. Note that ΔA is also a signed quantity.

The temperature- and field-dependence of the MCD spectrum carries information about the ground state magnetic parameters such as the spin, S, the g-value, and the zero-field splitting parameters in the case that S > 1/2. This information is collected in the form of a MCD magnetisation curve, constructed by plotting the relative intensity of the MCD signal at a fixed wavelength against the parameter $\beta B/2kT$, where β is the Bohr magneton, B the applied magnetic induction field, k Boltzmann's constant and T the absolute temperature (Fig. 11). In the simple case of a spin S = 1/2 ground state where no zero-field splittings are possible and when the g-value is isotropic the latter can be estimated from the so-called intercept value, I, where

$$I = \frac{\text{(asymptotic magnetisation value)}}{\text{(gradient of initial linear part of curve)}}$$

Then $g_{iso} = 1/I$. For an axially distorted system $g_{//} = 2/3I$ [23].

The power of the method for selectively probing the magnetic properties of individual metal centres within a complex protein can be illustrated by means of cytochrome c oxidase (CCo) [24]. Bovine CCO is the terminal oxidase of the respiratory chain which accepts electrons from cytochrome c, a soluble electron carrier, and passes them to O_2 for reduction to water. The enzyme CCO is a

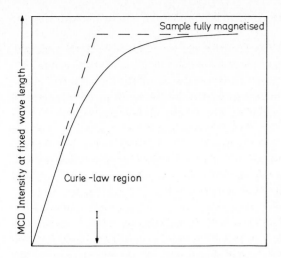

Fig. 11. The dependence of the MCD intensity on the applied magnetic field, B, and the absolute temperature, T, for the case of a ground state with a simple spin S = 1/2. This constitutes a MCD magnetisation curve[23]

membrane-spanning protein of great complexity both in structure and function. The protein comprises approx. 13 sub-units and at least four redox-active metal sites. The latter comprise two centres, cytochrome *a* and copper (Cu_A), which serve to transfer electrons to a special pair consisting of a haem group, cytochrome a_3, and a copper ion (Cu_B). In the oxidised, resting state of the enzyme the magnetic properties of cytochrome *a* and Cu_A^{2+} are detectable by means of EPR signals. Hence the ground state g-values of these two centres are known. By contrast the two metal ions thought to be closer together are not readily detectable by EPR spectroscopy. This pair is the site of di-oxygen reduction when

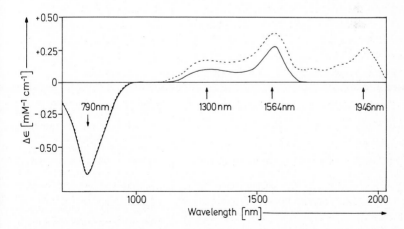

Fig. 12. Near infrared MCD spectra of bovine cytochrome c oxidase in the oxidised, resting state (——) and as the oxidised cyanide inhibited form (----) measured at 4.2 K and 5 Tesla. The *arrows* indicate the wavelengths at which MCD magnetisation curves were measured [Reproduced with permission from Biochem. J (1982) 207: 167]

the enzyme is fully reduced. Thus, there is considerable interest in knowing the structure of this centre.

Figure 12 shows the MCD spectrum at 4.2 K and 5 Tesla of the oxidised, resting and the CN^-, inhibited, forms of the enzyme in the wavelength range 700–2000 nm [25]. Clearly there are optical absorption bands throughout the region. These transitions belong to the metal centres and arise from the presence of unpaired electrons on the varied metals. The identity of these transitions is readily determined by monitoring the MCD magnetisation characteristics at the positions indicated by the arrows. Figure 13 shows the results. It is clear that the optical bands at 790 nm, 1564 nm and 1946 nm belong to paramagnets with different ground state properties because the magnetisation curves do not overlay. Furthermore, since the ground state g-values of cytochrome a and Cu_A are known from EPR studies the theoretical MCD magnetisation curves can be simulated. The results show that the band at 790 nm belongs to Cu_A^{2+} and that at 1564 nm to ferri-cytochrome a. This leaves the band at 1946 nm unassigned, and hence it must belong to the only other paramagnet in the protein, namely, the cytochrome a_3-Cu_B pair. This represents the first umabiguous identification of an optical band from this pair. Furthermore, the form of the magnetisation curve gives the magnetic properties of the ground state. The analysis shows that the centre in the CN^- inhibited form consists of a cyanide-bridged haem and Cu_B^{2+}

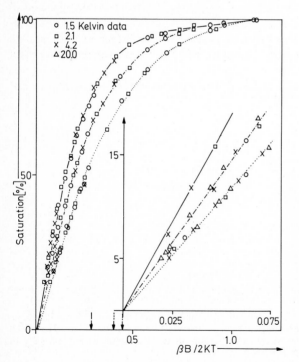

Fig. 13. A comparison of the MCD magnetisation curves of oxidised, cyanide inhibited bovine cytochrome c oxidase measured at 790 nm (••••), 1564 nm (-•-•) and 1946 nm (——) at temperatures of 1.5, 2.1, 4.2 and 20 K and fields between 0–5 Tesla. The *lines* are the theoretically simulated curves for Cu_A^{2+} (••••) and cytochrome a (-•-•). The *arrows* indicate the intercept values, I [Reproduced with permission from Biochem. J. (1982) 207: 162]

ion in a linear array, viz. $Fe^{III} - C \doteq N - Cu^{2+}$. The cyanide ion mediates ferromagnetic coupling of the spins on Fe^{III} and Cu^2. Since O_2 has virtually the same bond length as CN^- and the latter is a competitive inhibitor of di-oxygen reduction it can be assumed that di-oxygen becomes bound between a haem group and copper ion before it is reduced to water in a four-electron, four-proton process.

The use of an optical probe of paramagnetism greatly assists in deconvolution of the properties of individual metal centes and allows the ground state magnetic parameters to be determined without interference from impurities. This high degree of selectivity is a major advantage of the technique. It has now been applied to examples of metalloproteins containing iron-sulphur clusters, nickel, vanadium and molybdenum, as well as non-haem iron.

Recently a novel double resonance technique was reported in which the MCD signal is used to detect microwave paramagnetic resonance in the ground state [26]. The technique called paramagnetic resonance by optical detection (PROD) promises to allow even greater selectivity to be introduced into the technique.

4   References

1. Moscowitz A (1961) Tetrahedron 13: 48
2. Kuhn W (1930) Trans. Far. Soc. 46: 293
3. Weiss U, Ziffer H, Charney E (1965) Tetrahedron 21: 3105
4. Charney E (1965) Tetrahedron 21: 3127
5. Snätzke G (1979) Angew. Chem. Int. Ed. Engl. 18: 363
6. Moffitt W, Woodward RB, Moscowitz A, Klyne W, Djerassi C (1961) J. Amer. Chem. Soc. 80: 4013
7. Mason SF (1963) Quart. Rev. 17: 20
8. Harada N, Lai Chen S-M, Nakanishi K (1975) J. Amer. Chem. Soc. 97: 5354; Koreeda M, Harada N, Nakanishi K (1974) J. Amer. Chem. Soc. 95: 266
9. Kriebel AN, Albrecht AC (1976) J. Chem. Phys. 65: 4575
10. Moffitt W (1956) J. Chem. Phys. 25: 467
11. Compton LA, Johnson WC (1986) Anal. Biochem. 155: 155
12. Osborne GO, Cheng JC, Stephens PJ (1973) Rev. Sci. Inst. 44: 10
13. Nafie LA, Keiderling TA, Stephens PJ (1976) J. Amer. Chem. Soc. 98: 2715
14. Annamalai A, Keiderling TA (1984) J. Amer. Chem.Soc. 106: 6254
15. Stephens PJ, Lowe MA (1985) Ann. Rev. Phys. Chem. 36: 213
16. Nafie LA, Diem M (1979) Acc. Chem. Res. 12: 296
17. Hug W, Surbeck H (1979) Chem. Phys. Lett. 60: 186
18. Barron LD (1978) Tetrahedron 34: 607
19. Barron LD (1977) J. Chem. Soc., Perkin II: 1074
20. Piepho S, Schatz PN (1983) Group theory in spectroscopy, Wiley, New York
21. Schatz PN, McCaffery AJ (1969) Quart. Rev. 23: 552
22. Stephens PJ (1976) Adv. Chem. Phys. 35: 197
23. Thomson AJ, Johnson MK (1980) Biochem. J. 191: 411
24. Wikström M, Krab K, Saraste M (1981) 'Cytochrome c oxidase. A synthesis', Academic, London
25. Thomson AJ, Eglinton DG, Hill BC, Greenwood C (1982) Biochem. J. 207: 167
26. Barrett CP, Peterson J, Greenwood C, Thomson AJ (1986) J. Amer. Chem. Soc., 108: 3170

CHAPTER 11
Mass Spectrometry

C.S. Creaser and F.A. Mellon

1 History and Introduction

Mass spectrometry evolved out of researches into particle physics in the late 19th century. J.J. Thomson built the parabola mass spectrograph, generally considered to be the first mass spectrometer worthy of the name, in the first decade of the twentieth century. He went on to recognise negative ions, isotopes, "metastable" ions and ion-molecule reactions, and also predicted the analytical possibilities of the technique [1,2]. Aston built on this work and constructed instruments which helped to establish the presence of isotopes in most elements in the periodic table, at the same time measuring the "accurate masses" of many simple nuclei. [3]. Between 1940 and the early 1950s, the number of mass spectrometers throughout the world increased, according to A.O. Nier, a pioneer in the field, from "probably fewer than a dozen...[to] many hundreds" [4]. Many early applications of mass spectrometry were confined to petroleum chemistry (with a few biological, nuclear and geological applications). However, by the late 1950s the potential of the technique in general organic chemistry had been realised. A crucial development was the combination of gas chromatography and mass spectrometry (GC/MS), which took place in the mid 1960s. This produced a very rapid expansion in the application of mass spectrometry to chemical and biochemical studies. A further development was the routine use of computers to supervise the acquisition and processing of data, which GC/MS instruments produce in enormous amounts. The 1960s also saw the introduction of chemical ionisation (CI) followed in the 1970s and early 1980s by desorption chemical ionisation (DCI), field desorption (FD) and fast atom bombardment (FAB). These techniques helped to extend the range of applications of mass spectrometry, yielding information about the molecular weights and structures of thermally unstable and involatile compunds. Recent innovations include: coupled liquid chromatography/mass spectrometry, tandem mass spectrometry, supercritical fluid chromatography/mass spectrometry, laser desorption, and the development of a diverse range of instruments, from small benchtop GC/MS systems to large, multiple-analyser devices. This arsenal of technology is capable of tackling such a wide variety of problems that today virtually all types of material can be analysed with the aid of mass spectrometry.

Mass spectrometry is unlike the other forms of spectrometry described in this book, in that it is not concerned with non-destructive interactions between molecules and electromagnetic radiation. Instead, it involves the production and separation of ionised molecules and their ionic decomposition products. It is therefore a destructive technique; samples are consumed by the mass spectrometer and cannot be recovered at the end of the experiment. This is of little consequence in practice because mass spectrometry is one of the most sensitive of spectroscopic techniques and very little sample (usually nanomoles to picomoles) is required to obtain a spectrum. In favourable cases, for example if small quantities of a known compound are to be analysed by GC/MS, meticulous work can yield detection limits in the femtomole to attomole range.

2   Ionisation Methods

2.1   Electron Ionisation (EI)

In electron ionisation, the most common ionisation mode, the sample is vapourised under vacuum and the vapour introduced into the ion source, where it is bombarded by electrons accelerated from a tungsten or rhenium filament. Interactions between electrons and molecules result in the removal of an electron from the molecule to form a positively charged radical cation, the molecular ion. The efficiency of this process is very low ($< 0.1\%$),

$$M + e^- \rightarrow M^{+\cdot} + 2e^-$$
(70 eV)   Molecular ion

The relationship between the electron energy and the degree of ionisation is shown in Fig. 1. Electrons with translational energies below the first ionisation

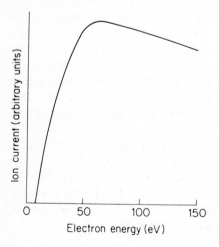

Fig. 1. Plot of Total Ion Current versus electro energy (reproduced from Chapman JR (1985) Practical Organic Mass Spectrometry, Wiley-Interscience Chichester, with permission)

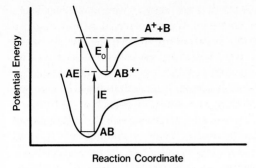

Fig. 2. Vertical transitions leading to ionisation and dissociation of a diatomic molecule AB

potential of the sample molecule (typically 8–12 eV for organic molecules) have insufficient energy to ionise the sample, but above the ionisation potential the ion current rises with increasing electron energy until a plateau region is reached. By convention, 70 eV electrons are normally used for electron ionisation, because of the better reproducibility and optimum ion yield at this energy. The formation of a radical cation is the dominant ionisation process, but multiply charged and negatively charged ions are also produced in low abundance.

Electron ionisation is very rapid ($< 10^{-15}$ s) compared to the time required for bond vibration, so the ionisation process obeys the Franck-Condon principle. This is illustrated for a diatomic molecule by the vertical transition between the potential energy curves of the ground state and the ionised state of the molecule in Fig. 2. These transitions lead to the formation of molecular ions in electronically and vibrationally excited states. In practice the electronic energy is rapidly converted to vibrational energy, thus producing a population of ions with a range internal energies immediately following ionisation. The probability, P(E), of an ion having an energy E, depends on the amount of energy transferred by each electron-molecule interaction and a typical distribution of internal energies is shown in Fig. 3. The maximum internal energy (E_{max}) which an ion can possess is given by:

$$E_{max} = E_{el} + E_{th} - IE \qquad (1)$$

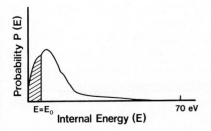

Fig. 3. Internal energy distribution of a molecular ion. The shaded portion indicates ions which have insufficient energy to decompose

where E_{el} is the electron energy, E_{th} is the thermal energy of the molecule prior to ionisation and IE is the ionisation energy of the molecule. It is important to emphasise that this is not a Boltzmann distribution, since at the low operating pressures in the ion source (10^{-5} to 10^{-6} Torr) very few collisions occur and all fragmentation reactions are therefore unimolecular. Ions with insufficient energy ($< E_0$) to undergo the lowest energy unimolecular fragmentation will be detected as molecular ions. These are indicated by the shaded area in Fig. 3. However, those ions with internal energies in excess of E_0 will dissociate to yield fragment ions whose abundance is determined by several factors, including their internal energy. The appearance energy (AE) for a particular ion in the mass spectrum is determined by the ionisation energy of the neutral molecule and the activation energy for bond dissociation (Fig. 2).

In addition to the thermochemical control of unimolecular fragmentation, kinetic factors must also be considered. The rate constant for fragmentation increases with internal energy to a limiting rate of $k_{max} \approx 10^{13}\,s^{-1}$, the time required for bond vibration. Since the ions spend approximately 10^{-6} s in the source prior to acceleration into the analyser region of a magnetic sector mass spectrometer (see Sec. 3) only those reactions with rate constants in the range 10^6–$10^{13}\,s^{-1}$ will be observed as normal ions in the mass spectrum. Ions which dissociate with rate constants in the range 10^4–$10^6\,s^{-1}$ (within a field-free region of the flight path) may be detected as weak, diffuse metastable ions (m^*) at a point on the mass scale determined by the masses of the parent (m_1) and daughter (m_2) ions according to the relationship

$$m^* = m_2^2/m_1 \tag{2}$$

Metastable ions therefore provide evidence for a specific precursor/product relationship. However, these ions are usually excluded during computer processing of the mass spectrometric data.

The combination of the thermochemical and kinetic constraints on the ionisation process results in a characteristic fragmentation pattern, or fingerprint, for the molecule under investigation. The relative abundance of the ions in the mass spectrum depends on the nature of the molecule being ionised and the stabilities of the molecular ion and fragment ions formed during ionisation. Thus, the mass spectrum of naphthalene, which forms a stable radical cation, shows an intense molecular ion at m/z 128 and a few weak fragments at lower mass, whereas the spectrum of n-decane, which produces a less stable radical cation, contains a small molecular ion at m/z 132 and an abundance of fragments. The presence of weak molecular ions for many molecules, or their complete absence, means that unambiguous molecular weight information may be difficult to obtain under high energy (70 eV) electron ionisation. This limitation may sometimes be overcome by lowering the electron energy and thus the internal energy distribution of the molecular ions, but this is at the expense of absolute ion intensity (Fig. 1). The preferred alternative is to use a 'soft' ionisation technique such as chemical ionisation.

2.2   Chemical Ionisation (CI)

In chemical ionisation a reagent gas such as methane, isobutane or ammonia, is introduced into a high pressure source (0.1 to 1 Torr as opposed to 10^{-5} Torr in EI) and ionised by electron bombardment. The primary ions produced undergo ion-molecule collisions to form a stable population of secondary reagent ions. For example, the formation of the reagent ions CH_5^+ and $C_2H_5^+$ from methane occurs as follows:

$$CH_4 + e^- \rightarrow CH_4^{+\cdot} + 2e^-$$

$$CH_4^{+\cdot} \rightarrow CH_3^+ + H\cdot$$

$$CH_4^{+\cdot} + CH_4 \rightarrow CH_5^+ + CH_3$$

$$CH_3^+ + CH_4 \rightarrow C_2H_5^+ + H_2$$

Similarly, NH_4^+ and $C_4H_9^+$ reagent ions may be produced from ammonia and isobutane respectively.

The introduction of a small amount of sample into the CI source results in reactions between the reagent ions and the sample molecules. The most important of these reactions is the transfer of a proton from the reagent ion to the neutral molecule, to give a protonated molecular ion, MH^+, for the analyte.

$$\begin{array}{cccc} CH_5^+ & +M & \rightarrow MH^+ & +CH_4 \\ \text{Reagent} & \text{Sample} & \text{Protonated} & \\ & \text{molecule} & \text{molecular ion} & \end{array}$$

Other ion-molecule reactions which can occur are electrophilic addition,

$$NH_4^+ + M \rightarrow [M + NH_4]^+$$

and hydride abstraction,

$$C_2H_5^+ + C_nH_{n+2} \rightarrow C_nH_{n+1}^+ + C_2H_6$$

Hydride abstraction is usually observed for saturated hydrocarbons, giving a characteristic ion at one mass unit below the molecular weight.

The transfer of a proton between two species A and B by the reaction

$$AH^+ + B \overset{K}{\rightleftharpoons} A + BH^+$$

depends on the proton affinities (PA) of the two species, where the proton affinity is defined by:

$$PA(M) = \Delta H_f(M) + \Delta H_f(H^+) - \Delta H_f(MH^+) \tag{3}$$

The reaction occurs if the proton affinity of B is greater than that of A, i.e. the reaction is exothermic. The equilibrium constant, K, is related to the free energy change (ΔG^0) by the relationship, $-\Delta G^0 = RT \ln K$ and since for large molecules proton transfer involves negligible entropy change, the exothermicity of the reaction is given by:

$$\Delta H^0 = -RT \ln K = PA(B) - PA(A) \tag{4}$$

The excess energy appears as internal energy in the products of the reaction and may lead to fragmentation in the same way as in electron ionisation. However, the internal energy resulting from chemical ionisation is generally much less than that obtained by 70 eV electron ionisation. For example, for the protonation of trimethylamine (PA 938 kJ mol^{-1}) by ammonia reagent gas (PA 857 kJ mol^{-1}) the difference in proton affinities (ΔPA) is equivalent to 0.8 eV. CI spectra are therefore characterised by strong protonated molecular ions and few fragments. A further consequence of the chemical ionisation process is the possibility of selectively ionising some components of a mixture in preference to others, by a careful choice of the reagent gas. If, in the example above, both trimethylamine and isobutane (PA 823 kJ mol^{-1}) are introduced into the source in the presence of ammonia reagent gas, only the trimethylamine would be protonated because of the low proton affinity of isobutane.

Negative chemical ionisation relies on the production of reagent ions, such as OH$^-$ (from water) and NH$_2$$^-$ (from ammonia) which subsequently ionise sample molecules by proton abstraction and nucleophilic addition, in a manner analogous to positive ion CI. Radical anions (M$^-$·) may also be produced in a chemical ionisation source by the resonance capture of low energy thermal electrons. These thermal electrons are produced as a result of interactions between the primary electrons and a buffer gas such as CH$_4$, Ar, or CO$_2$.

$$M + e^- \text{(thermal)} \rightarrow M^-$$

This process is very efficient for molecules with high electron affinities, for example, those containing halogen atoms, and detection limits can be up to a thousand-fold better than in positive ion electron ionisation.

2.3 Desorption Ionisation

Electron ionisation (EI) and chemical ionisation (CI) are well established methods for the mass spectrometric analysis of many classes of organic compound. These techniques usually require the sample to be present in the ion source in the vapour phase, although their range has been extended to some less volatile compounds by the introduction of methods such as in-beam EI and desorption chemical ionisation (DCI) [5]. Many classes of organic compounds are nevertheless too involatile or thermally unstable to meet the requirements of EI and CI. Considerable effort has therefore been directed towards the development of "desorption" ionisation methods for involatile and thermally labile compounds.

2.3.1   Fast Atom Bombardment

Fast atom bombardment (FAB) [6] has rapidly established itself as the principal condensed phase ionisation method because of its experimental simplicity and wide range of applications. The analyte is dissolved in a viscous liquid, typically glycerol, and ionisation is achieved by bombardment of the sample matrix by a beam of atoms (or atoms and ions) with translational energies in the range 2–10 keV (Fig. 4). Argon or xenon atom/ion beams are most widely used because of the ease of handling these gases. The range of liquid matrices which may be used for FAB ionisation is extensive and can include many common solvents and low melting solids, if a cooled or heated FAB target is used [7, 8].

The success of FAB ionisation depends on the solubility and the chemical state of the sample in the solvent, as well as processes occurring at the surface of the matrix. Ions are generally produced by two separate mechanisms. Ionisation arises initially as a result of the sputtering of a mixture of preformed ions, fragments, clusters and small droplets from the surface layers of the sample/matrix solution by the impacting atoms or ions. Preformed ions ejected from the liquid matrix by the bombardment process are desolvated and transferred directly into the gas phase. A variety of procedures, following the rules of solution chemistry, may be used to increase the concentration of preformed ions in the matrix. This in turn leads to an enhancement of the abundance of these ions in the FAB spectrum. Figure 5 shows the positive ion FAB spectrum of a phosphonium ion produced by direct expulsion from the glycerol matrix. In the high pressure region above the surface of the matrix (the selvedge), ejected species may also undergo ion-molecule interactions, similar to those which occur during chemical ionisation, to yield a variety of quasimolecular and adduct ions, such as $[MH]^+$, $[M + Na]^+$ and $[2M + H]^+$. For example, the FAB spectrum of the cyclic peptide gramicidin S shows a protonated molecular ion at m/z 1141 (Fig. 6), which is typical of the behaviour of involatile peptides. These FAB spectra also exhibit characteristic fragment ions and so provide both molecular weight and structural information.

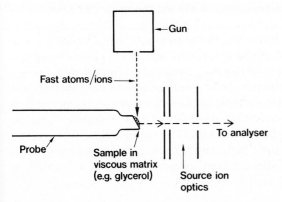

Fig. 4. Schematic representation of a Fast Atom Bombardment ion source

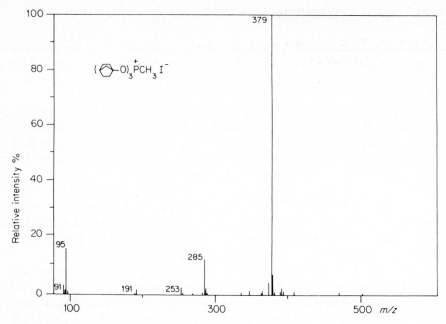

Fig. 5. Positive ion Fast Atom Bombardment spectrum of a Quasiphosphonium salt (reproduced from Org. Mass Spectrum. (1988) 23:148, with permission)

FAB has been successfully applied to the ionisation of high molecular weight samples of biological origin, and compounds with molecular weights well in excess of 5,000 Daltons may be routinely determined as a result of parallel developments in the available mass range of modern magnetic instruments. However, the technique is also widely used for the analysis of relatively low molecular weight (< 2000 Dalton) involatile, polar and ionic compounds, for which the sputtering processes appear to be highly efficient. At the other extreme, nutrient minerals including iron, calcium and zinc have been successfully determined by the FAB method [9].

A major drawback of conventional FAB ionisation is the problem of quantitative measurement. Since the primary FAB beam sputters only the surface layers of the sample matrix, differences in the surface activities of the various components which are present in real samples are reflected in the ion intensities observed in the FAB spectrum. FAB therefore samples the surface rather than the bulk concentration of the solutes present. In extreme cases, for example mixtures of hydrophilic and hydrophobic peptides, only the hydrophobic (surface active) peptides are observed. The difficulty of quantitative measurement may be overcome, in certain circumstances, by the use of internal standards, usually isotopically labelled compounds or closely related analogues [10]. An alternative approach is to carry out the FAB ionisation using a 'continuous flow' interface, in which the sample is carried to the probe tip, at flow rates of $1-10\,\mu$l per min, in

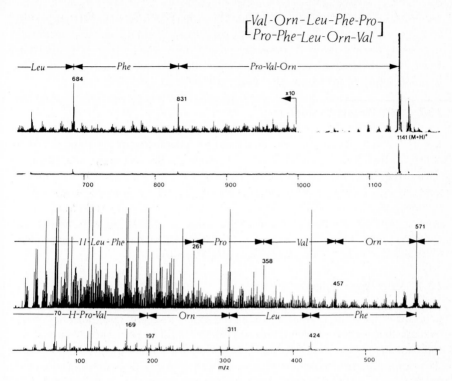

Fig. 6. Positive ion Fast Atom Bombardment spectrum of Gramicidin S (reproduced by permission of VG Analytical Ltd.)

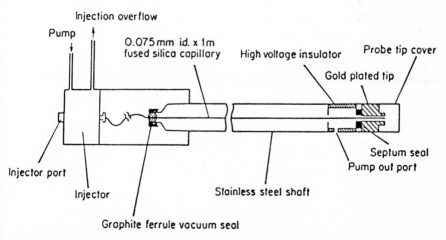

Fig. 7. Continuous flow Fast Atom Bombardment probe (reproduced by permission of Kratos Analytical Ltd.)

a suitable solvent (typically 5–10% glycerol in water) [11]. Surface effects appear to be less pronounced with continuous-flow FAB and good quantitative measurements may be carried out. A commercial continuous-flow FAB probe is shown in Fig. 7. This method of sample introduction has also been successfully used to interface liquid chromatography with FAB–MS (see Sect. 6.2).

2.3.2 ^{252}Cf Plasma Desorption

In ^{252}Cf plasma desorption (Fig. 8), the sample is deposited onto a thin foil which is held at high voltage ($\pm$ 10–20 kV) to accelerate any secondary ions formed. Fission fragments from a ^{252}Cf source, which have energies in the 100 MeV range, pass through the foil desorbing secondary ions. Secondary ions desorbed by this process are accelerated and mass measured, usually by a time-of-flight analyser (Sect. 3.4). The low energy of the desorption process, together with the absence of the mass range limit of a sector instrument, has yielded mass spectrometric data on molecules with masses in excess of 20,000 Dalton [12].

2.3.3   Laser Desorption

Laser desorption and ionisation is achieved by a high-power laser pulse, followed by mass analysis of the products using a time-of-flight or Fourier transform mass spectrometer. The pulsed operation of these instruments is ideally suited to the short ionisation timescale of laser desorption. High spatial resolution is possible because of the small diameter of the focused laser beam and the technique is

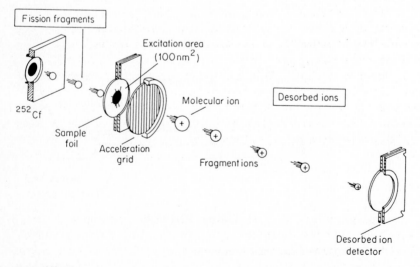

Fig. 8. ^{252}Cf Plasma Desorption probe (reproduced from Eur. Spectrosc. News (1988) 80: 13, with permission)

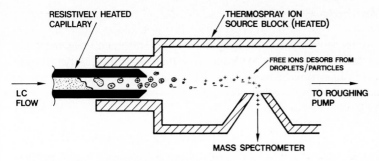

Fig. 9. Schematic of a Thermospray ion source

capable of providing information on selected areas of inhomogeneous samples. Multiphoton ionisation, using one or more lasers, has been shown to offer an unparalleled combination of selectivity and sensitivity [13, 14] (see Chap. 8).

2.3.4 Thermospray

The thermospray (TSP) ion source is usually associated with the interfacing of liquid chromatography to mass spectrometry, because of its ability to handle high solvent flow rates [15] (see Sec. 6.2.4). However, TSP is an ionisation technique in its own right. Sample dissolved in a flowing stream of aqueous (ionising) solvent is conducted through a resistively heated capillary tube into a specially designed ion source (Fig. 9). The temperature of the capillary is carefully controlled to achieve partial vapourisation of the solvent, producing a supersonic jet of vapour which contains an entrained mist of small, electrically charged, droplets. The charging arises statistically; no external electrical field is applied. The droplets shrink in size as they traverse the ion chamber (which is separately pumped to remove solvent vapour). Ionisation occurs by field-assisted ion evaporation from the small droplets. Ions leave the TSP source via a small orifice in a sampling cone and can be detected by either quadrupole or magnetic sector mass analysers. The ion evaporation effect is greatly enhanced by the presence of a volatile electrolyte (e.g. 0.1 M ammonium acetate) or if the sample itself is ionic. This type of ionisation often produces very stable molecular ions, and functions as a useful (positive or negative) soft ionisation technique.

3 Mass Analysis

There are four main types of mass analyser, magnetic, quadrupole, time-of-flight (ToF) and Fourier transform ion cyclotron resonance mass spectrometry (FTMS). Most commercially available instruments are either based on magnetic sectors or quadrupole mass filters. ToF and FTMS mass spectrometers are mainly confined to specialised applications.

3.1   Magnetic Sector Analyser

A schematic diagram of a single-focusing magnetic sector mass spectrometer is shown in Fig. 10. Ions are accelerated from the source by a voltage V (typically 2–10 kV) and gain translational energy equivalent to the drop in electrical potential energy experienced during acceleration:

$$zeV = \frac{1}{2}mv^2 \tag{5}$$

where v is the velocity of an ion of mass m, electronic charge e and number of charges z. The ions then coast through a 'field-free' region until they encounter a wedge-shaped magnetic field, whose force is exerted perpendicular to the direction of motion shown in the diagram. This causes the ion beam to curve through a circular path, in which the magnetic force balances the centrifugal forces:

$$Bzev = \frac{mv^2}{r} \tag{6}$$

(B is the magnetic induction field strength, r the radius of the ion path). Combining Eqs. (5) and (6) yields the relationship:

$$\frac{m}{z} = \frac{B^2r^2e}{2V} \tag{7}$$

Thus ions of different mass-to-charge ratio (m/z) can be induced to strike the collector by varying either B or V, i.e. by scanning either the magnetic field or the source accelerating voltage. Voltage scanning is generally used only for switching between peaks of known mass (for quantitative measurements), or for scanning

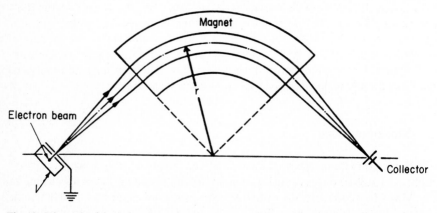

Fig. 10. Schematic of single-focusing magnetic sector mass spectrometer (reproduced from Chapman JR (1985) Practical Organic Mass Spectrometry, Wiley-Interscience, Chichester, with permission)

all the peaks in a limited range. Magnetic scanning is preferred because variation of the accelerating voltage changes the sensitivity and resolution of the mass spectrometer, whereas magnetic scanning yields a uniform, reproducible performance across the complete mass range.

In addition to acting as a mass separation device, the magnetic field also has a focusing action, akin to that of a light-focusing convex lens. Thus, any ions of the same mass which follow divergent paths when the leave the ion source, can be brought the same focal point at the collector (as shown in Fig. 10).

Mass spectrometric resolution (R) is defined by the minimum resolvable separation of two ion peaks of equal height and of masses m and m + Δm, and is given by:

$$R = \frac{m}{\Delta m} \tag{8}$$

where the overlapping peaks are separated by a valley which is 10% of the peak height. An alternative definition, which is used for quadrupole analysers, is based on the peak width measured at 50% peak height. In general, spectrometers are classified as low resolution (R < 2000), medium resolution (2000 < R < 6000) or high resolution (R > 6000).

The resolution of a single-focusing magnetic sector mass spectrometer is adjustable by varying the slit widths at the source and collector. The maximum resolution attainable is limited (in optimised designs) to about 6,000 (10% valley).

The main factor which limits the resolving power of single-focusing magnetic sector mass spectrometers is the spread in translational energies of ions of the

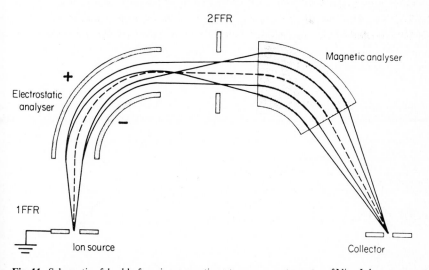

Fig. 11. Schematic of double-focusing magnetic sector mass spectrometer of Nier-Johnson geometry (reproduced from Chapman JR (1985) Practical Organic Mass Spectrometry, Wiley-Interscience, Chichester, with permission)

same mass. Increased resolution can be obtained by combining electric and magnetic sectors in "double-focusing" instruments, such as the Nier-Johnson design shown in Fig. 11 [16]. "Reverse geometry" instruments (in which the magnet precedes the electric sector) have also been developed successfully. The curved plates of the electric sector act as an energy filter, causing ions to traverse a circular path according to the equation:

$$R = 2V/E \tag{9}$$

where R is the radius of circle described by the ion beam, E the electric field strength and V the accelerating voltage.

The electric sector transmits all ions which have the same kinetic energy, irrespective of their mass. The electric and magnetic sectors are combined in such a way that the velocity dispersion in the two analysers is equal and opposite. This produces an instrument capable of yielding high resolution without an unacceptable loss of sensitivity. Maximum resolution of about 50,000 is typical of medium sized double-focusing instruments; larger designs can produce 100,000–200,000 resolution.

3.2   Quadrupole Analyser

The quadrupole mass analyser functions as a "mass filter" and is based on a completely different principle to magnetic sector instruments. Four parallel rods, with either circular or (ideally) hyperbolic cross-sections, are arranged in parallel and are connected to radiofrequency (RF) and direct current (DC) power supplies $V_0 \cos \omega t$ and U respectively. (Fig. 12). Opposing rods are electrically connected, and adjacent rods are oppositely charged. The motion of the ions inside the quadrupole filter is complex and the force acting on these ions is described by second order equations of the sort:

$$\frac{d^2 x}{dt^2} - \frac{2z}{mr_0^2} (U + V_0 \cos \omega t) x = 0 \tag{10}$$

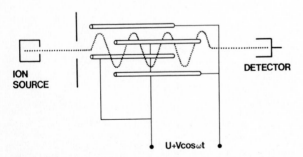

ION
SOURCE

DETECTOR

U+Vcosωt

Fig. 12. Schematic of a quadrupole mass spectrometer

Such equations are examples of the Mathieu type, for which the general form is

$$\frac{d^2x}{d\gamma^2} - (a_x + 2q_x \cos 2\gamma)x = 0 \qquad (11)$$

where γ is a time function $(\omega t/z)$. These have the following solutions for the coefficients a and q:

$$a = 8zeU/mr_0^2\omega^2 \qquad (12)$$

$$q = 4zeV/mr_0^2\omega^2 \qquad (13)$$

($2r_0$ is the spacing between rod faces). The parameters a and q are used to construct a stability diagram defining a space in which ions have a stable trajectory (Fig. 13). Masses inside the shaded area are transmitted and strike the detector; masses outside this region are lost, either striking the quadrupole rods or escaping between them.

Inside the stability region, the range of m/z values increases as a decreases, within the mass range of the particular quadrupole device, until all masses are transmitted at $a = 0$ (RF – only operation). Nearing the apex of the diagram (point X) fewer masses are transmitted, i.e. resolution effectively increases. An operating point close to the apex is chosen to give a good compromise of ion transmission and resolution. Under these conditions, resolving power is a simple

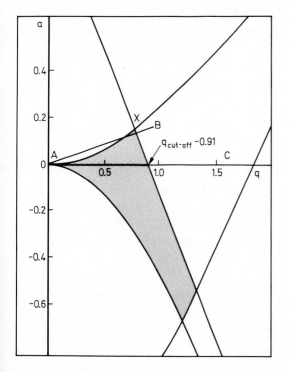

Fig. 13. Stability diagram for quadrupole and ion trap mass spectrometers (re-drawn from diagram supplied by Finnigan MAT Ltd., with permission)

multiple of mass (usually about 2 m) so the quadrupole filter is essentially a low resolution device. The mass spectrum is scanned along the line of mass-selective stability (scan line AB), sequentially bringing ions within the stable region near the apex of the stability diagram where they are transmitted to the detector, by varying U and V_0 so that the ratio U/V_0 remains constant. Because transmitted mass is linearly proportional to V_0, scan functions are easily determined.

Quadrupole analysers are relatively cheap to construct (compared to magnetic sector instruments), easy to use and can be made in very compact form, as in the "bench top" GC/MS instruments currently marketed by several manufacturers.

3.3   Ion Trap

Advances in ion trap technology have recently produced a compact and relatively inexpensive design of mass spectrometer [17]. The ion trap (Fig. 14) can be viewed as a solid of revolution of a two-dimensional quadrupole mass analyser. Its mode of operation can also be explained by referring to the a, q stability diagram (Fig. 13). In the simplest form of the device, an RF voltage is applied to the ring electrode and the end caps are earthed. Ions are formed inside the trap by a pulse of electrons from the filament and are all confined by operating in the stable region along the $a = 0$ line (Fig. 13). The mass spectrum is obtained by imposing conditions of mass-selective instability by increasing the q value of each ion to above the stability cut-off value $q = 0.91$ (scan line AC), so that

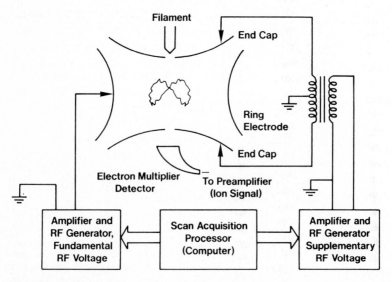

Fig. 14. Schematic of an Ion Trap mass spectrometer

ions are expelled sequentially through the end cap and are detected. This is in contrast to quadrupole operation, where ions following stable trajectories are detected. The presence of a damping gas at a pressure of about 10^{-3} Torr is essential for maintaining adequate mass resolution (the helium carrier gas from a capillary GC column is sufficient for this purpose). Chemical ionisation reactions are also possible [18]. Although CI reagent gases must be introduced at relatively low pressure, the CI plasma can be trapped for a lengthy period of time. This more than compensates for the reduced number of ion-molecule reactions, by increasing the plasma/sample contact time.

3.4   Time-of-Flight Analyser

Time-of-flight (ToF) mass spectrometry is based on measurement of the time required for an ion to travel from the source region to the detector (Fig. 15). A "packet" of ions, formed by a brief ionisation pulse, is directed by an accelerating voltage V into the field-free drift region. Since the ions have all gained the same kinetic energy ($1/2\, mv^2 = zeV$, where v is the ion velocity) during acceleration, they begin to separate according to mass as they drift along the flight tube, and ions of different mass arrive at the detector at different time, t, according to the equation

$$t = \left(\frac{m}{2zeV} \right)^{1/2} L \tag{14}$$

where L is the length of the drift region. Flight times are of the order of tens of μs in most ToF instruments, with mass separations of a few nanoseconds. Very fast detector electronics are essential in ToF mass spectrometry, and sophisticated data recording systems are also necessary. The ToF instrument has the capacity to record simultaneously all the ions produced in the mass

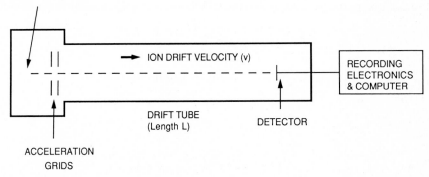

IONIZATION REGION
(Pulsed EI, FAB, SIMS,
Laser or ^{252}Cf)

ION DRIFT VELOCITY (v)

RECORDING
ELECTRONICS
& COMPUTER

DRIFT TUBE
(Length L)

DETECTOR

ACCELERATION
GRIDS

Fig. 15. Schematic of a Time-of-Flight mass spectrometer

spectrum, giving a "multichannel advantage". This contrasts with conventional scanning of quadrupole or magnetic sector instruments where only a fraction of the ions produced are detected at any one time. ToF spectrometers also have no inherent mass limit, in contrast to magnetic and quadrupole instruments.

The ToF mass analyser was once considered to be a rather esoteric device, with applications confined mainly to the study of fast reactions. It has recently gained a new lease of life because of its compatability with 'pulsed' ionisation techniques including pulsed lasers and ^{252}Cf plasma desorption sources. Early designs of ToF mass spectrometer were relatively low resolution instruments (typically about R ~ 500 at 50% valley). However, recent improvements comprising a flight tube bent into a "V" shape and an electrostatic mirror lens at the apex of the V to reflect ions, have produced resolutions of the order of 10,000 with laser desorption and ionisation methods [19]. These "reflection" or "reflectron" instruments have considerable potential in a variety of analytical applications (see Chap. 8).

3.5 Fourier Transform Mass Spectrometry

Fourier transform mass spectrometry (FTMS) [20] evolved from scanning ion cyclotron resonance (ICR) mass spectrometry and is often referred to in the literature as FT/ICR, a term which is interchangeable with FTMS. The principle of FTMS (Fig. 16) relies on the circular motion of ions perpendicular to a uniform magnetic field B. Each ion will have a characteristic frequency, ω, for its orbital motion, given by the cyclotron equation

$$\omega = \frac{zeB}{m} \tag{15}$$

Thus the cyclotron frequency of an ion is dependent on its m/z ratio. A pulse of ions is injected into the cell, which is usually cubic, located inside a high field

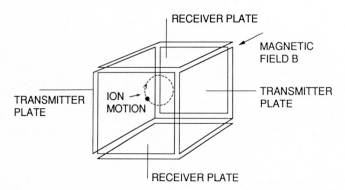

Fig. 16. Schematic of a Fourier Transform Ion Cyclotron Resonance mass spectrometer

conventional or superconducting magnet (fields of between 1 and 8 T are typical of most FTMS designs.) A small trapping voltage confines the ions inside the cell. An excitation "chirp", a broadband radio-frequency signal, is applied to the transmitter plates of the cell and excites the ions into larger orbits. The orbiting ions induce an "image current" (a small alternating current) in the receiver plates of the cell. This signal is converted into a voltage form and is digitised and stored by a data system. The complex FTMS signal contains frequency components characteristic of all the ions present in the cell. The "time domain" signal is processed by Fourier transformation and subsequently converted into a mass spectrum using the relationship given in Eq. (15).

FTMS instruments are capable of achieving very high resolving powers; resolutions well in excess of 1,000,000 have been attained (over a limited mass range). Resolution falls off with increasing mass, although a resolution of 150,000 has been reported at m/z 1130. The mass range of FTMS instruments is, according to Eq. (15), limited only by the ability to detect low frequency signals. In practice other instrumental factors impose limitations and the highest m/z value detected for an organic compound (at the time of writing) is approximately 12,384 Dalton [21].

FTMS is capable of recording tandem MS/MS spectra and sequential (MS^n) decompositions (see Sect. 7). It is also compatible with pulse ionisation techniques such as laser desorption and, like ToF under pulsed ionisation conditions, has a multichannel advantage. The main disadvantages are a result of the very high vacuum ($< 10^{-8}$ Torr) necessary for optimum sensitivity and resolution. This makes sample introduction difficult and a number of ingenious solutions (such as dual-cells and external sources) have been devised to overcome this problem; however these tend to increase the cost of an already expensive instrument.

4   Ion Detection

Most conventional organic mass spectrometers incorporate electron multipliers for detecting ions. There are two main types, discrete dynode or continuous dynode. Both make use of secondary electron emission to produce an amplified ion signal. Only the discrete dynode detector will be described here.

A series of Be–Cu alloy dynodes (usually 12–20) are arranged as shown in the diagram (Fig. 17). The ion beam strikes the first dynode and produces a shower of secondary electrons, which are accelerated towards the next dynode and so on, producing a cascade effect and enhancing greatly the original, very weak, ion current. The dynodes are held at an increasingly positive potential by a resistive network. The output current is converted to a voltage, which is amplified further and fed to suitable recording devices (analogue or computer). Electron multipliers can easily produce gains of 10^6 and, very importantly, have fast response times making them suitable for ToF and fast scanning GC/MS instruments.

Negative ion operation is achieved by directing the ion beam towards a

Ion beam

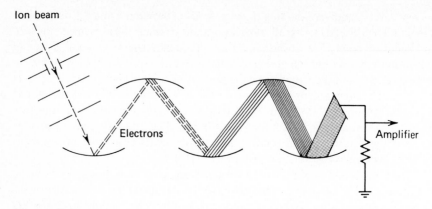

Electrons

Amplifier

Fig. 17. Schematic of an electron multiplier ion detector (reproduced from McFadden WH (1973) Techniques of Combined Gas Chromatography/Mass Spectrometry, Wiley-Interscience, New York, with permission)

"conversion dynode" which is usually held at a positive potential. Positive ions emitted by the conversion dynode are then detected by the electron multiplier in the usual manner.

5   Data Systems

5.1   Data Acquisition

Although mass spectra are still occasionally recorded using analogue devices, it is more usual to acquire and process mass spectrometric data using digital computers. The components of a typical mass spectrometer/data system are shown in Fig. 18. The analogue-to-digital converter (ADC) samples the fluctuating voltage output of the mass spectrometer detection system, producing discrete digital samples suitable for computer processing (conversion to masses and intensities). Some sort of edited output is usually produced on the visual display unit (VDU) during data acquisition, so that the progress of a sample run can be followed. This output can be either plot of total ion current or mass chromatogram (see below) against scan number, or abbreviated tabular or bar graph data. Low or high resolution mass spectra can be acquired (high resolution spectra, of course, require a high resolution mass spectrometer). If the spectra are recorded at high resolution, the computer can be instructed to calculate the empirical formula of each ion. A single GC/MS run is capable of generating many hundreds of mass spectra; consequently fast access disk systems are essential for storing mass spectrometric data for subsequent processing.

Most modern data systems are not simply passive receivers of information, they also help to control instrument operation via a digital-to-analogue converter (DAC) which passes instructions from the computer to the mass

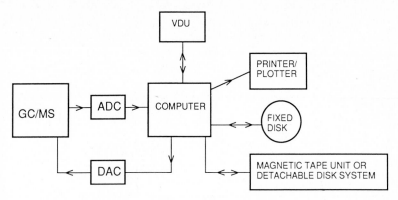

Fig. 18. Block diagram of a mas spectrometer data system

spectrometer. Scan speed, mass range, resolution, GC temperature ramps, ionisation modes, even focusing, are among some of the operational parameters which can be selected via the computer keyboard.

5.2   Data Processing

Mass spectra can be retrieved from the data system disk store in bar graph or tabular form. The mass spectrum is usually presented as a normalised bar-graph

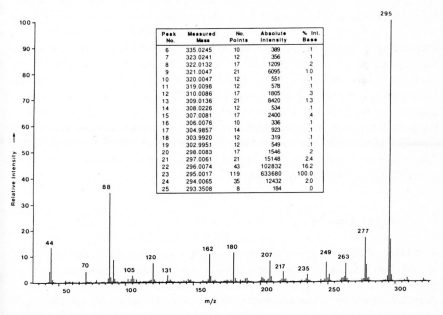

Peak No.	Measured Mass	No. Points	Absolute Intensity	% Int. Base
6	335.0245	10	389	.1
7	323.0241	12	356	.1
8	322.0132	17	1209	.2
9	321.0047	21	6095	1.0
10	320.0047	12	551	.1
11	319.0098	12	578	.1
12	310.0086	17	1805	.3
13	309.0136	21	8420	1.3
14	308.0226	12	534	.1
15	307.0081	17	2400	.4
16	306.0076	10	336	.1
17	304.9857	14	923	.1
18	303.9920	12	319	.1
19	302.9951	12	549	.1
20	298.0083	17	1546	.2
21	297.0061	21	15148	2.4
22	296.0074	43	102832	16.2
23	295.0017	119	633680	100.0
24	294.0065	35	12432	2.0
25	293.3508	8	184	.0

Fig. 19. Bar-graph and partial tabular output of the low resolution ammonia chemical ionisation spectrum of Aspartame

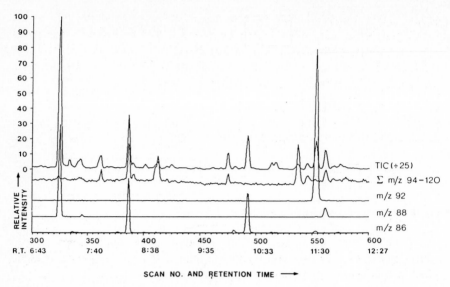

Fig. 20. Total and partial ion current and mass chromatograms of part of a GC/MS analysis

or table of % intensity, relative to the most abundant peak (the base peak), versus mass to charge ratio (m/z) (Fig. 19).

It is usual to acquire scans continuously when introducing samples, because this ensures that minor components are detected. In a typical GC/MS analysis, for example, spectra are recorded in less than a second, with an interval between scans of (typically) 0.2 s. By summing the ion intensities of all masses (or a specified mass range) in each scan and plotting these total or partial ion current values against scan number (and GC retention time), a reconstructed ion chromatogram is produced. Individual ion intensities may also be plotted against scan number (a mass chromatogram) (Fig. 20). This can assist in locating particular components in a complex chromatogram, if masses characteristic of particular structural features are chosen. Mass spectra often contain ions due to chemical interferents (e.g. GC column bleed or unresolved GC components). These may be removed by subtraction of "background" spectra or by more complex data processing techniques [22, 23].

5.3 Library Searching

A number of different data processing techniques are available as an aid to the identification of unknown analytes [24]. The most popular methods are based on some form of "library search". The unknown spectrum is compared with a large reference library (typically containing 50,000–100,000 entries) of mass spectra of known compounds. A suitable algorithm compares the analyte spectrum with the library entries and ranks these in order of similarity to the unknown. Different search procedures are available, depending on the particular type of analytical problem, purity of the spectrum etc. Some search procedures yield clues to the structure of the unknown, even when it is absent from the library, by matching against compounds which contain similar structural features.

6   Combined Chromatography/Mass Spectrometry

6.1   Gas Chromatography/Mass Spectrometry (GC/MS)

Gas chromatography and mass spectrometry are highly compatible analytical techniques, which are well matched in almost every respect save one. A gas chromatograph functions with the column outlet maintained at one atmosphere pressure, whereas a mass spectrometer operates (in EI mode) at source pressures of between 10^{-5} and 10^{-6} Torr ($0.1 - 1$ Torr in the case of CI), and typical analyser pressures of 10^{-6} to 10^{-7} Torr. Hence, high GC carrier gas flow rates (> 5 ml min^{-1}) must be reduced before the samples enter the ionisation chamber.

6.1.1   Molecular Separators

The interfacing of wide-bore capillary and packed GC columns with high carrier gas flow rates to MS can be accomplished by using a molecular separator, which selectively removes most of the carrier gas prior to introduction of samples into the source. Many different types of separator were developed when GC/MS was in its infancy but most modern instruments are now equipped with all-glass, single-stage jet separators (Fig. 21). These can accept flow rates in the range 20– 40 ml/min of hydrogen or helium carrier gas. The carrier gas and eluting organic material are pushed through the nozzle on the left and expand outwards into the vacuum-pumped region in a cone-shaped stream, as shown. Under these conditions the low molecular weight carrier gas diffuses outwards more rapidly

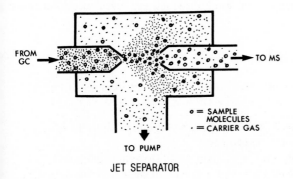

FROM
GC → → TO MS

TO PUMP

o = SAMPLE
      MOLECULES
· = CARRIER GAS

JET SEPARATOR

Fig. 21. Schematic of a single-stage jet separator

than the higher molecular weight organic molecules, which are concentrated in the centre of the cone. A "skimmer" nozzle, placed opposite the expansion nozzle, accepts this sample-enriched gas flow, and the bulk of the carrier gas is pumped away. About 40% of sample is transferred to the ion source, and the concentration of the sample in the carrier gas is increased by a factor of 30–100, while the flow rate is reduced to a level compatible with the mass spectrometer pumping system. Although some alternative designs of separator are capable of improving on these performance figures, none can match the inertness of the all-glass jet system. Contact between samples and surfaces is minimised by the rapid transit time of the column effluent through the separator, and sample decomposition is also minimised by the glass fabrication of the device. The separator and transfer lines are heated (typically to about 250°C) to prevent sample condensation.

6.1.2   Capillary Column Interface

The entire effluent of a capillary GC column can be introduced into the ion source if the mass spectrometer is equipped with high capacity diffusion or turbomolecular pumps. Flow rates of about 5 ml/min are acceptable without degrading mass spectrometer performance. Different techniques for coupling the capillary column to the mass spectrometer are available and the interested reader is referred to a recent review article for further information [25].

6.2   Liquid Chromatography/Mass Spectrometry (LC/MS)

One of the most challenging developments in analytical methodology has been the direct combination of high performance liquid chromatography (HPLC) with mass spectrometry. The impetus behind this development has been the lack of versatility of conventional HPLC detectors, and the realisation that LC/MS promises benefits even greater than those fulfilled by the implementation of routine GC/MS. Unfortunately, HPLC and mass spectrometry are not as compatible as GC and mass spectrometry; no truly "universal" LC/MS interface has been produced at the time of writing. Modern LC/MS systems tend to be based on a balance of compromises, some of which are more successful than others.

6.2.1   The ideal LC/MS system

Although such a device does not exist, one can draw up brief set of desirable specifications:

- No restriction of flow rate, solvents, buffers and gradient elution; chromatographic integrity should also be preserved.
- Mass spectrometer vacuum should be maintained (solvent should be removed preferentially, with concomitantly high transfer efficiency for the analytes).

- A variety of (positive and negative) ionisation modes should be accessible, allowing the analysis of a wide range of samples including involatile, thermally labile compounds.

There are, unfortunately, a number of obstacles to achieving such a specification. Firstly, sample concentration is considerably lower than in GC/MS; secondly vapour flow is considerably higher than in GC/MS (1 ml/min of liquid is approximately equivalent to 1,000 ml/min of vapour); thirdly, HPLC is often applied to samples too involatile or thermally labile for gas phase separation and finally, many HPLC separations entail the use of involatile buffer salts.

Because of the challenging nature of these problems, and because LC/MS is still an evolving technique, a range of different types of LC/MS interface has been developed. A representative selection of the main types of commercially available interface currently in use, is described here.

6.2.2 Direct Liquid Introduction

The simplest way to introduce HPLC effluent into a mass spectrometer is to split the flow and inject only that amount of solvent tolerable by the mass spectrometer pumping system. Although early experiments along these lines utilised a standard EI mass spectrometer, it was soon realised that a CI source, which accepts at least 1–2 orders of magnitude more solvent that EI, had greater potential [26, 27]. Since these early experiments, a number of direct liquid introduction (DL1) interfaces have been developed. Figure 22 is a schematic of one of the more successful designs [28]. The LC effluent flows past a pinhole orifice and a jet of liquid (about 1% of a 1 ml/min column flow) enters the mass spectrometer ion source. Cooling water is circulated through the probe to prevent premature vapourisation of the solvent. The CI plasma is produced by ionising the solvent vapour, consequently most spectra are somewhat dependent on the LC conditions used. Some DLI designs interpose a heated vapourisation chamber between probe and source; this yields improved performance with

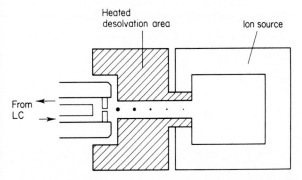

Fig. 22. Direct Liquid Introduction LC/MS interface (reproduced from Chapman JR (1985) Practical Organic Mass Spectrometry, Wiley-Interscience, Chichester, with permission)

difficult samples [29]. Microbore LC columns, which operate at low flow rates ($10–50\,\mu l/min$) are more compatible with DLI than are standard (4 mm bore) analytical columns because the entire effluent can be introduced into the ion source, without flow-splitting [30]. This results in greatly improved detection limits (a few ng in favourable cases).

The DLI method can work very well with certain compound classes, but difficulties are often encountered with labile, involatile, or high molecular weight solutes, where decomposition and adsorption can occur. The small size of the sampling orifice also means that complete or partial blockages can occur with unacceptable frequency.

6.2.3   Mechanical Transport Interfaces

The first mechanical transport interface was based on a "moving-wire" principle [31], which subsequently evolved into moving belt interface, the first commercial LC/MS system [32]. Sample is deposited onto a looped polyimide belt (Fig. 23), which passes under a focussed infra-red lamp where a large proportion of the solvent is vapourised. The belt then continues through a series of vacuum locks and pumping chambers which remove any residual solvent and help to maintain vacuum in the mass spectrometer. The sample is finally conducted into the ion source, where it is vapourised and ionised. A clean-up heater or solvent scrubber removes any residual sample adhering to the belt, preventing any "ghost" peaks which could result from re-cycled material. The solvent capacity of the belt interface can be as high as 1.5 ml/min for volatile solvents but is significantly lower for aqueous solvent systems. Both EI and CI mass spectra are obtainable, with best results from designs where the belt enters the ionisation chamber [33].

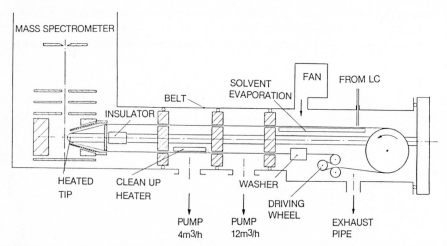

Fig. 23. Moving belt LC/MS interface for use with a magnetic sector mass spectrometer (reproduced by permission of Finnigan MAT Ltd.)

Improved performance (better sensitivity and less peak broadening) has also been obtained when column effluent is sprayed onto the belt [34] and when microbore HPLC columns are used [35].

Belt LC/MS systems have also been adapted so that samples can be ionised by FAB, by bombarding the belt surface directly [36, 37]. This has extended the range of compounds amenable to belt LC/MS enormously, with reports of applications to peptides and oligosaccharides [37]. Lasers have also shown some promise in desorbing/ionising analytes from the belt surface [38].

6.2.4 Thermospray Interface (TSP)

The TSP ion source has already been described in Sect. 2. TSP can accept solvent flow rates of up to approx. 2 ml/min. If ion production cannot be accomplished by ion evaporation (buffer ionisation), electron bombardment or a discharge electrode can be used to produce a CI plasma from the solvent vapour. Figure 24 demonstrates the analysis of a mixture of desulphoglucosinolates by TSP LC/MS [39]. These compounds, unusually, yield diagnostic fragment ions and proto-

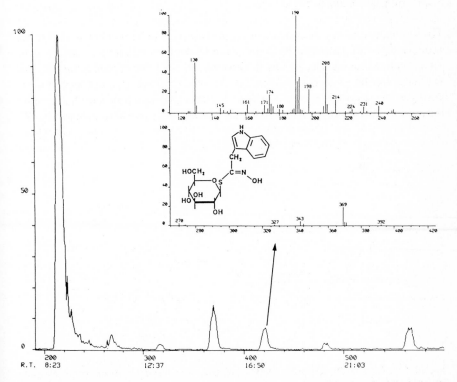

Fig. 24. Thermospray LC/MS chromatogram and mass spectrum of desulphoglucosinolates isolated from a variety of calabrese

nated molecular ions (fragmentation is often absent in TSP spectra, due to the mildness of the ionisation conditions).

6.2.5 Atmospheric Pressure Ionisation

The atmospheric pressure ionisation (API) source functions, in LC/MS mode, by vapourising the eluant at atmospheric pressure. Solute and solvent vapour are directed towards a chamber, where corona discharge produces a complex series of ion-molecule reactions. The ions generated in atmosphere are sampled through a small orifice, behind which a "curtain" of nitrogen gas is used to break up cluster ions. The ions are then directed into the mass analyser through a second orifice. It is possible to attain nanogram detection limits with this type of system. Several different sample vapourisation techniques have been used with API, and ionisation is sometimes effected by means other than an electrical discharge [40–42]. Because the API tends to yield stable molecular ions, it is often combined with MS/MS (Sect. 7) in order to generate diagnostic fragments.

6.2.6 Particle Beam Interface

The monodisperse aerosol generator for introduction of liquid chromatographic effluents (MAGIC) interface is an interesting recent approach to LC/MS which, apart from belt designs is the only practical system capable of producing EI mass spectra [43,44]. The LC effluent is directed through a nebuliser into a desolvation chamber, through which a dispersion gas flows (Fig. 25). This produces a mist of uniformly-sized droplets which are conducted into an aerosol beam separator, or momentum separator, where rotary-pumped chambers remove solvent. Sample molecules are routed to the source of the mass

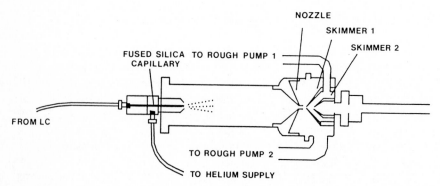

Fig. 25. Schematic of Particle-Beam LC/MS interface (reproduced by permission of Hewlett-Packard Ltd.)

spectrometer, and ionised by EI or CI. Liquid flows in the range 0.1–0.5 ml/min yield the best performance. The particle beam interfaces, and variations on this design, are currently available from three different commercial manufacturers of mass spectrometers.

6.2.7 Continuous FAB LC/MS

Continuous flow FAB (see Sect. 2) is suitable as a sample introduction system for LC/MS. Only very low flow rates (5–10 μl/min) are admissible with current source designs, so either microbore-LC Columns (0.32 mm i.d.) [45] or flow-splitting from larger bore columns is essential [46].

6.3 Supercritical Fluid Chromatography/Mass Spectrometry (SFC/MS)

Supercritical Fluid Chromatography (SFC) has some advantages over GC and HPLC in certain types of analysis and is an increasingly important separation technique [47]. The benefits of SFC derive from its combination of liquid-like solubility behaviour and gas-like diffusivity. These properties can produce narrower peak widths than in HPLC, together with the ability to handle high molecular weight compounds. Both packed (high flow rate) and capillary (low flow) systems are available. SFC is an attractive sample introduction technique for MS because CO_2 is much easier for the vacuum system to remove than are HPLC solvents. SFC capillary columns, which provide a low gas flow burden, can be combined directly with MS [48]. Packed column flows require the removal of the bulk of CO_2, and interfaces such as the belt [49] and modified thermospray [50] have been used.

6.4 Capillary Zone Electrophoresis/Mass Spectrometry (CZE/MS)

Capillary zone electrophoresis (CZE) is an ultra high-resolution type of zone electrophoresis carried out in small (< 100 μm i.d.) capillary tubes. Limitations of the technique largely centre on the sensitivity of detection techniques (typically UV and fluorescence) as injection volumes are, of necessity, very small (5–50 nl). An attempt to overcome these problems has been made by combining CZE with mass spectrometry [51, 52]. Samples are electrosprayed[53] into the mass spectrometer, and preliminary studies indicate that attomolar sensitivities are attainable. Other ionisation techniques, such as continuous flow FAB and thermospray, may also be compatible with CZE. CZE/MS is till in its early stages of development, but the methodology is promising and may have a significant impact if developed further.

7 Tandem Mass Spectrometry

Many of the 'soft' ionisation techniques, such as CI, FAB and thermospray ionisation, frequently yield intense molecular ions but little diagnostic fragmentation. The structural information which may be directly obtained from the mass spectrum is consequently limited. The widespread use of these ionisation methods has therefore been paralleled by the development of spectrometers capable of inducing fragmentation in "stable" ions and subjecting these to a second stage of mass analysis. These techniques are collectively described by the term 'tandem' mass spectrometry.

Tandem mass spectrometry has been described as the process of obtaining the mass spectrum of a particular ion in a mass spectrum [54]. This is achieved by combining two (or more) mass analysers in series, separated by an ion dissociation chamber. An ion selected by the first analyser is subjected to dissociation in the region between the analysers and the resulting fragments are then measured by the second analyser. The steps in a typical MS/MS experiment are shown schematically in Fig. 26. The principal uses of tandem mass spectrometry are (i) to obtain structural information on ions generated by 'soft' ionisation methods, by fragmenting the molecular or quasi-molecular ions formed, and (ii) direct analysis of complex mixtures by exploiting the improvement in selectivity possible with a multistage separation.

Almost every possible combination of magnetic (B), electric (E) and quadrupole (Q) analyser has been used in tandem experiments; some of the commoner permutations are listed in Table 1. The list includes the two 'ion trapping' techniques, Fourier transform (FTMS) and ion trap (ITMS) mass spectrometry, (Sect. 3) which have been used in tandem experiments. Configurations which

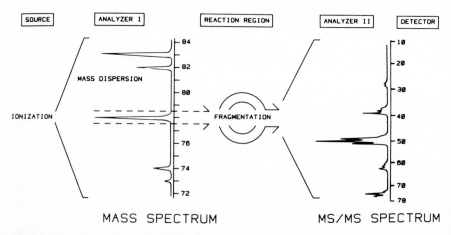

Fig. 26. Schematic representation of daughter ion analysis by tandem mass spectrometry (reproduced from McLafferty FW (ed) (1983) Tandem Mass Spectrometry, Wiley-Interscience, New York, with permission)

Table 1. Some Tandem MS configurations

B–(C)–E
EB–(C)–E
EB–(RFQ)–Q
EB–(C)–EB
Q–(RFQ)–Q
FTMS
Ion Trap

E = Electric Sector, B = Magnetic Sector, Q = Quadrupole, C = Collision cell, RFQ = RF-only quadrupole

employ a quadrupole as the final analyser (e.g. QQQ and EBQQ) also utilise an intermediate quadrupole, operated in an RF-only mode, which acts as a collision cell and efficiently transmits daughter ions to the analyser quadrupole.

The principal methods for achieving fragmentation of precursor ions are collisionally activated dissociation (CAD) and laser photodissociation (PD). CAD is the more commonly used and relies on collisions between the precursor ions, selected by the first analyser, and a suitable gas in the collision cell situated between the two analysers. The nature of the CAD spectrum obtained depends upon the energy transferred to the precursor ion during the collision process. This is usually sufficient to produce fragmentation similar to that obtained under electron ionisation, and the CAD spectrum therefore provides structural information on the selected precursor ion. In laser photodissociation, the laser is focused coaxially with, or perpendicular to, the ion beam. Absorption of photons transfers energies of a few electronvolts and is sufficient to induce fragmentation of the precursor ions [55]. The efficiency of PD is much greater in ion trap and FTMS instruments, where the precursor ions may be trapped and illuminated for

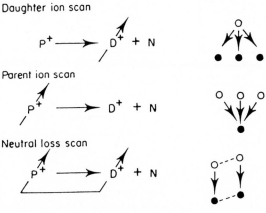

Daughter ion scan

Parent ion scan

Neutral loss scan

P^+ = parent ion(s), D^+ = daughter ion(s), N = neutrals

= scanning analyser

Fig. 27. Scan modes in tandem mass spectrometry (reproduced from (1988) Eur. Spectrosc. News, 80: 13, with permission)

a much longer time than is possible in conventional instruments, allowing the population of daughter ions to build up. Dissociation efficiencies in excess of 90% have been achieved in this way [56].

The three main types of scan function commonly used in tandem mass spectrometry are shown schematically in Fig. 27. A product ion spectrum is obtained by setting the first analyser to transmit ions of a chosen m/z value to the collision cell. These precursor ions are dissociated by CAD or PD and the daughter ion spectrum is then scanned by the second analyser (Fig. 26). A precursor ion scan is carried out by setting the second analyser to transmit a selected product ion to the detector, whilst the first analyser is scanned to pass each putative precursor to the collision cell sequentially. A signal is only observed when precursor ions which give rise to the selected daughter are transmitted by the first analyser. In the neutral loss scan, the first and second analysers are scanned simultaneously so as to maintain a constant mass difference between the ions transmitted by the two analysers. Under these conditions, only precursor ions which lose a neutral fragment of mass corresponding to the chosen mass difference in the transmission of the two analysers (for example members of a class of compounds which loose a common fragment) will give rise to detected product ions.

Ion trapping mass spectrometers, such as Fourier transform (FTMS) and ion trap (ITMS) instruments, rely on a temporal separation, rather than the spatial and temporal separation of ionisation, mass analysis and fragmentation processes of conventional tandem mass spectrometry. In a typical experiment the analyte is introduced into the trap, ionised and the resulting population of ions is confined in the cell using a strong magnetic field (FTMS) or by applying an RF

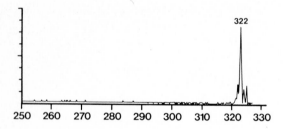

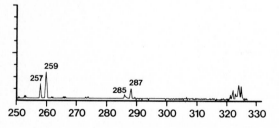

Fig. 28. Daughter ion mass spectrum of the m/z 322 ion of tetrachloro-dibenzo dioxin, obtained on an ion trap mass spectrometer

voltage (ITMS). All ions except those of a chosen mass are ejected from the cell and, after fragmentation by CAD or PD, the product ion spectrum is scanned. For CAD in an ion trap, the chosen ions are induced to perform oscillations of sufficient amplitude to cause energetic collisions with the buffer gas by applying an auxiliary RF voltage across the end caps at the secular frequency of the trapped ions. Collision-induced product ions are trapped and then detected in a simple but elegant MS/MS experiment. The product ion spectrum of the m/z 322 ion of tetrachlorodibenzo-p-dioxin (TCDD) is shown in Fig. 28 (lower spectrum). The upper trace shows the appearance of the spectrum prior to collisional activation. The advantage of this type of instrument is that multiple stage tandem experiments (e.g. MS/MS/MS or MS^n) may be carried out sequentially by trapping product ions of interest and subjecting these to further fragmentation and mass analysis. In conventional sector or quadrupole tandem spectrometers this type of analytical sequence is only possible by adding further analysers and collision cells to the instrument. Significant increases in selectivity are possible using these sequential MS^n procedures, as long as a measurable population of product ions remains in the trap.

8   References

1. Thomson JJ (1911) Phil. Mag. 21: 225
2. Thomson JJ (1913) Rays of positive electricity and their application to chemical analyses, Longmans Green, London
3. Aston FW (1942) Mass spectra and isotopes, Edward Arnold, London
4. Nier AO (1953) Scientific American 3: 256
5. Cotter RJ (1980) Anal. Chem. 52: 1589A
6. Barber M, Bordoli RS, Sedgwick RD, Tyler AN: J. Chem. Soc. Chem. Comm. 1981: 325
7. Heckles K, Johnstone RAW, Willby AH (1987) Tet. Lett. 28: 102
8. Ackermann BL, Watson JT, Holland JF (1985) Anal. Chem. 57: 2656
9. Pratt DE, Fairweather-Tait SJ, Eagles J, Symms LL (1988) In: Creaser CS, Davies AMC (eds) Analytical applications of spectrocopy, Roy. Soc. Chem., London, p 294
10. Holland J, Haskins NJ, Tibbetts PJC, Large R (1986) In: Todd JFJ (ed) Advances in mass spectrometry, Wiley, New York, p 1543
11. Caprioli RM, Fan T, Cottrell JS (1986) Anal. Chem. 58: 2949
12. McNeal CJ (ed) (1986) Mass spectrometry in the analysis of large molecules, Wiley, New York
13. Shinghal R, Ledingham K (1987) New Scientist 116: 52
14. Grotemeyer J, Boesl U, Walter K, Schlag EW (1986) Org. Mass Spectrom 21: 645
15. Vestal ML (1984) Science 226: 275
16. Johnson EG, Nier AO (1953) Phys. Rev. 9: 10
17. Stafford GC, Kelley PE, Syka JEP, Reynolds WE, Todd JFJ (1984) Int. J. Mass Spectrom. Ion Phys. 60: 35
18. Brodbelt JS, Louris JN, Cooks RG (1987) Anal. Chem. 59: 1677
19. Boesl U, Grotemeyer J, Walter K, Schlag EW (1986) Inst. Phys. Conf. Ser., 84: 151
20. Comisarow MB, Marshall AG (1974) Chem. Phys. Lett. 62: 293
21. Hunt DF, Shabanowitz J, Yates JR, Zhu N-Z, Russel DH, Castro M (1987) Proc. Natl. Acad. Sci. USA 34: 620
22. Biller JW, Biemann K (1974) Anal. Lett. 7: 515
23. Dromey RG, Stefik MJ, Rindfleisch TC, Duffield AM (1976) Anal. Chem 48: 1368
24. Chapman JR (1978) Computers in mass spectrometry, Academic Press, New York

25. Rose ME (1985) In: Rose ME (ed) Specialist periodical reports: Mass spectrometry, vol 8, Roy. Soc. Chem., London, p 216
26. Talrose VL, Karpov GV, Gordoetshu IG, Skurat VE (1968) Russ. J. Phys. Chem. 42: 1658
27. McLafferty FW, Baldwin MA (1973) Org. Mass. Spectrom., 8: 1111
28. Melera A (1980) In: Quayle A (ed) Advances in mass. spectrometry, vol 8B, Heyden, London, p 1597
29. Dedieu M, Juin C, Arpino P, Guiochon G (1982) Anal. Chem. 54: 2372
30. Lee ED, Henion JD (1985) J. Chromatogr. Sci. 23: 253
31. Scott RPW, Scott CG, Munroe M, Hess J (1974) J. Chromatogr. 99: 395
32. McFadden WH, Schwartz HL, Evans S (1976) J. Chromatogr. 122: 389
33. Games DE, McDowall MA, Levsen K, Schafer KH, Dobberstein P, Gower JL (1984) Biomed. Mass Spectrom. 11: 87
34. Hayes M, Lankmayer EP, Vouros P, Karger BL, McGuire JM (1983) Anal. Chem. 55: 1745
35. Games DE, Lant MS, Westwood SA, Cocksedge MJ, Evans N, Williamson J, Woodhall BJ (1982) Biomed. Mass Spectrom. 9: 215
36. Dobberstein P, Korte E, Meyerhoff G, Pesch R (1983) Int. J. Mass Spectrom. Ion Phys. 46: 185
37. Arpino P (1989) Mass Spectrom. Rec. 8: 35
38. Hardin ED, Fan TP, Blakley CR, Vestal ML (1984) Anal. Chem. 56: 2
39. Mellon FA, Fenwick GR, Lewis JA, Spinks GA (1988) In: Creaser CS, Davies AMC (eds) Analytical applications of spectroscopy, Roy. Soc. Chem., London, p 301
40. Covey TR, Lee ED, Bruins AP, Henion JD (1986) Anal. Chem. 58: 1451A
41. Bruins AP, Covey TR, Henion JD (1987) Anal. Chem. 59: 2642
42. Covey TR, Bruins AP, Henion JD (1988) Org. Mass. Spectrom. 23: 178
43. Willoughby RC, Browner RF (1984) Anal. Chem. 56: 2626
44. Winkler PC, Perkins DD, Williams WK, Browner RF (1988) Anal, Chem., 60: 489
45. Ashcroft AE, Chapman JR, Cottrell JS (1987) J. Chromatogr. 394: 15
46. Games DE, Pleasance S, Ramsey ED, McDowall MA (1988) Biomed. Env. Mass Spectrom. 15: 179
47. Smith RM (ed) (1988) Supercritical fluid chromatography, Roy. Soc. Chem., London
48. Smith RD, Udseth HR (1987) Anal. Chem. 59: 13
49. Berry AJ, Games DE, Perkins JR (1986) J. Chromatogr. 363: 147
50. Berry AJ, Games DE, Mylchreest IC, Perkins JR, Pleasance S (1988) Biomed. Env. Mass Spectrom 15: 105
51. Olivares JA, Nguyen NT, Yonker CR, Smith RD (1987) Anal. Chem. 59: 1230
52. Smith RD, Olivares JA, Nguyen NT, Udseth HR (1988) Anal. Chem. 60: 436
53. Whitehouse CM, Dreyer RN, Yamashita M, Fenn JB (1985) Anal. Chem. 57: 675
54. McLafferty FW (ed) (1983) Tandem mass spectrometry, Wiley, New York
55. Levsen K (1986) In: Todd JFG (ed) Advances in mass spectrometry, vol 10, Wiley, New York, p 350
56. Louris JN, Brodbelt JS, Cooks RG (1987) Int. J. Mass Spectrom, Ion Proc., 75: 345

CHAPTER 12

Electron Paramagnetic Resonance and Electron Nuclear Double Resonance Spectroscopy

A.J. Thomson

1 Introduction

An unpaired electron can be aligned by an applied magnetic field so that its spin is precessing about the field. The direction of spin precession can be either clock- or anticlockwise corresponding to the two energy states of the electron (Fig. 1). Irradiation with an electromagnetic beam of the correct frequency can induce transitions of the unpaired electron from one spin direction to the other. The transition is induced by the oscillating magnetic field component of the electromagnetic radiation. Thus the transition is a magnetic dipole process and consequently the selection rule for the transition is that the spin component along the direction of the applied field, given by M_S, must change by ± 1 unit. The energy equation for resonance is the well-known expression

$$h\nu_0 = g\mu_B B_0$$

where g is the splitting factor, μ_B the Bohr magneton and B_0 the applied field. For a magnetic induction field of 0.33 T and a g-factor of 2.0 the microwave energy, $h\nu_0$, is $0.3\,cm^{-1}$. A resonance spectrum is obtained by placing the sample in a microwave cavity tuned to the frequency of the microwave radiation. An external magnetic field is applied and swept until the resonance condition is found.

The technique provides a rather sensitive method of detecting the presence of unpaired electrons in molecules. This chapter illustrates, by means of examples, the wide range of problems which have been investigated by EPR spectroscopy. These include the study of radical species in polymers, the use of spin-labels to follow molecular motion in membranes and on surfaces, spin-traps to stabilise reactive radical intermediates for mechanistic investigation and the spin-labelling of proteins to determine molecular structure. The study of transition-metal ion compounds with unpaired electrons is also illustrated. Finally, a brief section describes electron nuclear double resonance (ENDOR) spectroscopy. This technique is becoming more routine and provides a method of obtaining the nuclear magnetic resonance properties from paramagnetic species.

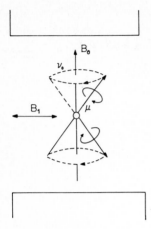

Fig. 1. The ESR Experiment. Precession of a spinning electron with magnetic dipole moment, μ, in an applied magnetic induction field, B_0, with a frequency ν_0. A spin-1/2 particle has two quantised orientations with respect to B_0. An oscillating magnetic field, B_1, perpendicular to the B_0 field, will induce a flip of the spin from one precession direction to the other provided that the frequency of oscillation of B_1 matches that of ν_0

2 EPR Spectrometer

A block diagram of a typical EPR spectrometer is shown in Fig. 2a. The source of microwaves may be either a Klystron (vacuum tube) or a Gunn diode (a solid-state device) either of which provides a monochromatic, coherent source of electromagnetic radiation. Commonly spectrometers operate in the X- or Q-band frequency regions corresponding to $\sim 9\,\mathrm{GHz}$ (wavelength $\sim 3.0\,\mathrm{cm}$) or to $35\,\mathrm{GHz}$ (wavelength $1.1\,\mathrm{cm}$). The source frequency can be swept either side of the centre frequency by about $\pm 100\,\mathrm{MHz}$ for the purpose of tuning the microwave circuit and cavity. The sample is placed inside a resonant cavity (Fig. 2b). This constitutes a hollow conducting box either cuboid or cylindrical with dimensions designed to support a standing microwave pattern. The cavity has two functions. First, it stores microwave energy so that the intensity of the radiation field at the sample is maximised. Since magnetic dipole transitions are weak this function of the cavity plays a vital role in giving the EPR technique good sensitivity. Secondly, the cavity design ensures a standing wave pattern which maximises the magnetic field component of the radiation field and minimises the electric field component at the sample. In this way dielectric loss by the sample via electric dipole absorption is avoided. This would lead to the cavity failing to resonate and sensitivity being lost.

As the source has a fixed frequency the external applied magnetic field, B_0, is swept in order to bring the Zeeman separation, $g\mu_B B_0$, into resonance with the energy of the microwave field, $h\nu_0$. Superimposed upon the field B_0 is a second magnetic field of a few gauss $(10^{-4}\,\mathrm{T})$ which is modulated plus and minus at a frequency of, say, $100\,\mathrm{kHz}$. This modulates the intensity of the reflected microwave at this frequency. Electronic detection of the reflected microwave beam is carried out by phase-sensitive detection locked to the magnetic field modulation frequency. As a consequence the EPR spectrum is displayed as the

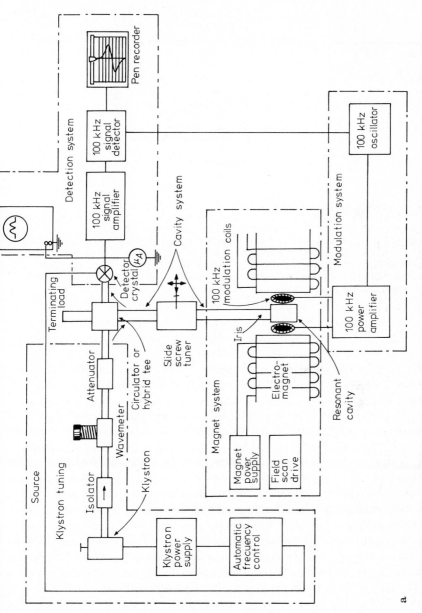

Fig. 2a. Block diagram of a typical X-band ESR spectrometer employing 100 kHz phase-sensitive detection

a

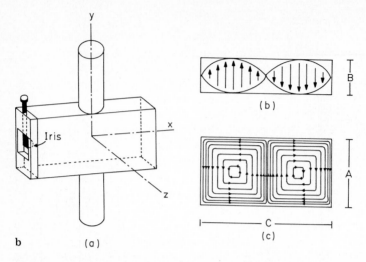

Fig. 2b. A rectangular parallelepiped TE_{102} microwave cavity. (a) Cylindrical extensions above and below the cavity prevent excessive leakage of microwave radiation out of the cavity and act as positioning guides for a sample. The microwave energy is coupled into the cavity through the iris hole at the left. This coupling may be varied by means of the iris screw. (b) The electric field contours in the xz plane. One half-wavelength in the x direction corresponds to the shortest distance between points of equal field intensity but of opposite phase. (c) Magnetic field flux in the xy plane. A is approximately one half-wavelength and C is exactly two half-wavelengths. The B dimension is not critical but should be less than one half-length
[Reproduced with permission from Wertz JE, Bolton JR (1986) Electron spin resonance, Chapman and Hall, London]

first derivative with respect to the field of the intensity of magnetisation. occasionally the second derivative is plotted to improve the resolution of closely lying lines.

Facilities for temperature control of the sample may be required. Rapidly relaxing paramagnets, typically transition metal ions with d^n configuration where $n = 2$–8, require cooling to liquid helium temperature (< 77 K) to be observed. All radical species if sufficiently stable can be observed readily at room temperature. The sample can be a liquid, frozen glass, solid powder, single crystal or amorphous polymer. However, the sample must be magnetically dilute, that is, the electron spins must be sufficiently isolated from one another to minimize spin-spin coupling otherwise this may broaden the spectrum beyond detection.

The sensitivity of the technique depends upon the number of spins which can be packed into the cavity whilst still retaining the need for magnetic dilution. Therefore it depends upon the cavity size and thus upon the microwave frequency of the source. Typically a modern X-band spectrometer can detect 10^{15}–10^{16} spins whereas a Q-band spectrometer can achieve a factor of 10 better. This implies a sensitivity of 10^{-7}–10^{-8} moles at X-band and 10^{-9} moles at Q-band. In practice 1 ml of a solution which is 10^{-4}–10^{-5} M in spins is ideal at X-band.

3   Spectral Parameters

The spectral parameters which can be determined in favourable circumstances are the electronic g-factor, and the nuclear hyperfine coupling parameters which measure the unpaired electron spin density at a given nucleus with a spin, I_N. The origin of these parameters is illustrated by means of examples.

The EPR spectrum of the methyl radical is shown in Fig. 3.[1]. Here $\cdot CH_3$ was generated in liquid methane at $-176\,°C$ by irradiation with 2.8 MeV electron radiolysis during observation. The spectrum is a quartet of lines with components separated by 23.0 gauss (0.0023 T) and an intensity ratio 1:3:3:1. The energy level diagram shows the electron spin levels arising from the $S = \frac{1}{2}$ electron. In the absence of nuclear hyperfine coupling to the protons $(I = \frac{1}{2})$ a single line would be observed. In the case of an organic radical the electron will have no orbital magnetic moment: hence $g = 2.0023$ to a good level of approximation. Thus for organic radicals the interest lies in the nuclear hyperfine structure of the EPR transition. Figure 3 shows the energy level diagrams and the allowed microwave transitions according to the selection rules $\Delta M_S = \pm 1, \Delta M_I = 0$. These imply that no nuclear spin changes may take place simultaneously with changes in electron spin. The nuclear spin degeneracies of the levels determine the relative intensities of the transition as shown.

The methyl radical spectrum is a simple one. At the temperature shown, $-176\,°C$, the radical is tumbling rapidly enough to average out the anisotropic components in the proton coupling to the electron as well as the dipole–dipole interaction with nuclei of the CH_4 solvent molecules. The nuclear hyperfine coupling constant is the same for all 3 protons of the radical and has a value of 23.0 gauss. For a free hydrogen atom $(S = \frac{1}{2}, I = \frac{1}{2})$ the hyperfine coupling constant $a^H = 508$ gauss. The unpaired electron spin density in the 1s orbital of each H atom in $\cdot CH_3$ is given by the ratio $(23/508) = 0.046$. This provides an accurate evaluation of the spin polarisation of the C—H σ bond by an unpaired electron entirely in the $2p_\pi$ orbital of an α-carbon atom.

The ethyl radical $\cdot CH_2$—CH_3 in liquid ethane at $-180\,°C$, also produced by 2.8 MeV electron bombardment of the liquid, is shown in Fig. 4(a) [2]. The two fragments $C_\alpha H_2$ and $CH_\beta H_3$ have nuclear spins of $I = 1$ and $3/2$ respectively arising from the protons. The coupling constants are $a_\alpha^H = 22.4$ gauss and $a_\beta^H = 26.9$ gauss. The stick diagram in Fig. 4(a) indicates how the splitting pattern arises. Note that the hyperfine coupling constant a_α^H is close to the 23.0 gauss value for the methyl radical. Also shown in this figure is the EPR spectrum of the ethyl radical at 4.2 K immobilized in an argon matrix. The spectrum is more complex because the rotational tumbling of the radical has been frozen out. The anisotropies in the proton couplings which were averaged out in the tumbling spectrum (see later) are not evident. Table 1 gives the results of a complete analysis of the spectrum of the immobilized ethyl radical and quotes the anisotropies in the proton hyperfine coupling constants. The latter arise from the dipole–dipole coupling of the electron and proton spins.

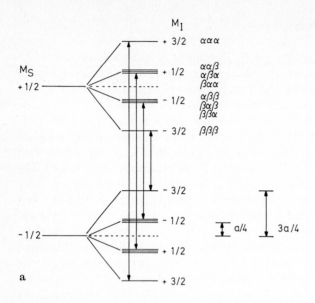

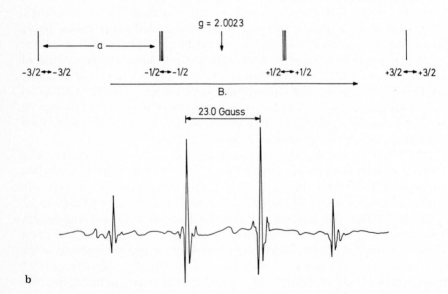

Fig. 3. The ESR spectrum of the methyl radical in liquid methane at − 176 °C produced by 2.8-MeV electron radiolysis of the liquid during observation
[Reproduced with permission from J. Chem. Phys. (1963) 39: 2417]

Table 1. ESR parameters of $CH_3CH_2^.$ radicals isolated in the argon matrix at 4.2 K

g_{av}	Coupling group	Proton couplings (gauss)		
		Principal value		a_{iso}
	CH$_2$	$a_\parallel^\alpha$	29.4	22.6
		$a_\perp^\alpha$	19.3	
2.002 ± 0.0003				
	CH$_3$	$a_\parallel^\beta$	28.5	26.6
		$a_\perp^\beta$	25.7	

[Reproduced with permission from J. Chem. Phys. (1974) 58: 114]

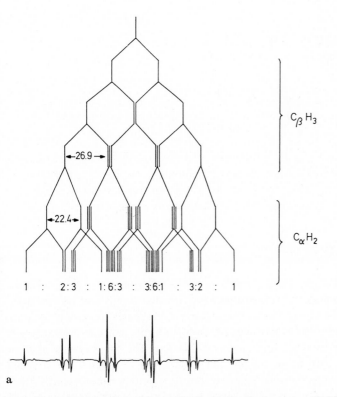

Fig. 4(a). The ESR spectrum of the ethyl radical in liquid ethane at − 180 °C produced by 2.8 MeV electron-beam bombardment of the liquid during observation
[Reproduced with permission from J. Chem. Phys. (1963) 39: 2147]

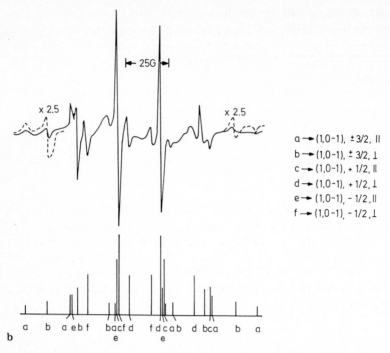

Fig. 4(b). The ESR spectrum of trapped ethyl radicals obtained by UV photolysis of C_2H_4 + HI + Ar (ratios 1:1:500) at 4.2 K. Broken curves at the extremities correspond to an increased amplification of 2.5 times. The stick diagram at the base represents the assigned spectrum as indicated [Reproduced with permission from J. Chem. Phys. (1974) 58:114]

The ability to observe the properties of radicals gives EPR spectroscopy a valuable utility in a wide variety of areas. The literature on the subject is large (see bibliography). In this chapter a few simple examples are given which are illustrative of the way in which useful information can be obtained.

4 Radicals in Polymers

EPR has a long history of applications in polymer chemistry starting with the experiments of Schneider and co-workers on X-irradiated poly(methyl methacrylate) in 1951 [3]. Radical species are present in polymers as a result of irradiation with high-energy photons or particles, photodegradation, mechanical or thermal degradation, oxidative or other chemical damage. The mechanisms of these various processes need to be studied and understood in order to control and prevent their occurrence because polymers are often used in circumstances where any one of these processes may take place. Radical species are also important in the processes of polymerization itself. Chain growth may proceed by radical initiation and propagation just as chain scission can take place by radical damage. Thus details of initiation, propagation and termination reactions inaccessible to other methods can often be studied with EPR. The range of work can be followed in the sources given in the bibliography. Two well-studied examples, polyethylene and polyfluoroethylene, are chosen to illustrate the methods used.

On irradiation at 77 K polyethylene forms free radicals by breakage of a C—H bond;

$$
\begin{array}{ccccc}
H & H & H & H & H \\
| & | & | & | & | \\
-C_\gamma - C_\beta - C_\alpha - C_\beta - C_\gamma - \\
| & | & \cdot & | & | \\
H & H & & H & H
\end{array}
$$

The g anisotropy of this radical is small and the EPR spectrum appears to be a sextet arising from five equally coupled protons. Apparently the couplings to the α and β protons are equal to one another whereas the coupling to the γ protons is too small to produce an observable splitting. This radical is not stable at room temperature. On warming the sample the EPR signal changes to that of another radical species not securely identified. It may be that the predominant radical species is of the alkene type:

$$
\begin{array}{ccccc}
H & H & H & H & H \\
| & | & | & | & | \\
-C - C - C = C - C - \\
| & \cdot & & & | \\
H & & & & H
\end{array}
$$

The X-irradiation of polytetrafluoroethylene (Teflon) in the total absence of oxygen leads to a free radical analogous to that found in polyethylene, namely

$$
\begin{array}{ccccc}
F & F & F & F & F \\
| & | & | & | & | \\
-C - C_\beta - C_\alpha - C_\beta - C - \\
| & | & \cdot & | & | \\
F & F & & F & F
\end{array}
$$

At 35 GHz (Q-band) the $^{19}F(I = \frac{1}{2})$ hyperfine structure is clearly visible in the EPR spectrum, Fig. 5 [4]. The spectrum is a doublet of quintets. The four equivalent β-fluorine atoms give rise to a quintet of lines and all have an equivalent coupling of 33 gauss (92 MHz). The α-fluorine splits the quintets into two overlapping sets with a coupling of 92 gauss (258 MHz). In the centre of the spectrum can be observed a second radical species, corresponding to about 10% of the total radical content. This is a triplet arising from a —CF_2 radical with a hyperfine coupling constant of 16 gauss. Because this radical is present only at a chain end it has been called a propagating radical. The intensity of this type of radical is much increased upon UV irradiation in the presence of oxygen.

The structures of the species formed in irradiated polytetrafluoethylene after combination with oxygen have been elucidated by reaction with $^{16}O_2$ isotopi-

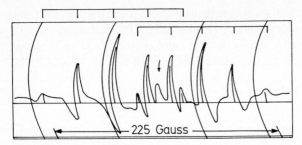

Fig. 5. Electron paramagnetic resonance of a γ-irradiated sample of Teflon pre-evacuated for 24 hours. The *curve* represents the second derivative of the actual resonance curve. The *bars* represent the theoretical hyperfine components for a free radical with 5 coupling F nuclei – one with coupling of 92 gauss and four with equivalent couplings of 33 gauss each. The *arrow* pointing downward marks the position for the resonance of DPPH, g = 2.0036. The small peak appearing at this point is thought to arise from a second radical. The observations were made at room temperature at a frequency of 9 GHz. A dosage of 5×10^6 rad was given with a kilocurie cobalt-60 source
[Reproduced with permission from J. Chem. Phys (1959) 30: 399]

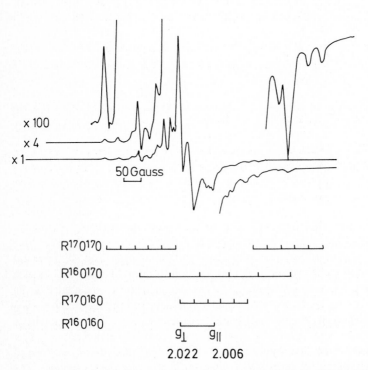

Fig. 6. EPR spectrum of the peroxy chain radical $-CF_2CF(O_1O_2)-CF_2-$ at 300 K and 35 GHz. Only the outlying features of the $C(^{17}O^{17}O)$ spectrum are marked
[Reproduced with permission from J. Chem. Phys. (1976) 64: 2370]

cally enriched with $^{17}O_2 (I = \frac{5}{2})$. The chain radical reacts with molecular oxygen to form the relatively stable peroxide radical;

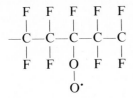

Figure 6 shows the EPR spectrum of the peroxy chain radical at 300 K and 35 GHz [5]. The unpaired spin density is localised on the oxygen atoms and does not interact with the F nuclei sufficiently strongly to produce fluorine hyperfine structure. By using a $^{16}O_2/^{17}O_2$ mixture the structure of the radical can be established. The g-value is axial and slightly anisotropic, with $g_\perp = 2.0220$ and $g_\parallel = 2.0060$. Two sets of six equally spaced lines are clearly resolved. This confirms the presence of two oxygen atoms in the radical. For the mixed isotope species $R^{16}O^{17}O$ and $R^{17}O^{16}O$ the hyperfine coupling is different, being a = 40 and a = 89 gauss respectively. This shows the inequivalence of the two oxygen atoms and justifies the structure as end-on. The corresponding peroxy radical of the propagating radical has been shown by similar techniques to be $CF_2-OO^\cdot$.

This study has provided a benchmark for similar work establishing the presence of peroxide radicals in irradiated polyethylene, polypropylene and numerous other polymers.

5   Molecular Motion and Spin Labels

Stable free radicals in liquid solutions of low viscosity give lines in their EPR spectra which are sharp because the anisotropies in both the g tensor and the nuclear hyperfine couplings are averaged out by the rapidly tumbling motions. Since the average of dipole–dipole coupling over all directions in space is zero the EPR hyperfine structure for radicals in liquid solutions gives only the Fermi contact interactions. This occurs if the mean tumbling rates, v_t, are much greater than the magnetic resonance frequency, v_0. In the other limit, that is when $v_t \ll v_0$, the tumbling motions may be neglected and the magnetic axes considered as fixed in space. The EPR spectrum obtained in this case is the same as that for randomly oriented radicals in powders, glasses or rigid polymers. An estimate of the magnitude of the correlation time, τ_c, which is required to sharpen up the EPR spectrum of a tumbling radical can be obtained from the modified Debye expression;

$$\tau_c = 3 \times 10^{23} \frac{\eta r^3}{T}$$

where r is the effective radius of the molecules treated as spheres, T is the absolute temperature and η the viscosity of the liquid, each expressed in SI units. At 300 K water has a viscosity of $\eta = 1.0 \times 10^{-3}$ Pa s and r = 1.5×10^{-10} m indicating that $\tau_c = 3.4 \times 10^{-12}$ s. This is three orders of magnitude less than $(1/v_0) = 10^{-9}$ s for X-band EPR frequencies. Hence the effect of motion is to broaden the narrow rapid-motion spectrum with a correlation time of faster than 10^{-9} s until the rigid limit is reached for any motion slower than 10^{-7} s [6].

5.1 Membranes

These principles have been put to use by McConnell and co-workers [7] to study the motions of radicals in biological membranes. The nitroxide radical, NO·, can be stabilized by incorporating it into a sterically hindered organic framework:

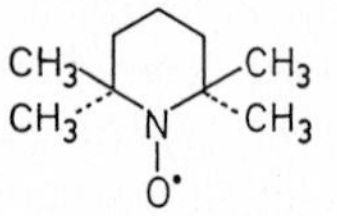

The g-values of this radical are close to isotropic, ($g_x = 2.0088$, $g_y = 2.0058$, $g_z = 2.0022$). However, the odd electron couples to the nuclear spin of ^{14}N (I = 1) to give a three-line nuclear hyperfine pattern which is highly anisotropic ($a_x = 5.8$ gauss, $a_y = 5.8$ gauss, $a_z = 32$ gauss). Figure 7 shows the hyperfine structure for an oriented specimen of the radical with the externally applied magnetic field along

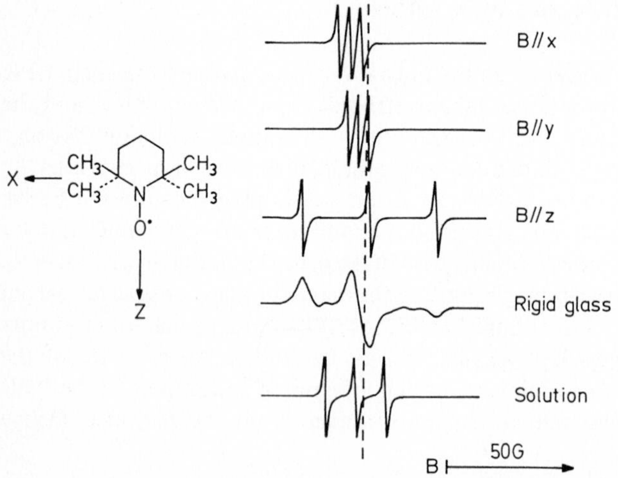

Fig. 7. EPR spectra (9 GHz: 0.3T) of nitroxide spin-label showing anisotropy of the hyperfine interaction given by the line separation when the magnetic induction field, B, is along each of the three principal nitroxide axes.
[Reproduced with permission from Biochem. Soc. Trans (1987) 13: 588]

Table 2. A representative number of different nitroxide spin-labels used for studying biochemical problems

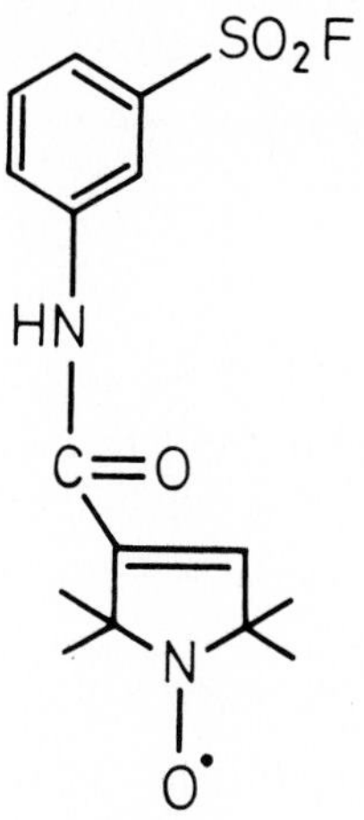

Sulphonylating label

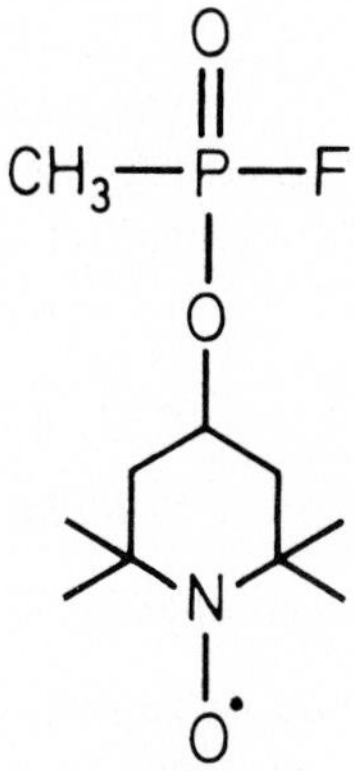

Phosphonylating label

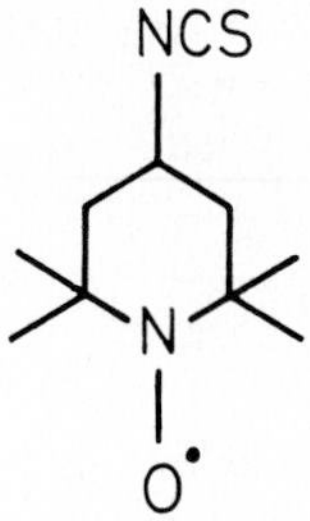

Alkylating label for proteins and nucleic acids

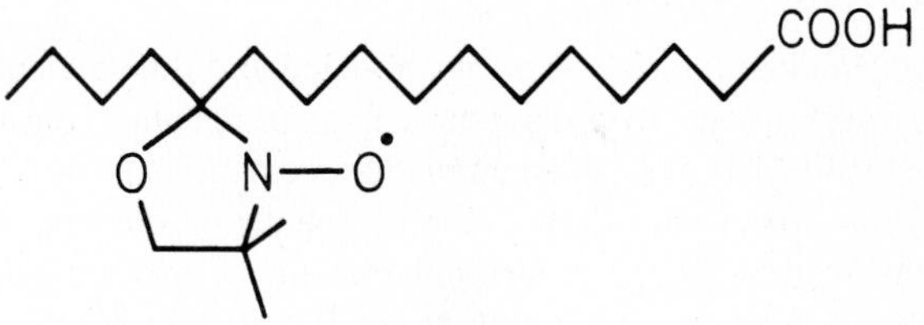

Lipid label

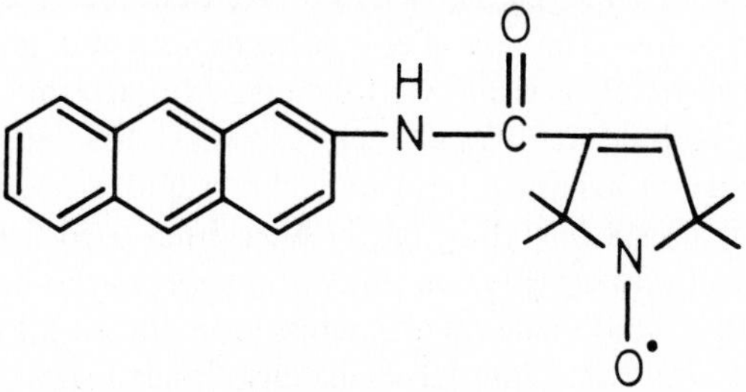

DNA intercalating label

[Reproduced with permission from Biochem. Soc. Trans. (1985) 13: 588]

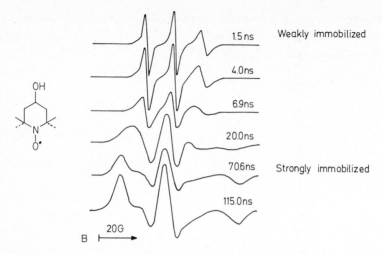

Fig. 8. EPR spectra (9 GHz: 0.3T) of nitroxide spin-labels to demonstrate the effect of mobility on spin-label spectra
[Reproduced with permission from Biochem. Soc. Trans. (1987) 13: 588]

each of three orthogonal axes. Also shown is the EPR spectrum of the radical rapidly tumbling in fluid solution. In this case the three a values are averaged to yield an isotropic hyperfine splitting $a_{iso}(= 14.5$ gauss) where

$$a_{iso} = \tfrac{1}{3}(a_x + a_y + a_z)$$

In a frozen glass the radical is immobilized and the spectrum is broadened by the overlap and superposition of spectra from molecules lying at various angles to the applied field. However, in an environment with a viscosity between the fluid and the glassy state the shape of the spectrum changes depending upon the correlation time of the rotational motion of the radical. Figure 8 shows this clearly and lists the correlation times for the various states.

Spin-labels covalently attached to suitable chains (see Table 2) can be incorporated into a membrane [6]. The temperature-dependence of the spectrum can indicate melting points. The rate of spin-label motion and hence membrane fluidity can be determined. This is a powerful method for determining rotational motions of liquids in fluid biological membranes. It has been applied, for example, to the study of synaptosomal and chromaffin granule membranes. Sulphonylating and phosphonylating labels have been used for study of the active-site geometry of serine proteases. Alkylating agents can be used to label non-specifically proteins and nucleic acids, while aromatic amine nitroxides can intercalate into ordered DNA. Lipid labels usually closely resemble their parent analogues with the nitroxide covalently attached to either one acyl-chain or, for immunological studies, to the lipid head-group.

The electron density distribution across the nitroxide bond is determined by the polarity of the local environment. This is reflected in a change in the average isotropic hyperfine splitting parameter which is close to 17 gauss in water and 13.0 gauss in an organic solvent. Using spin labels with chains of different lengths the nitroxide radical may be moved across a membrane, and by monitoring the variation in the value of a_{iso} the polarity profile of the membrane can be obtained.

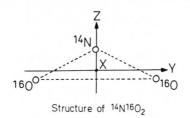

Structure of $^{14}N^{16}O_2$

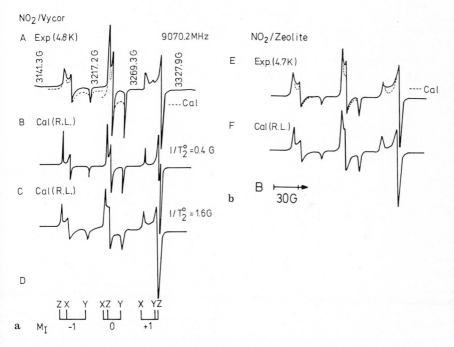

Fig. 9. (a) Experimental spectrum of NO_2 adsorbed on Vycor at 4.8 K (A) and spectra (B and C) calculated by the rigid-limit simulation program in order to determine the a and g tensor components precisely. Spectra B and C best fit the observed one at the band corresponding to $M_1 = 0$ and x, z components of $M_1 = \pm 1$, respectively, with parameters $g_x = 2.0051$, $g_y = 1.9913$, $g_z = 2.0017$, $a_x = 50.0\,G$, $a_y = 46.0\,G$, $a_z = 65.5\,G$, and Lorentzian line shape. Line widths of $1/T_2^0 = 0.4$ and $1.6\,G$ were used for B and C, respectively. The dotted lines in A correspond to the calculated line shape of B at $M_1 = 0$ and of C at the $M_1 = \pm 1$ bands
(b) Experimental spectrum of NO_2 in zeolite (Na-Zeolon) at 4.7 K (E) and best fitting one (F) calculated by using the rigid limit program. The a and g tensor components used are the same as those of NO_2-Vycor, but with a Lorentzian linewidth of $1/T_2^0 = 1.2\,G$
[Reproduced with permission from J. Phys. Chem. (1981) 85: 3873]

There have been many applications of the nitroxide spin-label technique in biological problems including the study of melting of regions of nucleic acids, and the probing of protein conformational changes on ligand binding by attachment of a spin lable. Nitroxides used as haptens have been used to probe hapten-antibody binding.

5.2   Surfaces

Spin labels have also been put to use to investigate anisotropic motion on surfaces. The work of Shiotani and Freed [8] provided a pioneering study with use of the stable paramagnetic gaseous molecule NO_2 to examine its motion when absorbed on crushed porous Vycor glass and on a Zeolite. NO_2 is a bent radical, (Fig. 9), which gives a highly resolved EPR spectrum at very low temperature, 4.8 K, corresponding to an immobilized molecule. The g-value is slightly anisotropic; $g_z = 2.0017$, $g_y = 1.9913$ and $g_x = 2.0051$. Because ^{14}N has I $= 1$ there are three hyperfine lines for each g-value, giving a nine-line spectrum. The rigid limit spectra are very similar on Vycor and on Zeolite (Na-Zeolon or mordenite, $Na_{8.7}(AlO_2)_{8.7}(SiO_2)_{39.3} \cdot 24H_2O$).

On warming the sample, (Fig. 10) the spectrum broadens. The line shapes

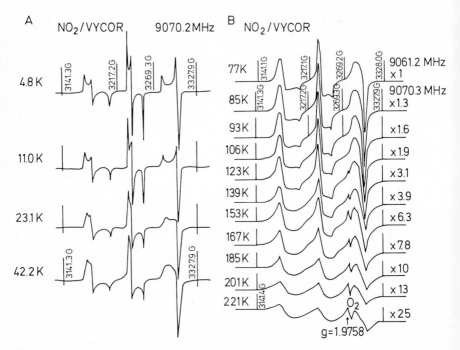

Fig. 10. Temperature-dependent ESR spectra of NO_2-Vycor: (**A**) below 42.4 K; (**B**) above 77 K [Reproduced with permission from J. Phys. Chem. (1981) 85: 3873]

depend strongly on temperature. By using simulation techniques Freed and co-workers were able to arrive at the following conclusions. The low temperature motion, below 77 K, is dominated by anisotropic rotation of the NO_2 at a single surface site and the preferred axis of rotation is the y-axis (0...0 direction). An activation energy of $210 \pm 65\,\mathrm{kJ\,mol^{-1}}$ was estimated. Above 77 K the rotational motion is isotropic and dominated by translational diffusion of NO_2 to surface sites of different orientation. The activation energy is $2.1\,\mathrm{kJ\,mol^{-1}}$. Such a low activation energy indicated that NO_2 is weakly physisorbed to the Vycor surface.

6 Spin Traps

The detection and identification of short-lived free radicals to investigate mechanisms of reactions involving such intermediates can be pursued in a number of ways. For example, the rapid freezing of products has been used in order to stabilise a radical intermediate. However, the technique of spin trapping involves reaction of a short-lived radical with an exogenous compound to produce a relatively stable radical. The EPR spectrum of the product can then be used to identify and quantify the initial short-lived product. With an efficient spin-trap the concentration of the spin adduct should increase with time to a maximum value, dependent upon the relative rates of formation and decomposition. A good spin trap should react rapidly with radicals to form an EPR spectrum which will allow the unambiguous identification of the radical trapped [9].

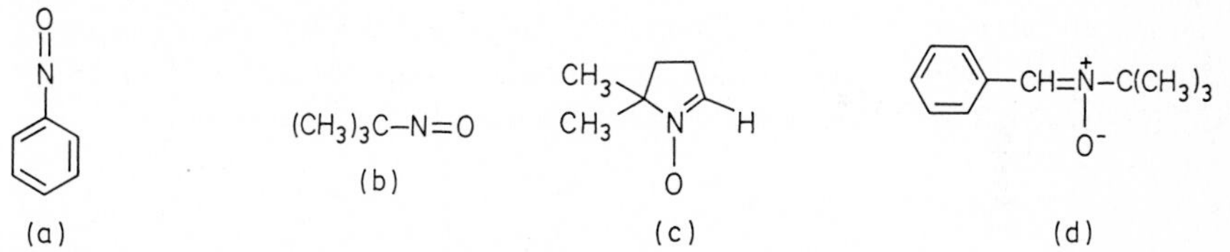

Fig. 11. Some examples of spin traps (a) Nitrosobenzene; (b) methyl-2-nitrosopropane; (c) DMPO; (d) PBN

Unfortunately, no spin traps available at present come close to satisfying these criteria. Two groups of compounds are most useful, namely, nitrosoarenes or nitrosoalkanes such as nitrosobenzene or 2-methyl-2-nitrosopropane, (Fig. 11), and nitrones such as DMPO (5,5-dimethylpyrroline 1-oxide) and phenyl-N-t-butylnitrone (PBN). DMPO will react with a radical such as OH˙ to give DMPO—OH˙,

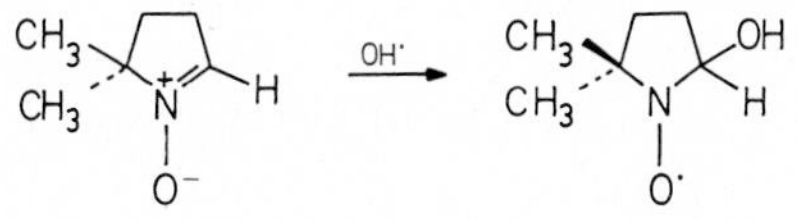

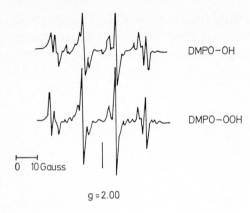

DMPO–OH

DMPO–OOH

0   10 Gauss

g = 2.00

Fig. 12. EPR spectra of DMPO–OH and DMPO–OOH
[Reproduced with permission from Biochem. Biophys. Acta (1982) 715: 116]

Table 3. Hyperfine splittings of the hydroperoxyl and hydroxyl radical adducts of DMPO

DMPO O_2H	DMPO OH
$a_N = 14.3$ G	$a_N = 14.8$ G
$a_b^H = 11.7$ G	$a_H = 14.8$ G
$a_g^H = 1.25$ G	

and superoxide, O_2^-, to give DMPO—OOH. The hyperfine splittings of these species are shown in Table 3. These two traps have different EPR spectra, (Fig. 12) [10]. The intensities of the spectra of the two species can be used to assess the relative yields of these two radicals in certain biological reactions.

7   Transition-Metal Ions in Proteins

EPR spectroscopy is an important method for detecting the presence of transition-metal ions in proteins and for investigating their structures and functions. All metal ions with an odd number of unpaired electrons should give EPR spectra. Since a transition-metal ion can usually be brought into a variety of oxidation levels it is often possible to arrange conditions to elicit an EPR signal from a metal ion. There are two conditions under which is not possible to obtain and EPR signal. If the metal ion has an even number of unpaired electrons the EPR signal is sometimes not detectable because zero field-splitting arising from low symmetry crystal fields moves the resonance transition out of the frequency range of most spectrometers. If the metal ions occur in pairs, then the EPR signals are often undetectable unless the pair of metal ions together possesses an odd number of electrons.

The detection of EPR signals from metal ions in proteins can provide information of the following type:

(a) Identification of the presence of the metal ion. Confirmation that the EPR signal belongs to a given type of metal can sometimes be obtained if the metal nucleus has a spin and if nuclear hyperfine structure is present in the spectrum.

(b) A means of following changes in the oxidation state of a metal ion. This enables redox potentials to be measured.

(c) Identification of ligands of the metal may be possible. If the unpaired electron of a metal ion is delocalised, by covalence, onto the nuclei of surrounding ligands which have nuclear spins, it can be possible to observe ligand nuclear hyperfine structure in the EPR spectrum and hence to identify the type of nucleus in the ligand.

One example will illustrate these points using a protein which contains nickel ion. This is an enzyme called hydrogenase which is present in a bacterium which reduces sulphate to H_2S. Hydrogenase catalyses the reduction of protons to hydrogen or vice versa;

$$2H^+ + 2e^- \rightleftharpoons H_2.$$

Some bacteria use protons as a terminal oxidant rather than oxygen as warm-blooded animals do. The hydrogenase gives an EPR signal that is quite unusual, (Fig. 13). Various lines of evidence suggested that the signal might arise from nickel ion. In order to prove this bacteria were grown on a medium enriched with ^{61}Ni, an isotope with a nuclear spin $I = \frac{3}{2}$. An electron spin coupled to a nucleus of spin I yields an EPR spectrum split into $(2I + 1)$ hyperfine lines. Some of the features in the EPR spectrum of the hydrogenase were indeed split into 4 lines when the bacterium had been grown on ^{61}Ni.

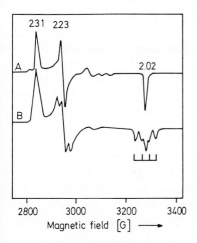

Fig. 13. EPR spectra of *D.gigas* hydrogenase-enriched (B) and unenriched (A) ^{61}Ni enzyme [Reproduced with permission from Biochem. Biophys. Res. Commun. (1982) 108: 1388]

This provided unambiguous proof that the signal arises fron nickel. Since the bacteria will not grow when deprived of a source of nickel this provides strong evidence of the essentiality of this metal ion and indeed points to a likely functional role for the nickel in hydrogenase. The EPR signal comes from the Ni(III) oxidation state and the redox potential for the reduction of Ni(III) to Ni(II) could be determined by observing the disappearance of the EPR signal on titration with a reductant. The mid-point potential is $-220\,mV$ which is remarkably low for the couple Ni(III)/Ni(II). Many inorganic complexes of Ni(III) have high, positive potentials of $> +800\,mV$. This attests to a most unusual environment for the nickel(III) ion in hydrogenase.

8   Electron Nuclear Double Resonance (ENDOR)

ENDOR spectroscopy measures the flip of a nuclear spin by a radiofrequency field, detected by a change in the intensity of the EPR signal arising from the electron to which that nucleus is coupled. The origin of the effect is seen most clearly by means of an energy level diagram for an effective electron spin of $\frac{1}{2}$ and a nuclear spin of 1 (e.g. ^{14}N) (Fig. 14). To simplify the example we take the case in which the external applied magnetic induction field B is applied along the principle g-tensor direction g_{zz} which is, in turn, parallel to the principle z-axis of the nuclear hyperfine tensor. The Hamiltonian for the electronic Zeeman and nuclear terms can be written

$$H = g_{\parallel}\beta_e S_z B_z \quad + a_{zz}I_z S_z + P_{zz}[I_z^2 - \tfrac{1}{3}I(I+1)] \quad - g_n\beta_n B_z I_z.$$

electronic   nuclear hyperfine interaction   nuclear

Zeeman   Zeeman

interaction   interaction

Here a_{zz} is the z-component of the magnetic hyperfine interaction and P_{zz} is the z component of the quadrupole interaction. The energy scheme in Fig. 14 shows the effects of the various terms in the spin Hamiltonian. Also given are the three allowed EPR transitions corresponding to the selection rules $\Delta M_S = 1, \Delta M_I = 0$. In the ENDOR experiment the intensity of one of the EPR transitions is measured whilst the radiofrequency is swept over several tens of megahertz in order to induce nuclear transitions according to the selection rule $\Delta M_I = \pm 1$. For an I = 1 nucleus four ENDOR transitions should be observed with energies given by

$$h\nu_{ENDOR} = \tfrac{1}{2}|A_{zz}| \pm |P_{zz}| \pm |g_n\beta_n B_z|$$

These correspond to the four transition A′–B′, C–B and B–A, B′–C′, occurring as two pairs separated by $2g_n\beta_n B$, that is, twice the nuclear Zeeman energy. The resulting ENDOR spectrum is shown at the bottom of Fig. 14. For an $I = \frac{1}{2}$ nucleus (where the quadrupolar interaction, P_{zz}, is zero) two ENDOR transitions

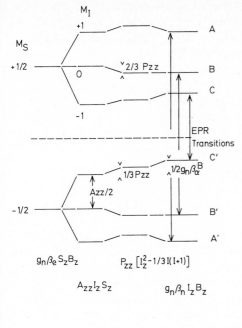

$$g_n \beta_e S_z B_z \qquad P_{zz}\left[I_z^2 - 1/3\, I(I+1)\right]$$

$$A_{zz} I_z S_z \qquad g_n \beta_n I_z B_z$$

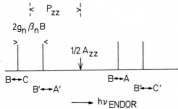

Fig. 14. Energy levels (top) and ENDOR frequencies (bottom) of a system with an effective electron spin of $\frac{1}{2}$ and a nuclear spin of 1 (e.g. ^{14}N), with the magnetic field along the haem axial symmetry axis. The contributions of electron Zeeman, electron-nuclear hyperfine, nuclear quadrupole, and nuclear Zeeman interactions are shown. In ENDOR, the EPR transitions are monitored while the nuclear transitions are induced. The energies of the four nuclear transitions are found to occur in pairs separated by $2g_n \beta_n B$. The energy levels are not drawn to scale; the EPR frequency is approximately three orders of magnitude larger than the ENDOR frequencies

are observed at

$$h\nu_{ENDOR} = \tfrac{1}{2}|A_{zz}| \pm g_n \beta_n B$$

The first example of an ENDOR spectrum is that of the low-spin cobalt(II) Schiff base complex, Co (acacen), diluted into the diamagnetic single crystal host lattice of Ni(acacen). $\frac{1}{2}$H$_2$O. (Fig. 15). The upper part of the figure shows the EPR spectrum at X-band of this crystal. This spectrum is complicated by the presence

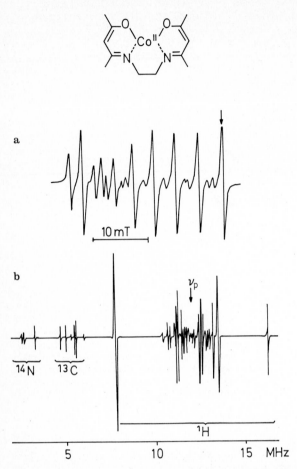

Fig. 15. EPR and ENDOR spectrum of the low spin Co(II) Schiff base complex Co(acacen) diluted into a Ni(acacen) · $\frac{1}{2}$ H$_2$O single crystal, temperature 8 K. (**a**) EPR spectrum; the two magnetically non-equivalent sites coincide for this particular orientation (EPR observer is marked by an *arrow*); (**b**) ENDOR spectrum of ^{1}H, ^{13}C (enriched) and ^{14}N ligand nuclei; ν_p denotes the free proton frequency, asterisks denote the $\Delta m_N = \pm 2$ nitrogen ENDOR transitions
[Reproduced with Permission from Struct. and Bonding (1982) 51: 3]

of an anisotropic g-value, that is $g_\parallel \neq g_\perp$, and the fine structure which arises predominantly from hyperfine coupling of the unpaired cobalt electron to the ^{57}Co nucleus ($I = \frac{7}{2}$). The lower part of the figure shows, by contrast, the ENDOR spectrum between 0–17 MHz. Lines due to ^{14}N, ^{13}C and ^{1}H couplings are clearly resolved and well separated from one another. This ENDOR spectrum was measured by monitoring the intensity change in the EPR spectrum indicated by the arrow in Fig. 15(a). ENDOR spectra can be collected by monitoring other lines in the EPR spectrum, thus giving rise to a virtually complete characteris-

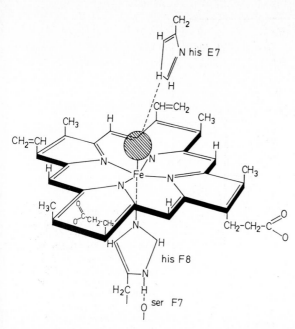

Fig. 16. Haem is the planar molecule in the centre of this figure, and it is depicted here as it is found in myoglobin and hemoglobin. Note the four haem nitrogens attached to the haem iron. A proximal histidine (F8) nitrogen is also attached to the haem iron. The *shaded area* represents the binding site for the sixth ligand. Distal histidine (E7) interacts with the sixth ligand but not with the haem iron. The histidines F8 and E7 are amino acids connected to the protein and are not integral parts of the haem. If the haem is removed from the protein, then the histidines in this figure disappear, but the shaded area can still be occupied by an anion
[Reproduced with permission from J. Amer. Chem. Soc. (1982) 104: 2724]

ation of all the hyperfine coupling contsants between the unpaired cobalt electrons and the ligand nuclei which experience unpaired electron spin density.

As a second illustration of ENDOR spectra we take as an example the metalloprotein myoglobin. This protein, responsible for reversibly binding O_2 and storing it in muscle, contains the haem group bound to a histidine residue, (Fig. 16). The Fe atom is coordinated by 4 nitrogen atoms in the plane of the porphyrin ring and 1 nitrogen atom from the histidine residue (His F8). In the oxidized form the iron is in the ferric state, and possesses five unpaired electrons $(S = \frac{5}{2})$ which give intense EPR signals with g values of 6.0 and 2.0 when the sample is cooled to the temperature of liquid helium. By monitoring the intensity of the EPR signal at g = 2.0 the ^{14}N ENDOR spectrum can be obtained, (Fig. 17). Four ENDOR lines from the equivalent in-plane haem nitrogen atoms and four lines from the histidine nitrogen were expected. The results from protein containing ^{14}N haem and ^{15}N $(I = \frac{1}{2})$ haem showed that the haem nitrogen atoms were slightly inequivalent. The lines from the histidine ligand were also clearly visible.

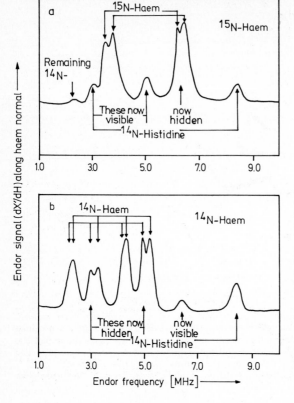

Fig. 17. Nitrogen ENDOR signals from haem and histidine nitrogens taken with the magnetic field along the haem normal, T = 2.1 K. Spectra were taken at $g_z = 2.00$ and had EPR frequencies of 9.09 GHz for (a) and 9.08 GHz for (b). Spectrum (a) is from a single crystal of [^{15}N] haem metmyoglobin. Spectrum (b) is from a single crystal of [^{14}N] haem metmyoglobin. Note that the combination of spectra (a) and (b) reveals all four [^{14}N] histidine ENDOR lines [Reproduced with permission from J. Amer. Chem. Soc. (1982) 104: 2724]

The ENDOR technique is difficult and complex. However, the sensitivity and ease of carrying out experiments is improving rapidly. New designs of microwave cavity, called loop-gap resonators, are being introduced. They require much smaller amounts of sample, can easily be wound with external radiofrequency coils, and can be placed inside a cryostat. These developments are now making ENDOR a useful technique for solving problems of structure in metalloproteins where sample volume is limited.

ENDOR provides a means of obtaining nuclear magnetic resonance signals from paramagnetic samples and from frozen solid solutions. For example, the observation of ^{57}Fe nuclear magnetic resonances in paramagnetic samples has only been successfully carried out by ENDOR spectroscopy; ^{57}Fe NMR even on diamagnetic samples requires large quantities of material and the sensitivity is poor. Although ENDOR spectroscopy will always remain a highly specialised technique its application in metalloproteins and in supported catalyst work will undoubtedly increase.

9   Pulsed EPR and Electron Spin Echo Envelope Modulation (ESEEM)

The advent of powerful computers in the field of NMR spectroscopy enabled the introduction of pulsed sources followed by Fourier transform of the nuclear echoes. The FT-NMR method has revolutionized the field. Will such methods ever become routine in the field of EPR spectroscopy? There is a fundamental difficulty, namely, the speed with which phase coherence of electron spins is lost. Because an electron is coupled much more strongly to its chemical environment than a nucleus is, the electron relaxes on a timescale orders of magnitude shorter than a nucleus does. High-power microwave sources are available and can be designed to pulse into cavities with rapid decay times, so that detection within a few tens of nanoseconds of a pulse is possible. This is sufficient to enable the detection of an echo from a set of electron spins refocusing after a 90° pulse. Therefore a free induction decay can be built up stepwise using varying time delays between pulses. The envelope of the train of spin echoes is not a smooth exponential decay but is modulated by coupling between the relaxing electron and neighbouring nuclei with non-zero spin to which the electron is coupled. The modulated decay was called the electron-spin echo envelope modulation (ESEEM) by its inventor Mims [13]. Fourier transform of the ESEEM expresses the results in the frequency domain and the resulting peaks give directly the hyperfine coupling frequencies of neighbouring nuclei. Nuclei which are dipolar coupled to the electron and which have nuclear quadrupole moments yield the best spectra. Hence the technique is in many ways complementary to that of ENDOR, detecting more distant nuclei with hyperfine couplings in the low-frequency range, 0–15 MHz, where ENDOR transitions are very weak [14]. Pulsed ENDOR methods have also been reported. There is little doubt that with new design of microwave cavities which are transparent to radiofrequency radiation the area of pulsed EPR will grow in the future.

10   References

Alkyl radicals
 1. Fessenden RW, Schuler RH (1963) J. Chem. Phys. 39: 2147
 2. McDowell A, Raghunathan P, Skimokoshi K (1974) J. Chem. Phys. 58: 114
Radicals in polymers
 3. Schneider EE, Day MJ, Stein G (1951) Nature (London) 168: 645
 4. Rexroad HN, Gordy W (1959) J. Chem. Phys. 30: 399
 5. Che M, Tench AJ (1976) J. Chem. Phys. 64: 2370

Molecular motion and spin labels
6. Watts A (1985) Biochem. Soc. Trans. 13: 588
7. Stone TJ, Buckman T, Nordic PL, McConnell H (1965) Proc. Natl. Acad. Sci. USA 54: 1010
8. Shiotani M, Freed JH (1981) J. Phys. Chem 85: 3873
Spin Traps
9. Jansen EG (1971) Acc. Chem. Res. 4: 31
10. Bannister JV, Bannister WH, Hill HAO, Thornally PJ (1982) Biochem. Biophys. Acta. 715: 116
ENDOR
11. Scholes CP, Lapidot A, Mascarenhas R, Inubushi T, Isaacson RA, Feher G (1982) J. Amer. Chem. Soc. 104: 2724
12. Scholes CP (1979) In: Dorio MM, Freed JH (eds) Multiple electron resonance spectroscopy, Plenum, New York, p 297
ESEEM
13. Mims WB, Peisach J (1981) Biol. Magnetic Reson. 3: 213
14. Ichikawa T, Kevan L (1982) J. Amer. Chem. Soc. 104: 1481

11 Bibliography

Abragam A, Bleaney B (1970) Electron paramagnetic resonance of transition ions, Oxford University Press,
Atherton NM (1973) Electron spin resonance, Halsted, London
Carrington A, McLachlan (1967) Introduction to magnetic resonance, Harper and Row, New York
Electron spin resonance, vols 1–8, Specialist Periodical Reports (Royal Society of Chemistry, London) Senior Reporter P.B. Ayscough
Gordy W (1980) Theory and applications of ESR, Wiley, New York
Ranby B, Rabek JF (1977) E.S.R. spectroscopy in polymer research, Springer Berlin Heidelberg New York
Symons M (1978) Chemical and biochemical aspects of electron-spin resonance, Van Nostrand Reinhold,
Wertz JE, Bolton JR (1972) Electron spin resonance, McGraw Hill, New York

Index